Statistical
Mechanics

STATISTICAL MECHANICS

DONALD RAPP
The University of Texas at Dallas

HOLT, RINEHART AND WINSTON, INC.
New York   Chicago   San Francisco   Atlanta
Dallas   Montreal   Toronto   London   Sydney

PREFACE

This book is intended as a general introduction to the subject of Statistical Mechanics for both chemists and physicists. It should be useful for senior undergraduates and first year graduate students. I have chosen not to devote the first part of the book to reviews of quantum mechanics, classical mechanics, mathematics and thermodynamics. Only a very brief review of thermodynamics is given. Some of the special features of the book are treatments of the gaseous, liquid and solid states, theory of reaction rates, the Darwin-Fowler method, classical statistical mechanics, symmetry numbers, chemical equilibrium, imperfect gases and isotope effects. Topics on solutions, magnetic properties, the grand canonical ensemble, and radiation are not included. I have tried to strike a good balance between communicability and rigor. I hope it will be of value to young people meeting the subject for the first time.

DONALD RAPP

Dallas, Texas
June 1972

CONTENTS

INTRODUCTION: A REVIEW OF THERMODYNAMICS

1 THE FIRST LAW OF THERMODYNAMICS

Consider a system and surroundings which can exchange heat energy and can do work on each other, but which cannot exchange matter. The work *done on* the surroundings is designated W and the heat *received by* the system from the surroundings is called Q. In any process involving work and heat exchange between the system and surroundings, conservation of energy requires that the difference $Q - W$ is the energy increase of the system.

The first law of thermodynamics is a postulate that states the following: *When a system undergoes a change in state, the quantity $Q - W$ depends only on the initial and final states of the system, and is independent of the path taken.* Therefore, a quantity, E, may be defined, representing the total energy of the system. This energy depends only on the state of the system, not its past history. Conservation of energy requires that for any actual process, the change in E, ΔE, is given by

$$\Delta E = Q - W \tag{1}$$

The nature of E, as a function that depends only on initial and final states, requires that the quantity dE is a perfect differential in the sense that

$$\Delta E = E_2 - E_1 = \int_1^2 dE \tag{2}$$

regardless of the path. This is not true for small amounts δQ and δW, although it is true[1] for

$$dE = \delta Q - \delta W \tag{3}$$

At this point, the concepts of reversible and irreversible processes

---

[1]Note that the functions $x dy$ and $-y dx$ are *not* perfect differentials because $x dy$ and $-y dx$ are dependent on the paths $x(y)$ and $y(x)$. Note, however, that the difference between these, $x dy - (-y dx) = d(xy)$ is a perfect differential because $\int_1^2 d(xy) = (xy)_2 - (xy)_1$.

1

must be introduced.   A reversible process is defined as a process carried out sufficiently slowly that all intermediate states of the system are essentially equilibrium states, each homogeneous portion of the system remains essentially uniform, and it is possible to slowly return to the initial state without a change in the surroundings.   All actual processes are irreversible, but a slow real process may be dealt with in the limit as it approaches reversibility.

2   THE SECOND LAW OF THERMODYNAMICS

The second law of thermodynamics is a set of postulates which may be stated as follows:

(a) For a reversible process, a property of the system, T, may be defined which depends only on the state of the system, such that

$$\frac{\delta Q_{rev}}{T} \tag{4}$$

is an exact differential.   That is, a quantity S may be defined, such that changes in S are given by

$$\Delta S = \int_1^2 dS = \int_1^2 \frac{\delta Q_{rev}}{T} \tag{5}$$

The quantity T is called the absolute temperature, and S is the entropy.

(b) For all natural processes (irreversible)

$$\Delta S = S_2 - S_1 > \int_1^2 \frac{\delta Q}{T} \tag{6}$$

where T is the temperature of the surroundings.

These two postulates can be used to prove a number of important corollaries.   Alternatively, any pair of independent corollaries can be defined as the second law, and postulates (a) and (b) above can be proven.

Thermodynamic temperature can be defined in a number of ways. With any method, a particular system must be defined as some standard value.   Usually, the temperature of ice is chosen at 1 atm. pressure in equilibrium with liquid water, T_i, to be 273.16° absolute.   Then the temperature, T, of any other system can be defined as T_i multiplied by the limiting ratio $(P/P_i)_V$ of pressures as $P_i \rightarrow 0$ with the volumes of each gas constant and equal.   Another method involves setting T/T_i equal to the ratio, Q/Q_i, of efficiencies of *Carnot cycles* carried out on systems in equilibrium with a system at temperature T and a system at temperature T_i.   Regardless of what procedure is used, it can be shown that the definition is commensurate with the second law.

3   ENTROPY OF MIXING OF IDEAL GASES

Suppose there are two bulbs containing different gases A and B as illustrated in Fig. 1. There are n_i molecules of type i ($i = A$ or B) in volume V_i, and p and T are the same for both gases. Now let the stopcock be opened. If the gases are assumed to be ideal, the final pressure and temperature are p and T. The entropy change is the same as the sum of entropy changes for expanding n_A molecules in volume V_A to $V_A + V_B$, plus n_B molecules in V_B expanded to $V_A + V_B$. The result[2] is

$$\Delta S = -n_A k \ln x_A - n_B k \ln x_B \qquad (7)$$

where $x_A = n_A/(n_A + n_B)$, and $x_B = n_B/(n_A + n_B)$, and k is the Boltzmann constant. This entropy change can be viewed as due to expansion of each gas into a larger container of volume $V_A + V_B$. Another interpretation can also be given which may at first sight appear very different, but which is really equivalent. Before the stopcock is opened, it is *known* that all the A molecules are in the left bulb, and all the B molecules are in the right bulb. After opening the stopcock, the gases mix, and *information* about the system is lost. It can be shown that all increases in entropy can be correlated with decreases in information about the system. The amount of information known about a system varies inversely with the number of possible equivalent arrangements of the system. For example, if there is a gas of molecules, all permutations of the molecules are equivalent arrangements. In the case of the two bulbs discussed above, the information on the location of gas A after mixing is x_A of the information before mixing, and similarly for B. This is because the location of gas i was known to within volume V_i before mixing, but the possible range of locations expanded to $V_A + V_B$ after mixing. The information on a substance I_i can be defined as the inverse of the number of equivalent arrangements, W_i:

$$I_i = W_i^{-1} \qquad i = A, B \qquad (8)$$

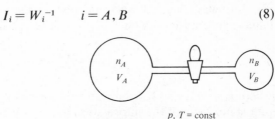

$p, T = \text{const}$

FIGURE 1   Two gas bulbs at constant pressure and temperature, n_A molecules of A are in volume V_A, and n_B molecules of B are in V_B, before mixing.

---

[2]F. T. Wall, *Thermodynamics* (San Francisco, Calif.: Freeman Book Co., 1965).

Since entropy varies inversely with I, it must vary monotonically with W. If there is a system composed of two subsystems (for example, the bulbs of gas with unopened stopcock), any arbitrary particular microscopic arrangement of the A molecules can be associated with any arbitrary microscopic arrangement of the B molecules. Therefore, the total number of arrangements for the system is the product

$$W_T = W_A W_B \tag{9}$$

The total entropy of the system, however, is

$$S_T = S_A + S_B \tag{10}$$

Since S_T varies monotonically with W_T, it follows that

$$S_i = k \ln W_i \qquad i = A, B, \text{ or } T \tag{11}$$

where k is a constant.[3] Since $W_i = I_i^{-1}$, the relation between entropy and information is

$$S_i = -k \ln I_i \tag{12}$$

Now return to the mixing of gases A and B. The ratio of information on gas i $(i = A, B)$, after mixing to before mixing, is x_i. Hence, Eq. (7) follows immediately. Equation (12) can be used for all real substances, although Eq. (7) will only be true for noninteracting particles.

4   THE THIRD LAW OF THERMODYNAMICS

The entropy of any substance composed of non-interacting particles can be regarded as the entropy of random mixing of particles in different quantum states. Suppose each particle can be in three quantum states. Then at equilibrium, at temperature T, n_i are in state i $(i = 1, 2, \text{ or } 3)$. The entropy of the substance can then be regarded as due to the decrease in information caused by randomly mixing these states if they were initially isolated, as shown in Fig. 2. After mixing, the entropy increase is

$$\Delta S = -\sum_{i=1}^{3} n_i k \ln x_i \tag{13}$$

where $x_i = n_i/\Sigma n_i$. If a substance has all its particles in the same quantum state, then all the information that can possibly be known is known about the substance. In this case, $I = 1$ since $I \leq 1$ [see Eq. (8)]. There-fore, the entropy is zero. The third law of thermodynamics is based on the assumption that at absolute zero on the thermodynamic temperature

---

[3]It will later be shown that k is the Boltzmann constant (gas constant/Avogadro's number).

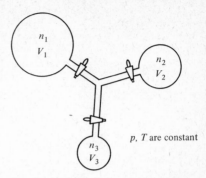

FIGURE 2   Hypothetical separation of n_1 molecules in state 1 into volume V_1, n_2 molecules in state 2 in V_2, and n_3 in state 3 in V_3, before mixing.

scale, all particles go into the lowest energy level.   If this level corresponds to a single quantum state, the entropy is zero.   If there are g quantum states with this energy, $1/g$ of the particles will be in each quantum state and the entropy of mixing remaining at $T = 0$ is

$$S_0 = Nk \ln g \tag{14}$$

where N is the total number of particles.   Equation (14) is the third law.

chapter 1

INDEPENDENT LOCALIZED SYSTEMS (Boltzmann Statistics)

1.1 INTRODUCTION

A goal of equilibrium statistical mechanics is to evaluate the properties of macroscopic matter in terms of the (presumed) known properties
of a single particle. It is assumed that the macroscopic material is
simply a collection of a very large number (say, 10^{20}) of particles in some
enclosure. One could, in principle, write the microscopic equations of
motion for the particles. This would yield $\sim 10^{20}$ coupled differential
equations to be solved and would be impossibly complicated. Instead,
we deal with the *average* properties of the entire collection of particles,
and apply statistical methods. For example, one may calculate the fraction of the total number of particles that are in some particular state.
However, it is impossible to identify *which* particles are in this state.
The total energy of a collection of particles can be calculated, but the
energies of each of the particles cannot.

It will be assumed that each particle exists in a quantum state with
an energy defined by the laws of quantum mechanics. That is, if the
particles under consideration are harmonic oscillators, then the appropriate wave functions and energy levels of an harmonic oscillator should be
used in the resulting formulae. However, it is not necessary to discuss
these microscopic properties of a single particle in detail at this point.
There is a general quantum-mechanical property that is very important,
namely, the principle of indistinguishability. According to quantum
mechanics, no representation of the state of a collection of particles is
valid unless it treats all identical particles in the collection as completely
*in*distinguishable. It would be improper to say, for example, that particle
A is in state 1 and particle *B* is in state 2, if *A* and *B* are both equivalent
particles. A correct description would be that one indistinguishable
particle is in state 1 and the other is in state 2, without labeling the
particles.

In this chapter, only collections of *independent localized systems*
are considered. Such a collection or *ensemble* is described in terms of a

7

crystal lattice, which is the major physical application of this section.   A crystal lattice is a collection of sites in a regular spatial array, with particles (atoms or molecules) at each of the sites.   A crystal composed of a large number of sites will be considered, with identical particles at each of the sites.   Although the particles are indeed indistinguishable, the lattice sites are identifiable and may be labeled, as in Fig. 1-1.   Therefore, one may utilize the phrase "the particle at site *b*" without violating the principle of indistinguishability.   It is understood that it is impossible to specify *which* particle is occupying site *b*.   In this chapter a loose figure of speech, "particle *a*," will be used to mean the particle at site *a*.   This in no way implies that labels have been attached to the particles.   Nevertheless, indistinguishable particles located at distinguishable lattice sites are referred to loosely as "distinguishable particles."

It is postulated that the particles in the crystal lattice interact with very weak forces, causing energy to flow slowly from particle to particle, creating an equilibration of the total energy of the lattice.   If the lattice was initially formed with a nonequilibrium distribution of states of the particles in the lattice, energy would be transferred between the particles until an equilibrium distribution of states was reached, corresponding to thermal equilibrium between the lattice and its surroundings.

Another postulate is made that may at first appear to be mutually exclusive with the previous postulate.   It is assumed that the forces between particles are so weak that the state of each particle is independent of the states of adjacent particles in the lattice.   Thus, the *localized* particles are assumed to be *independent*.   There really is no conflict with the assumption of equilibrium, because extremely weak forces can be postulated which bring the system very slowly to equilibrium, with the forces neglected over the time span of interest in determining the properties of the system at equilibrium.

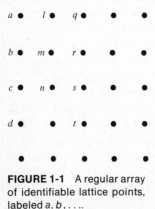

FIGURE 1-1   A regular array
of identifiable lattice points,
labeled *a*, *b*, ...

1.2 MICROSTATES AND MACROSTATES FOR INDEPENDENT LOCALIZED SYSTEMS

The following postulates are made:

(a) A thermodynamic system consists of an ensemble (or collection) of identical particles. The particles may be treated as if they were distinguishable because they occupy distinguishable lattice sites. The particles will, therefore, be labeled $a, b, c, \ldots$.

(b) Each of the labeled particles $a, b, c, \ldots$ can exist in any one of a number of quantum states $1, 2, 3, \ldots$, which are called *cells*.

(c) A *microstate* is a specification of exactly *which* particles are in the various cells (for example, particle a in state 7, b in state 4, and so forth).

(d) A *macrostate* is a specification of *how many* of the particles are in each cell (for example, 174 particles in state 3, 1,629 in state 8, and so forth).

(e) It is assumed that all microstates are equally *a priori* probable.

The methods of statistical mechanics of independent localized systems can be applied to any collection of distinguishable objects which can be in identifiable cells. For example, suppose there is a collection of distinguishable billiard balls labeled $a, b, c, \ldots$, which can be distributed in various ways among pool-table pockets labeled $1, 2, 3, \ldots$. A microstate is a specification of which balls are in which pockets. An arbitrary particular microstate is

$$\boxed{a\,e\,p}\ \boxed{b\,c}\ \boxed{f}\ \cdots \qquad (1\text{-}1)$$
$$\quad 1 \qquad 2 \quad 3$$

meaning that balls a, e, and p are in pocket 1, b and c are in pocket 2, f is in pocket 3, and so forth. A macrostate is a specification of how many balls are in each pocket, without regard to which particular balls are in which pockets. For example, a macrostate can be specified by writing $n_1 = 3$, $n_2 = 2$, $n_3 = 1$, $\ldots$, meaning three balls in pocket 1, 2 balls in pocket 2, 1 ball in pocket 3, and so forth. A symbolic representation of this macrostate is

$$\boxed{\cdots}\ \boxed{\cdot\,\cdot}\ \boxed{\cdot}\ \cdots \qquad (1\text{-}2)$$

The microstate previously given in (1-1) evidently corresponds to this. There are many conceivable microstates corresponding to each macrostate. For example, the microstate

$$\boxed{f\,g\,h}\ \boxed{d\,m}\ \boxed{r}\ \cdots \qquad (1\text{-}3)$$

also corresponds to the above macrostate. The analogy to postulate (e) is obtained by imagining a pool table with randomly scrambled billiard balls. The balls randomly fall into pool-table pockets. Each microstate is equally probable of being formed.

Another useful analogy is in terms of bridge hands dealt from a standard deck of cards. Each particular hand, say,

$$
\begin{array}{cccc}
S & H & D & C \\
A & K & 6 & Q \\
K & 10 & 4 & 3 \\
J & 4 & 2 & \\
3 & 2 & &
\end{array}
\tag{1-4}
$$

is a microstate. A macrostate is denoted by giving the number of cards in each suit. Thus,

$$
\begin{array}{cccc}
S & H & D & C \\
\cdot & \cdot & \cdot & \cdot \\
\cdot & \cdot & \cdot & \cdot \\
\cdot & \cdot & \cdot & \\
\cdot & \cdot & &
\end{array}
\tag{1-5}
$$

is the macrostate to which microstate (1-4) corresponds. According to postulate (e), the microstate

$$
\begin{array}{cccc}
S & H & D & C \\
A & & & \\
K & & & \\
Q & & & \\
J & & & \\
\cdot & & & \\
\cdot & & & \\
\cdot & & & \\
2 & & &
\end{array}
\tag{1-6}
$$

is just as probable of being dealt as (1-4). A bridge player will be dealt hand (1-6) just as often as hand (1-4) from a well shuffled deck. However, a 4-4-3-2 hand will occur far more frequently than a 13-0-0-0 hand. This is because there are a very large number of possible microstates corresponding to the 4-4-3-2 macrostate, and only one microstate associated with the 13-0-0-0 macrostate.

The total number of microstates corresponding to macrostate j is called the *thermodynamic degeneracy*, W_j.[1] Since all microstates are equally probable, the thermodynamic degeneracy of a macrostate will be a

---

[1]The term *degeneracy* is generally used to denote states which have the same values of the energy.

measure of the relative probability of that macrostate occurring in nature. For example, suppose one has four particles, a, b, c, and d, and two cells, 1 and 2. A macrostate is specified by giving n_1 and n_2, the number of particles in cells 1 and 2. There are five possible macrostates: $(n_1, n_2) = (4, 0), (3, 1), (2, 2), (1, 3),$ and $(0, 4)$. These macrostates will be denoted as 1, 2, 3, 4, and 5, respectively. To determine the thermodynamic degeneracy of the macrostates, we must determine the possible microstates. For macrostate 1, there is only one microstate,

$$\boxed{a\,b\,c\,d}\quad\boxed{}$$

$$\quad 1 \qquad\quad 2$$

Thus $W_1 = 1$. For macrostate 2, the possible microstates are

cell 1	$a\,b\,c$	$a\,b\,d$	$a\,c\,d$	$b\,c\,d$
cell 2	d	c	b	a

microstate 1   microstate 2   microstate 3   microstate 4

Thus, $W_2 = 4$.

PROBLEM: Show that $W_3 = 6$, $W_4 = 4$, and $W_5 = 1$.

Thus, if the particles are randomly placed in the cells, the probabilities of the macrostates occurring are in the ratio $1:4:6:4:1$, or $\frac{1}{16}:\frac{1}{4}:\frac{3}{8}:\frac{1}{4}:\frac{1}{16}$.

These results can be obtained in general for any number of particles in any number of cells. Consider the following combination and probability theorems:

(a) The number of ways W that N distinguishable objects can be permuted is $N!$. A permutation is defined as a linear array of the symbols representing the objects, and two arrangements are different if they are not identical. For example, if there are the objects a, b, and c, two of the possible permutations are abc and cba. To prove the theorem, set up N cells, and place one object in each cell.

$$\boxed{N \text{ ways}}\ \boxed{\begin{array}{c}N-1\\ \text{ways}\end{array}}\ \boxed{\begin{array}{c}N-2\\ \text{ways}\end{array}}\ \cdots\ \boxed{\begin{array}{c}1\\ \text{way}\end{array}}$$

$$\quad 1 \qquad 2 \qquad 3 \quad \cdots \quad N$$

The object that goes into cell 1 may be chosen N ways. Once having chosen this, the number of ways to pick the object in cell 2 is $N-1$ since there are $N-1$ objects remaining to choose from. This may be continued until there is only one object (and therefore one way) to fill cell N. Since each possible selection in one cell may be combined with any of the selections in the other cells, it follows that

$$W = N(N-1)(N-2) \ldots \qquad 1 = N!$$

(b) The number of ways that N distinguishable objects can be placed into r distinguishable cells without regard to order within a cell, such that n_1 are in cell 1, n_2 are in cell 2, ..., n_r are in cell r, with $n_1 + n_2 + \cdots n_r = N$, is

$$W = \frac{N!}{n_1!n_2!\ldots, n_r!} = \frac{N!}{\prod\limits_{j=1}^{r} n_j!}$$

The proof of this is obtained by using the symbolism

$$| \, ab \, | \, cde \, | \, fg \, | \, h \, | \, ij \, | \cdots | \, |$$
$$1 \quad 2 \quad 3 \quad 4 \quad 5 \qquad r$$

to denote an arrangement with objects a and b in cell 1, c, d, and e in cell 2, f and g in cell 3, and so forth. The vertical lines denote cell boundaries. The number of permutations of the letters is $N!$. Many of these permutations correspond to different arrangements of the particles in the cells. However, some correspond only to permutations within a cell, and these should not be included as different because the order of placement in a cell is irrelevant. For example,

$$| \, ab \, | \, cde \, | \, fg \, | \ldots$$

and

$$| \, ba \, | \, dce \, | \, gf \, | \ldots$$

are considered as equivalent. But there are $n_i!$ ways to permute the objects in cell i. Therefore,

$$W = \frac{N!}{n_1! \, n_2! \ldots n_r!} = \frac{N!}{\prod\limits_{j=1}^{r} n_j!} \tag{1-7}$$

PROBLEM: Apply Eq. (1-7) to the problem of four particles and two cells, and show that W for the various macrostates is the same as previously calculated.

1.3 THE MOST PROBABLE DISTRIBUTION

In this section, we shall consider an ensemble[2] of N particles, each located at a well defined lattice site, where N is very large (say $\sim 10^{20}$).

---

[2] An "ensemble" merely means a collection of particles as an isolated system in the thermodynamic sense.

Each particle can exist in any of states $1, 2, 3, \ldots, i, \ldots$, the energy levels[3] of these states being $\epsilon_1, \epsilon_2, \epsilon_3, \ldots, \epsilon_i, \ldots$. The number of particles in state i is n_i, and

$$n_1 + n_2 + \cdots = \sum_{\substack{i=\text{all} \\ \text{states}}} n_i = N \tag{1-8}$$

if the total number of particles does not change with time. The total energy of the ensemble is

$$n_1\epsilon_1 + n_2\epsilon_2 + \cdots = \sum_{\substack{i=\text{all} \\ \text{states}}} n_i\epsilon_i = E \tag{1-9}$$

Suppose there is some arbitrary distribution of the particles among the various states, such that Eqs. (1-8) and (1-9) are satisfied. This macrostate can be formed in many different ways because a large number of microstates will correspond to the macrostate. The number of microstates per macrostate is simply the number of ways of placing N objects in cells such that n_1 are in the first, n_2 are in the second, and so forth. According to Eq. (1-7),

$$W = \frac{N!}{\displaystyle\prod_{\substack{i=\text{all} \\ \text{states}}} n_i!} \tag{1-10}$$

Since W is generally a large number if N is large, it is convenient to consider the natural log of W:

$$\ln W = \ln N! - \ln \left[\prod_i n_i! \right]$$

$$= \ln N! - \sum_i \ln n_i! \tag{1-11}$$

According to Stirling's approximation (see Appendix VII) the natural log of a large factorial can be approximated closely by the function

$$\ln N! \cong N \ln N - N \tag{1-12}$$

At this point, a crucial assumption is made, that even though N is partitioned into the n_i, each of the n_i is nevertheless sufficiently large that Stirling's approximation can be used for $\ln n_i!$. For those n_i that are not large enough to justify this assumption, it does not really matter, because the term $\ln n_i!$ in Eq. (1-11) will be very small for such a state. When Stirling's approximation is used in Eq. (1-11), the result is

$$\ln W = N \ln N - N - \sum_i \left[n_i \ln n_i - n_i \right] \tag{1-13}$$

---

[3]It is possible that energy levels of several of the states may be equal, but this is immaterial to the discussion. Each state, regardless of energy, is treated as a cell.

When Eq. (1-8) is used to simplify, this becomes

$$\ln W \cong N \ln N - \sum_i n_i \ln n_i \qquad (1\text{-}14)$$

The *thermodynamic degeneracy*, or relative probability, of any set n_1, n_2, . . . of n_i is determined from Eq. (1-14). According to the basic postulates given in Sec. 1.2, particles exchange energy and change states over a long period of time. Although these energy exchanges leave the total energy E unchanged, they usually result in a different distribution of the n_i. For example, suppose that at some initial time, 10,000 particles are in state 6 with energy ξ, 10,000 are in state 7 with energy 2ξ, and 20,000 are in state 10 with energy 3ξ.[4] After a passage of time, it is found that, say, 12,500 are in state 6, 5,000 are in state 7, and 22,500 are in state 10. The total energy is unchanged since 2,500 particles increased their energy by ξ and 2,500 particles lost ξ. However, the set $n_1 = 10^4$, $n_2 = 10^4$, $n_3 = 2 \times 10^4$ has been changed to $n_1 = 1.25 \times 10^4$, $n_2 = 0.5 \times 10^4$, $n_3 = 2.25 \times 10^4$. As these random energy transfer processes take place, the n_i's vary with time. Over a period of time, the probability of a particular set of n_i occurring will be proportional to the thermodynamic degeneracy of that macrostate. It is the goal here to determine which macrostates (that is, which sets of n_i) have the highest probability of occurring.

There are many possible distributions of particles among the cells (states) which conserve particles and energy. Each distribution $(n_1, n_2, \ldots)$ is a macrostate to which many microstates correspond. Since all microstates are equally probable, the macrostate with the greatest number of microstates will be the most probable macrostate. Therefore, we can find the most probable set $n_1{}^*, n_2{}^*, \ldots$, by requiring that this set of n_i maximize $W(n_1, n_2, \ldots)$. If the n_i were continuous mathematical variables (instead of large integers) the most probable set could be obtained by allowing the n_i to vary, and setting $dW(n_1, n_2, \ldots) = 0$. Since $\ln W$ varies monotonically with W, the set of n_i that maximizes W also maximizes $\ln W$. Therefore, a maximum in W would be obtained by solving

$$d \ln W(n_1, n_2, \ldots) = 0 \qquad (1\text{-}15)$$

Since

$$d \ln W = \frac{\partial \ln W}{\partial n_1} dn_1 + \frac{\partial \ln W_2}{\partial n_2} dn_2 + \cdots \qquad (1\text{-}16)$$

and $\ln W$ is given in Eq. (1-14), so that

$$\frac{\partial \ln W}{\partial n_j} dn_j = -n_j \frac{dn_j}{n_j} - \ln n_j dn_j$$

---

[4]The selection of equispaced levels is entirely arbitrary.

it follows that

$$0 = d \ln W = -\sum_j dn_j - \sum_j \ln n_j dn_j \qquad (1\text{-}17)$$

Because variations in the n_j are only being taken subject to the restrictive condition

$$\sum_j n_j = N$$

or

$$\sum_j dn_j = 0$$

the first sum in Eq. (1-17) is equal to zero. Hence, if the n_j were continuous variables, it would follow that the most probable set of n_i are obtained from the equation

$$\sum_j \ln n_j{}^* dn_j = 0 \qquad (1\text{-}18)$$

However, the n_j are, in actuality, integers. Nevertheless, if these integers are large enough, and the changes dn_j are integers which, though large, are still small compared to n_j, then the n_j may be regarded as continuous variables to a high degree of approximation. For example, suppose n_j is an integer like 1.0×10^{14} and dn_j is an integer like 1.0×10^8; then the use of differentiation formulas for functions of n_j will be a good approximation. *To this approximation then, the most probable set of $n_i{}^*$* satisfy the equation

$$\ln n_1{}^* dn_1 + \ln n_2{}^* dn_2 + \cdots = 0 \qquad (1\text{-}19)$$

where the possible variations dn_i are subject to the restrictive conditions

$$dn_1 + dn_2 + \cdots = 0 \qquad (1\text{-}20)$$
$$\epsilon_1 dn_1 + \epsilon_2 dn_2 + \cdots = 0 \qquad (1\text{-}21)$$

Equation (1-19) may be written in the form

$$Ax + By + Cz + \cdots = 0 \qquad (1\text{-}22)$$

where $x, y, z, \dots$ represent the mathematical variables $dn_1, dn_2, dn_3, \dots$. If these variables were totally independent, there would only be one solution possible, namely $A = B = C = \cdots = 0$. This is because $x, y, z, \dots$ can take on any values, and the equality still holds. The only possible way for $Ax + By + \cdots$ to be zero for all values of $x, y, z, \dots$ is for $A = B = C = \cdots = 0$. However, if there are dependent relationships between the variables $x, y, z, \dots$, then it is not obvious that $A = B = \cdots = 0$, because the sum $Ax + By + \cdots$ could conceivably be zero due to these dependent relationships. Since the dependent relationships are of the form

$$a_1 x + b_1 y + \cdots = 0 \qquad (1\text{-}23)$$
$$a_2 x + b_2 y + \cdots = 0 \qquad (1\text{-}24)$$

a new equation can be formed by multiplying Eqs. (1-23) and (1-24) by respective constants K_1 and K_2, and adding to Eq. (1-22). The resulting equation is

$$(A + K_1 a_1 + K_2 a_2)x + (B + K_1 b_1 + K_2 b_2)y + \cdots = 0 \qquad (1\text{-}25)$$

The variables in this equation may be regarded as being independent because there are no external relationships between $x, y, z, \ldots$. Equation (1-25) is the most general linear combination that can be formed from Eqs. (1-22), (1-23), and (1-24). By including the external conditions in our main equation, the variables became effectively independent. Therefore, the solution of Eq. (1-25) is that each coefficient of the variables $x, y, \ldots$ is equal to zero. When applied to Eqs. (1-19), (1-20) and (1-21), this method[5] gives

$$\sum_j \{\ln n_i{}^* - \ln \alpha + \beta \epsilon_i\} \, dn_i = 0 \qquad (1.26)$$

and

$$\ln n_i{}^* - \ln \alpha + \beta \epsilon_i = 0 \qquad (1\text{-}27)$$

for all i, where $K_1 = -\ln \alpha$, $K_2 = \beta$, and α and β are undetermined multiplier constants. Equation (1-27) may be rewritten

$$n_i{}^* = \alpha e^{-\beta \epsilon_i} \qquad (1\text{-}28)$$

Therefore, the most probable number of particles in the ith state is proportional to $e^{-\beta \epsilon_i}$, showing that states with high energy are less populated.

The constant α may be evaluated by summing the $n_i{}^*$ over all i and setting the result equal to N. Thus,

$$N = \sum_{\substack{i=\text{all} \\ \text{states}}} \alpha \, e^{-\beta \epsilon_i} = \alpha Q$$

where

$$Q = \sum_{\substack{i=\text{all} \\ \text{states}}} e^{-\beta \epsilon_i} \qquad (1\text{-}29)$$

is defined as the *partition function* of a particle. Hence,

$$\alpha = N/Q \qquad (1\text{-}30)$$

and

$$\frac{n_i{}^*}{N} = \frac{e^{-\beta \epsilon_i}}{\sum_i e^{-\beta \epsilon_i}} \qquad (1\text{-}31)$$

The name *partition function* derives from Eq. (1-31), from which it may

---

[5]This procedure is known as Lagrange's method of undetermined multipliers. In making the application to Eqs. (1-19), (1-20), and (1-21), we take $dn_1, dn_2, \ldots$ as the independent variables $x, y, \ldots$.

be seen that the total number of particles N is partitioned among the possible quantum states in proportion to $e^{-\beta\epsilon_i}$.

If several quantum states have the same energy, these states are said to comprise a degenerate energy level. Instead of dealing with numbers of particles in various states, we may alternatively consider the numbers of particles in various energy levels. The degeneracy of the nth level is called g_n. Suppose, for example, that some particular level is threefold degenerate so that $g = 3$. Then $e^{-\beta\epsilon_i}$ is the same for the three states comprising the level, and the terms in the partition function for these states are

$$e^{-\beta\epsilon_i} + e^{-\beta\epsilon_i} + e^{-\beta\epsilon_i} = 3\,e^{-\beta\epsilon_i} = g_n\,e^{-\beta\epsilon_n}$$

in which ϵ_i for the three degenerate states is the same as ϵ_n for the nth level. Therefore,

$$Q = \sum_{\substack{i=\text{all}\\ \text{states}}} e^{-\beta\epsilon_i} = \sum_{\substack{n=\text{all}\\ \text{levels}}} g_n\,e^{-\beta\epsilon_n} \tag{1-32}$$

and the number of particles in the nth energy level is

$$\frac{n_n}{N} = \frac{g_n\,e^{-\beta\epsilon_n}}{Q} \tag{1-33}$$

1.4 ENTROPY AND RANDOM DISORDER

In this section, the statistical mechanical ensemble will be arranged to be consistent with the laws of thermodynamics. According to the second law of thermodynamics, the entropy change dS accompanying any physical process is the reversible heat absorbed δq_{rev}, divided by the temperature T. Thus

$$\delta q_{\text{rev}} = T\,dS \tag{1-34}$$

But the first law of thermodynamics requires that[6]

$$\delta q_{\text{rev}} = dE + P\,dV \tag{1-35}$$

and hence

$$dS = \frac{1}{T}\,dE + \frac{P}{T}\,dV \tag{1-36}$$

But if E and V are treated as independent variables on which S depends, we may write

$$dS = \left(\frac{\partial S}{\partial E}\right)_V dE + \left(\frac{\partial S}{\partial V}\right)_E dV \tag{1-37}$$

---

[6]Provided that only pV work is done.

Therefore,[7]

$$\left(\frac{\partial S}{\partial E}\right)_V = \frac{1}{T} \tag{1-38}$$

In order to make statistical mechanics consistent with these results, a statistical mechanical entropy σ must be defined. To do this, it must be noted that there is a close relationship between the thermodynamic entropy of a system and the random disorder of the system. *Random disorder* is a term used to describe the nonuniformity of a system due to random mixing of different particles. Two particles are different if they are different substances, or if they are the same substance but in different states. For example, the entropy of a perfect crystal, at absolute zero, where all the particles are in the ground energy level, is $R \ln g_0$ per mole, where R is the gas constant, and g_0 is the degeneracy of the ground level. If $g_0 = 1$, all the particles are the same, and $S = 0$. If $g_0 = 2$, there are effectively two different kinds of particles, and S is $R \ln 2$, the entropy of mixing $\frac{1}{2}$ mole of one kind with $\frac{1}{2}$ mole of the other. The mixing has to be random to make this valid. If all the atoms in one state were on one side of the lattice, and all the atoms in the other state were on the other side, S would still be 0. Only a random mixture throughout the lattice results in $S = R \ln 2$. Any spontaneous process can be analyzed as a process resulting in an increase in random disorder of the system. For example, consider a pair of bulbs containing gas A and gas B, connected by a stopcock. When the stopcock is opened, random mixing of the molecules in both bulbs produces an increase in entropy $\Delta S = NK \ln 2$, if $N/2$ molecules were in each bulb initially.

According to the postulates of statistical mechanics, a measure of the random disorder of a system is the number *a priori* probable microstates that conserve particles and energy. If there are many such microstates, then random disorder will be large, and the system will randomly shift from microstate to microstate as time proceeds. For any particular values of E and N, the total number of microstates is

$$\Omega(E, N) = \sum_{\substack{j=\text{all} \\ \text{macrostates}}} W_j(n_1{}^j, n_2{}^j \ldots) \tag{1-39}$$

where $n_i{}^j$ is the number of particles in state i when the system is in macrostate j corresponding to a total energy E and N particles.

The statistical mechanical entropy σ varies monotonically with Ω

---

[7]Since these equations were written for constant N, Eq. (1-38) should more rigorously be written as

$$\left(\frac{\partial S}{\partial E}\right)_{V,N} = \frac{1}{T}$$

because an increase in entropy is accompanied by an increase in the random disorder of a system. For a system composed of two parts,

$$\sigma_{\text{tot}} = \sigma_1 + \sigma_2 \tag{1-40}$$

whereas

$$\Omega_{\text{tot}} = \Omega_1 \Omega_2 \tag{1-41}$$

since each microstate of part 1 can be combined with any of the microstates of part 2. Therefore, the relation between σ and Ω for a system is

$$\sigma = k \ln \Omega \tag{1-42}$$

where k is a constant to be evaluated later. In order to make this entropy equivalent to S from thermodynamics, set

$$\left(\frac{\partial \sigma}{\partial E} \right)_V = \frac{1}{T} \tag{1-43}$$

where E is the total (internal) energy of the particles. Before doing this, an assumption will be made, which will be justified in Sec. 1.7. It will be shown that if N is very large (say 10^{20} particles), $\ln \Omega$ is very nearly equal to $\ln W^*$, where W^* is W corresponding to the most probable macrostate. That is, if Eq. (1-39) is written as

$$\ln \Omega = \ln \sum_j W_j \tag{1-44}$$

then the sum can be replaced by the single term W^* for which W_j is a maximum. It is important to realize that replacing the sum by W^* in Eq. (1-44) is less stringent a requirement than if it were replaced by W^* in Eq. (1-39). The reason for this will become clear in Sec. 1.7. Write

$$\sigma^* = k \ln W^*$$

With Stirling's approximation used for the factorials, this becomes

$$\sigma^* = k \left[N \ln N - N - \sum_i n^*_i \ln n^*_i + \sum_i n^*_i \right]$$

$$\sigma^* = k \left[N \ln N - \sum_i n^*_i \ln n^*_i \right]$$

$$\sigma^* = k \left[N \ln N - \frac{N}{Q} \sum_i e^{-\beta \epsilon_i} (\ln N - \beta \epsilon_i - \ln Q) \right]$$

$$\sigma^* = k \left[N \ln N - N \ln N + \beta \sum_i n_i \epsilon_i + N \ln Q \right]$$

$$\sigma^* = k \beta E + N k \ln Q \tag{1-45}$$

In order to make the statistical mechanical entropy consistent with the laws of thermodynamics, choose

$$\left(\frac{\partial \sigma^*}{\partial E}\right)_V = \frac{1}{T} \qquad (1\text{-}46)$$

If Eq. (1-45) is differentiated with respect to E at constant volume, the following is obtained:

$$\left(\frac{\partial \sigma^*}{\partial E}\right)_V = k\beta + kE\left(\frac{\partial \beta}{\partial E}\right)_V + \frac{Nk}{Q}\left(\frac{\partial Q}{\partial \beta}\right)_V\left(\frac{\partial \beta}{\partial E}\right)_V$$

But

$$\left(\frac{\partial Q}{\partial \beta}\right)_V = \frac{\partial}{\partial \beta}\sum_i e^{-\beta\epsilon_i} = \sum_i -\epsilon_i\, e^{-\beta\epsilon_i} \qquad (1\text{-}47)$$

$$= -\frac{Q}{N}\sum_i n_i\epsilon_i = -\frac{QE}{N} \qquad (1\text{-}48)$$

Therefore,

$$\left(\frac{\partial \sigma^*}{\partial E}\right)_V = k\beta \qquad (1\text{-}49)$$

Hence,

$$\beta = \frac{1}{kT} \qquad (1\text{-}50)$$

and k is still not defined.   It will be shown later, when statistical mechanics is applied to independent nonlocalized particles, that in order to obtain the proper expression for the equation of state of an ideal gas, k must be chosen equal to the Boltzmann constant (that is, the gas constant per molecule R/N_{AV}, where N_{AV} is Avogadro's number).

With this choice of β, σ^* may be set equal to S, and statistical mechanical formulae will be in conformity with thermodynamics.

1.5   THERMODYNAMIC FUNCTIONS AND THE PARTITION FUNCTION

The thermodynamic functions of an ensemble of independent localized particles may now be evaluated in terms of the partition function.

The most probable distribution of particles among the states is

$$n_i^* = \frac{N}{Q}e^{-\epsilon_i/kT} \qquad (1\text{-}51)$$

where

$$Q = \sum_{\substack{i=\text{all} \\ \text{states}}} e^{-\epsilon_i/kT} \qquad (1\text{-}52)$$

The entropy is determined by Eqs. (1-45) and (1-50) to be

$$S = Nk\ln Q + E/T \qquad (1\text{-}53)$$

and the internal energy is

$$E = \sum_i n_i^* \epsilon_i = \frac{N}{Q} \sum_i \epsilon_i e^{-\epsilon_i/kT} \tag{1-54}$$

$$E = \frac{N}{Q} kT^2 \frac{dQ}{dT} = NkT^2 \frac{d \ln Q}{dT} \tag{1-55}$$

This expression[8] can be substituted for E in Eq. (1-53) to obtain S entirely as a function of Q and T. All the other thermodynamic functions can be expressed in terms of S and E, and therefore they may be expressed in terms of Q and T. For example, the Helmholtz free energy is

$$A = E - TS = -NkT \ln Q \tag{1-56}$$

Since the partition function depends only on the energy levels of the substance, the thermodynamic properties are determined in terms of these levels and the temperature.

1.6 SYSTEMS WITH ONLY TWO QUANTUM STATES

In this section, the formulas of preceding sections will be illustrated in terms of particles with only two quantum states. Although this is mainly an academic problem, it does clearly illustrate the behavior of more complicated systems. There are a few cases involving particles with spin in magnetic fields for which there are only two quantum states per particle, in which case this section has practical as well as academic significance.

As a first example, suppose each particle has two states 1 and 2, with equal energies (that is, $\epsilon_1 = \epsilon_2 = $ "ϵ") where ϵ is a constant. First calculate the partition function

$$Q = \sum_{i=1}^{2} e^{-\epsilon_i/kT} = e^{-\epsilon/kT} + e^{-\epsilon/kT} = 2e^{-\epsilon/kT}$$

The distribution functions are

$$n_1^* = n_2^* = \frac{N}{Q} e^{-\epsilon/kT} = \frac{N}{2}$$

and the particles in the lattice are 50% in state 1, and 50% in state 2, at all temperatures. This is a special case of a more general result, that all

---

[8]Since the system is held at constant volume and the number of particles is implicitly constant, a more proper expression is

$$E = NkT^2 \left(\frac{\partial \ln Q}{\partial T} \right)_{V,N}$$

quantum states with the same energy are equally populated. The energy is

$$E = \sum_{i=1}^{2} n_i^* \epsilon_i = \frac{N}{2}\epsilon + \frac{N}{2}\epsilon = N\epsilon$$

corresponding to N particles, each with energy ϵ. The entropy is

$$S = Nk \ln Q + E/T = Nk\left[\ln 2 - \frac{\epsilon}{kT}\right] + \frac{N\epsilon}{T}$$

$$S = Nk \ln 2$$

which is the entropy of randomly mixing $N/2$ particles in state 1 with $N/2$ particles in state 2. A lattice of $N/2$ particles in state 1 has $S = 0$ because $W = 1$. It is only on forming a random mixture of particles in different states that S becomes nonzero.

Next, consider the more interesting case where the two states have different energies:

$$
\begin{array}{l}
\text{———————} \epsilon_2 = \text{``}\epsilon\text{''} \\
\uparrow \\
\epsilon \\
\downarrow \\
\text{———————} \epsilon_1 = 0
\end{array}
$$

Let $\theta = \epsilon/k$, which has the dimensions of temperature. The partition function for a particle with these two states is

$$Q = \sum_{i=1}^{2} e^{-\epsilon_i/kT} = 1 + e^{-\epsilon/kT} = 1 + e^{-\theta/T}$$

The numbers of particles in each of the states is

$$n_1 = \frac{N}{Q}e^0 = \frac{N}{Q} = \frac{N}{1 + e^{-\theta/T}}$$

$$n_2 = \frac{Ne^{-\theta/T}}{1 + e^{-\theta/T}}$$

In the limit of low T, if $T \ll \theta$, $e^{-\theta/T} \cong 0$, and $n_1 \cong N$ and $n_2 \cong 0$. At very high T, $n_1 \cong N/2$ and $n_2 \cong N/2$. For T comparable to θ, n_1 and n_2 vary between these limits. In Figs. 1-2 and 1-3, the partition function and n_1 and n_2 are plotted vs T. It can be seen that at low T, all the particles go into the lowest energy level. At very high T, all energy levels are equally populated. This result holds more generally for systems with many energy levels. The partition function is a measure of the effective number of states populated if the ground energy level is denoted as the 0 of energy. Thus, Q varies from 1 at low T to 2 at high T.

The energy of the substance is

$$E = n_1\epsilon_1 + n_2\epsilon_2 = 0 + \frac{N\epsilon\, e^{-\theta/T}}{1 + e^{-\theta/T}}$$

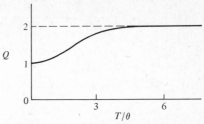

FIGURE 1-2 Partition function of a system composed of particles with only two energy levels. The quantity $\theta = \epsilon/k$, where ϵ is the spacing between levels.

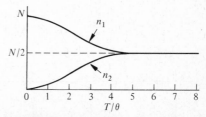

FIGURE 1-3 Number of particles in each state of a system composed of particles with only two energy levels. The quantity $\theta = \epsilon/k$, where ϵ is the spacing between levels.

and $E \to N\epsilon/2$ as $T \to \infty$, and $E \to 0$ as $T \to 0$. The entropy is

$$S = Nk \ln Q + E/T = Nk \left[\ln (1 + e^{-\theta/T}) + \frac{\theta/T}{e^{\theta/T} + 1} \right]$$

As $T \to 0$, $S \to 0$, and as T becomes large, $S \to Nk[\ln Q + \theta/2T] \to Nk \ln 2$. Thus, the entropy of the substance may be regarded as the entropy of mixing of n_1 particles in state 1 with n_2 particles in state 2. For a system with a total of g quantum states, the entropy approaches $Nk \ln g$ as $T \to \infty$. If the degeneracy of the lowest energy level is g_0, the entropy at $T = 0$ is $Nk \ln g_0$. The entropy and energy are plotted in Figs. 1-4 and 1-5. The specific heat is simply the slope of the energy curve [that is, $c_v = (\partial E/\partial T)_v$]. The specific heat is plotted in Fig. 1-6. The specific heat approaches 0 for $T \gg \theta$ because there are no new energy levels to take up additional energy, and a very large change in T will produce a very small change in the energy.

PROBLEM: Construct qualitative plots of Q, n_1, n_2, n_3, S, E and c_v vs T/θ for the following systems. (Note: In this case n_1, n_2, and n_3 refer to the numbers of particles in energy levels 1, 2, and 3, not quantum states.) Let $\theta = \epsilon/k$.

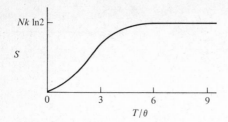

FIGURE 1-4 Entropy of a system composed of particles with only two energy levels. The quantity $\theta = \epsilon/k$, where ϵ is the spacing between levels.

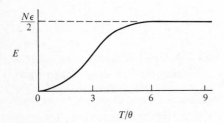

FIGURE 1-5 Energy of a system composed of particles with only two energy levels. The quantity $\theta = \epsilon/k$, where ϵ is the spacing between levels.

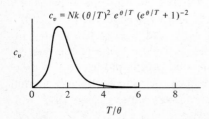

$$c_v = Nk \, (\theta/T)^2 \, e^{\theta/T} \, (e^{\theta/T} + 1)^{-2}$$

FIGURE 1-6 Specific heat of a system composed of particles with only two energy levels. The quantity $\theta = \epsilon/k$, where ϵ is the spacing between the levels.

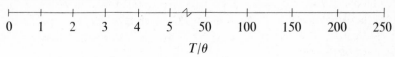

Use as a horizontal scale:

$$T/\theta$$

1.7   JUSTIFICATION OF THE REPLACEMENT OF ln Ω BY ln W^*

In the previous sections, it has been assumed that

$$\ln \Omega \cong \ln W^* \qquad (1\text{-}57)$$

so that the entropy corresponding to the most probable macrostate can be used for the entropy of a system. This implies that the most probable macrostate is so highly probable that it overwhelms all other macrostates in taking

$$\ln \Omega = \ln \left\{ \sum_{\substack{j=\text{all} \\ \text{macrostates}}} W_j \right\} \qquad (1\text{-}58)$$

This assumption will be examined for the special case where each particle has two quantum states of equal energy. According to Sec. 1.6,

$$n_1{}^* = n_2{}^* = N/2$$

Therefore,

$$W^* = \frac{N!}{n_1{}^*! n_2{}^*!} = \frac{N!}{[(N/2)!]^2} \qquad (1\text{-}59)$$

According to Stirling's approximation,

$$\ln N! \cong (N + \tfrac{1}{2}) \ln N - N + \tfrac{1}{2} \ln (2\pi) + O(1/N) \qquad (1\text{-}60)$$

where $O(1/N)$ means terms of the order of $1/N$. Then

$$\ln W^* \cong N \ln 2 + \tfrac{1}{2} \ln (2/\pi N) \qquad (1\text{-}61)$$

If $N \cong 10^{20}$, the second term is ≈ -20, while the first term is $\approx 10^{20}$, so

$$\ln W^* \cong N \ln 2 \qquad (1\text{-}62)$$

However, it is improper to write

$$W^* \cong 2^N \tag{1-63}$$

because according to Eq. (1-61),

$$W^* = 2^N \cdot \left[\frac{2}{(2\pi N)^{1/2}} \right] \tag{1-64}$$

The term in brackets in Eq. (1-64) is $\approx 10^{-10}$. Thus, despite the fact that Eq. (1-62) is good to about 1 part in 10^{19}, Eq. (1-63) is incorrect by a direct factor of 10^{10}. The reason for this is simply that N appears as an exponent in 2^N, and therefore 2^N is so much larger than $N^{1/2}$ that $\ln (2^N)$ is much larger than $\ln (N^{1/2})$.

To calculate Ω, we merely note that each particle can be in either of two states, and each state of one particle can be combined with each state of the others. Therefore, the total number of microstates is

$$\Omega = \underbrace{2 \cdot 2 \cdot 2 \cdot 2 \cdots 2}_{N\text{-fold}} = 2^N \tag{1-65}$$

It is evident from Eqs. (1-62), (1-64), and (1-65), that although $W^* \neq \Omega$, nevertheless

$$\ln \Omega \cong \ln W^* \tag{1-66}$$

According to Eqs. (1-61) and (1-65),

$$\ln \Omega - \ln W^* = \ln [(2\pi N)^{1/2}/2] \tag{1-67}$$

Since $\ln \Omega$ is of the order of 10^{20} and the term on the right side of Eq. (1-67) is ≈ 20, it is evident that Eq. (1-66) is good to about 1 part in 10^{19}. Thus, the use of $\ln W^*$ in place of $\ln \Omega$ is justified by this illustration.

Next, $W(n_1, n_2)$ is calculated for $n_1 \neq n_1^*$. Assume that n_i is almost (but not quite) equal to n_i^*. Thus, write

$$n_1 = \frac{N}{2} + x \qquad n_2 = \frac{N}{2} - x$$

with $N \gg x$. Then, for this macrostate,

$$W = \frac{N!}{n_1! n_2!}$$

and using Stirling's approximation,

$$\ln W \cong N \ln N - N - \left(\frac{N}{2} + x \right) \ln \left(\frac{N}{2} + x \right)$$
$$+ \left(\frac{N}{2} + x \right) - \left(\frac{N}{2} - x \right) \ln \left(\frac{N}{2} - x \right) + \left(\frac{N}{2} - x \right) + \frac{1}{2} \ln (2/\pi N)$$

$$\ln W \cong N \ln N - \left(\frac{N}{2}+x\right) \ln \left(\frac{N}{2}+x\right) - \left(\frac{N}{2}-x\right) \ln \left(\frac{N}{2}-x\right) + \frac{1}{2} \ln (2/\pi N)$$

But

$$\ln \left(\frac{N}{2}\pm x\right) = \ln \left[\frac{N}{2}\left(1\pm\frac{2x}{N}\right)\right] = \ln \frac{N}{2} + \ln \left(1\pm\frac{2x}{N}\right) \cong \ln \frac{N}{2} \pm \frac{2x}{N}$$

if $N \gg x$. Thus,

$$\ln W \cong N \ln N - \left(\frac{N}{2}+x\right)\left(\ln \frac{N}{2}+\frac{2x}{N}\right)$$

$$- \left(\frac{N}{2}-x\right)\left(\ln \frac{N}{2}-\frac{2x}{N}\right) + \frac{1}{2} \ln (2/\pi N)$$

$$\ln W \cong N \ln 2 - \frac{4x^2}{N} + \frac{1}{2} \ln (2/\pi N)$$

$$W \cong W^* \, e^{-4x^2/N}$$

A plot of W vs x is given in Fig. 1-7. When $x = \sqrt{N}/2$, $W = e^{-1}W^*$. As x increases beyond this value, W rapidly decreases. A typical value of N is 4×10^{20} particles, in which case $\sqrt{N}/2 = 10^{10}$. The macrostate for which W is e^{-1} of W^* has $x = 10^{10}$ (or in other words n_1 is $2 \times 10^{20} + 10^{10}$). When x is as large as 10^{15}, $W = W^* \exp(-10^{10})$. Therefore, the thermodynamic degeneracies of macrostates with n_i substantially different than n_i^*, are very small.

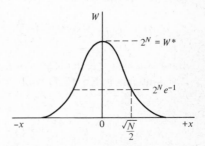

FIGURE 1-7 The number of ways, W, to arrange N particles in two degenerate states with $N/2+x$ in one state, and $N/2-x$ in the other state.

1.8 THE EINSTEIN TREATMENT OF THE SPECIFIC HEAT OF A METAL

Metals form crystals consisting of metal atoms in a regular three-dimensional lattice. Each atom is bound to its equilibrium position by forces from neighboring atoms. The further an atom is displaced from equilibrium, the greater is the restoring force tending to bring it back. In 1905, Einstein proposed a simple model of such a substance which gives most of the essence of the physical reality. Each particle is treated as a three-dimensional isotropic harmonic oscillator, uncoupled from the other particles. According to quantum mechanics, the energy levels of a three-dimensional isotropic oscillator are the same as for three one-dimensional oscillators. Therefore, $3N$ one-dimensional oscillators will be dealt with. If the crystal is uniform (neglect surface effects), the energy levels of the oscillators are

$$\epsilon_j = (j+\tfrac{1}{2})h\nu = (j+\tfrac{1}{2})\hbar\omega \qquad (1\text{-}68)$$

where j is the vibrational quantum number, ν is the frequency given by

$$\nu = (2\pi)^{-1}(f/m)^{1/2} \qquad (1\text{-}69)$$

f is the force constant, m is the particle mass, and $\omega = 2\pi\nu$. Each energy level is singly degenerate. A particle can be in any of the levels $n = 0, 1, 2, \ldots$.

To calculate the specific heat of this substance, it is necessary to evaluate E and take $c_v = (\partial E/\partial T)_V$. E will be evaluated by taking

$$E = \sum_{j=0}^{\infty} n_j^* \epsilon_j \qquad (1\text{-}70)$$

First, the partition function is calculated.

$$Q = \sum_{j=0}^{\infty} e^{-\epsilon_j/kT} = \sum_{j=0}^{\infty} e^{-(j+1/2)h\nu/kT} = e^{-\theta/2T} \sum_{j=0}^{\infty} e^{-j\theta/T} \qquad (1\text{-}71)$$

where $\theta = h\nu/k$ has the dimensions of temperature.

The sum in Eq. (1-71) may be expanded to

$$Q = e^{-\theta/2T}\{1 + e^{-\theta/T} + e^{-2\theta/T} + \cdots\} \qquad (1\text{-}72)$$

$$Q = e^{-\theta/2T}\{1 + u + u^2 + \cdots\} \qquad (1\text{-}73)$$

where $u = e^{-\theta/T}$. The sum in braces in Eq. (1-73) is a geometric progression and it is equal to

$$1 + u + u^2 + \cdots = \frac{1}{1-u} \qquad (1\text{-}74)$$

Thus,

$$Q = \frac{e^{-\theta/2T}}{1 - e^{-\theta/T}} \qquad (1\text{-}75)$$

Since there are an infinite number of energy levels for each oscillator, the partition function increases without limit as T increases. The number of oscillators in the jth energy level is

$$n_j = \frac{3N}{Q} e^{-\theta/2T} e^{-j\theta/T} \tag{1-76}$$

where $3N$ is the total number of oscillators.   As j increases, n_j decreases. However, the rate of fall-off of n_j decreases at higher T.   The energy of the lattice is

$$E = \sum_{j=0}^{\infty} n_j \epsilon_j = \sum_{j=0}^{\infty} \left(\frac{3N e^{-\theta/2T}}{Q} \right) e^{-j\theta/T} \left([j+\tfrac{1}{2}] h\nu \right)$$

$$E = \frac{3N e^{-\theta/2T}}{Q} h\nu \left\{ \sum_{j=0}^{\infty} j e^{-j\theta/T} + \frac{1}{2} \sum_{j=0}^{\infty} e^{-j\theta/T} \right\} \tag{1-77}$$

The first sum in Eq. (1-77) can be evaluated in the following way.   The partial of $\sum e^{-j\theta/T}$ with respect to T is

$$\frac{\partial}{\partial T} \left\{ \sum_{j=0}^{\infty} e^{-j\theta/T} \right\} = \frac{\theta}{T^2} \sum_{j=0}^{\infty} j e^{-j\theta/T}$$

Therefore,

$$\text{sum} = \sum_{j=0}^{\infty} j e^{-j\theta/T} = \frac{T^2}{\theta} \frac{\partial}{\partial T} \left(\sum e^{-j\theta/T} \right) \tag{1-78}$$

The sum that is differentiated in Eq. (1-78) has been evaluated previously when Q was calculated.   Thus,

$$\text{sum} = \frac{T^2}{\theta} \frac{\partial}{\partial T} \left(\frac{1}{1-e^{-\theta/T}} \right) = \frac{e^{-\theta/T}}{(1-e^{-\theta/T})^2} \tag{1-79}$$

The second sum in Eq. (1-77) is, as before, $(1-e^{-\theta/T})^{-1}$, so

$$E = \frac{3Nh\nu}{2} + \frac{3Nh\nu \, e^{-\theta/T}}{1-e^{-\theta/T}} = 3Nh\nu \left\{ \frac{1}{2} + \frac{1}{e^{\theta/T}-1} \right\} \tag{1-80}$$

The first term in Eq. (1-80) represents the zero point energy $3N$ oscillators would have if each oscillator was in the ground state.   The second term represents the extra energy due to oscillators in excited states.   A plot of E vs T is given in Fig. 1-8.   At high T, such that $\theta/T \ll 1$, the denominator in Eq. (1-80) can be expanded in a power series

$$1 - e^{-\theta/T} \cong 1 - \left(1 - \frac{\theta}{T} + \frac{\theta^2}{T^2} + \cdots \right) = \frac{\theta}{T} + \cdots \tag{1-81}$$

Thus, Eq. (1-80) becomes

$$E \cong 3NkT$$

at high T.   In this, $e^{-\theta/T}$ has been set equal to 1, and the first term in Eq. (1-80) has been neglected compared to the second term.

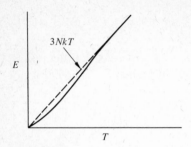

FIGURE 1-8   Energy of a collection of $3N$ harmonic oscillators according to the Einstein model.

The specific heat is

$$c_v = \left(\frac{\partial E}{\partial T}\right)_V = \frac{3Nk(\theta/T)^2\, e^{\theta/T}}{(e^{\theta/T}-1)^2} \tag{1-82}$$

A plot of c_v vs T is given in Fig. 1-9.   It is seen that c_v goes to zero as $T \to 0$, and approaches $3Nk$ as $T \to \infty$.   For 1 mole of material, N is Avogadro's number, and $Nk = R$.

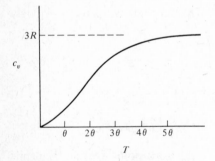

FIGURE 1-9   Specific heat of a crystalline solid according to the Einstein model vs T/θ where $\theta \approx h\nu/k$ and ν is the frequency of vibration of a particle in the lattice.

According to Eq. (1-69), the vibrational frequencies of metals tend to vary inversely with mass.   The effect is to make θ vary inversely with mass.   Therefore, a plot of c_v vs T for various metals is as shown in Fig. 1-10.

The entropy of a metal, according to this model, is

$$S = 3Nk \ln Q + E/T = 3Nk \ln\left(\frac{e^{-\theta/2T}}{1-e^{-\theta/T}}\right) + \frac{3Nk\theta}{2T} + \frac{3Nk\theta}{T(e^{\theta/T}-1)}$$

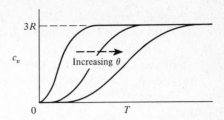

FIGURE 1-10 Specific heat functions according to the Einstein model for various solids with differing values of θ.

As $T \rightarrow 0$,

$$S \rightarrow 3Nk \ln (e^{-\theta/2T}) + \frac{3Nk\theta}{2T} + 0$$

and this is

$$-3Nk\theta/2T + 3Nk\theta/2T = 0$$

PROBLEM: Plot S vs T with T in units of θ.

Finally, a remark should be made on the distribution of oscillators among the various states.   The ratio n_j/n_k is

$$\frac{n_j}{n_k} = e^{-(j-k)\theta/T}$$

If $T \gg \theta$, this ratio approaches unity.   If $T \ll \theta$, the ratio is large if $j < k$, and small if $j > k$.   The lower states tend to be equally populated if $T \gg \theta$, and the population falls off rapidly if $T \ll \theta$.

The Einstein model gives a good semi-quantitative description of the energy and specific heat of a solid, and was an important step in the development of quantum theory.   Before Einstein's paper in 1905, there was no way to explain why c_v decreased at low T from the result $3Nk$ predicted by classical mechanics.   However, the Einstein model has defects, because the expression for c_v can only be approximately fitted to c_v data for substances over wide ranges of temperature.   To build a critique on the Einstein model, it is necessary to go back to the model itself.   It is assumed that the particles of the crystal lattice are bound by harmonic restoring forces to their equilibrium positions, and that interactions between neighboring particles can be neglected.   At the same time, it is assumed that very weak interactions do in fact occur, causing interchange of energy between the various lattice sites so that equilibrium is eventually established.   The problem is that the idea of each atom being bound to its equilibrium position by a mysterious restoring force is a fiction.   The only forces that actually act on an atom in the lattice are the forces due to the interactions with neighboring atoms. The atoms tend to resist stretching and compression of the inter-atomic distances, and this is the source of the harmonic restoring force used in the Einstein calculation. Since this force is due to interactions with neighboring atoms, it could hardly be correct to treat the atoms as independent.

The next section deals with the interactions between atoms in a lattice.

1.9   THE DEBYE MODEL FOR THE SPECIFIC HEAT OF A CRYSTAL

The forces that act on a particle in a crystal lattice are due to the adjacent particles pushing or pulling in order to strive to keep the inter-atomic distances as close to equilibrium as possible. At a temperature of absolute zero, all the particles would be regularly ordered with only the zero-point motion remaining. At moderate temperature, the thermal energy of the particles enables them to partially overcome the interaction forces and to undergo various kinds of vibrations. Consider the elementary one-dimensional model of a crystal lattice shown in Fig. 1-11.

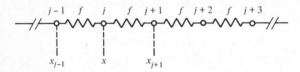

FIGURE 1-11 One-dimensional crystal lattice modeled by particles coupled by springs.

Each particle is represented as a mass point, and the interactions are represented as springs with force constant f connecting the particles. The equilibrium position of the jth particle is denoted as x_j^0, and the deviation from equilibrium is $\xi_j = x_j - x_j^0$. For a lattice of N particles, the force on a typical particle j (except for particles 1 and N), exerted by particle $j+1$, is $f(\xi_{j+1} - \xi_j)$ in the direction to the right. The force on particle j to the left due to particle $j-1$ is $f(\xi_j - \xi_{j-1})$. Thus, the net force to the right on particle j is

$$f(\xi_{j+1} - 2\xi_j + \xi_{j-1})$$

The classical equation of motion for this particle is

$$m\ddot{\xi}_j = f(\xi_{j+1} - 2\xi_j + \xi_{j-1}) \tag{1-83}$$

where m is the particle mass, and $\ddot{\xi}_j$ is $\partial^2 \xi_j / \partial t^2$. The atoms at the ends of the chain are special cases. In a real crystal lattice they are simply cantilevered out. However, they are much easier to deal with in this model if it is assumed that the end atoms are bound by springs to fixed walls, as shown in Fig. 1-12. The equations of motion for the end atoms are then the same as Eq. (1-83) if we set $\xi_0 = \xi_{N+1} = 0$. Equations (1-83) are a set of coupled differential equations in $N+2$ coordinates. (i varies from 0 to $N+1$) with two coordinates (ξ_0 and ξ_{N+1}) being trivially zero.

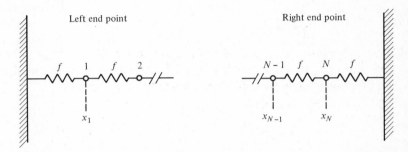

FIGURE 1-12   End points of one-dimensional lattice.

There are, therefore, only N actual physically significant coordinates. A system of N particles coupled by springs has N normal modes. A normal mode is a pattern of motion that repeats at regular intervals, in which the position of each particle varies sinusoidally with time. Normal modes are determined by seeking solutions of Eq. (1-83) which vary sinusoidally with time. Temporarily assume that it is possible to write the position of particle j in normal mode n as

$$\xi_{jn} = \sin(jk_n) \sin(\omega_n t + \alpha_n) \tag{1-84}$$

where k_n is a quantity to be determined, and ω_n is the angular frequency of the nth normal mode. The phase factor α_n merely defines the zero of time. It is clear that this function repeats at regular intervals with period $\tau = 2\pi/\omega_n$. The value of k_n is found from the boundary conditions. For $j = 0$, $\xi_{0n} = 0$, so the left boundary condition is automatically satisfied. The right boundary condition is $\xi_{N+1,n} = 0$, which requires that

$$\sin[(N+1)k_n] = 0 \tag{1-85}$$

Therefore,

$$(N+1)k_n = n\pi \qquad (n = 1, 2, 3, \ldots N) \tag{1-86}$$

If $n > N$, $\sin(jk_n)$ is either zero or the same as a previous solution. For example, for $n = N+1$, $\sin(jk_n) = \sin j\pi = 0$. For $n = N+2$, $\sin(jk_n) = \sin[j\pi + j\pi/(N+1)] = \cos(jn) \sin[j\pi/(N+1)] = \pm\sin(jk_1)$. Thus, there are only N distinct sets of function ξ_{jn} for $n = 1, 2, \ldots N$.

Now verify that Eq. (1-84) is a solution of Eq. (1-83) by direct substitution.   It is found that

$$\omega_n^2 = -\frac{f}{m}\left\{\frac{\sin[(j+1)k_n] - 2\sin[jk_n] + \sin[(j-1)k_n]}{\sin[jk_n]}\right\}$$

$$\omega_n^2 = \frac{2f}{m}\{1 - \cos(k_n)\} = \frac{4f}{m}\sin^2\left[\frac{n\pi}{2(N+1)}\right] \tag{1-87}$$

$$\omega_n = 2\sqrt{\frac{f}{m}}\sin\left[\frac{n\pi}{2(N+1)}\right] \tag{1-88}$$

For such values of ω_n, Eq. (1-84) is a solution of Eq. (1-83).   A qualitative plot of ω_n vs n is shown in Fig. 1-13.   If N is a very large number, the distribution of ω_n in n can be approximated by a continuum.

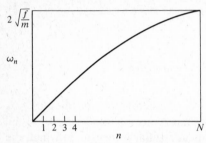

FIGURE 1-13 Radial frequency ω_n of the nth mode of vibration of the one-dimensional lattice.

Thus, write

$$\omega(n) = 2\omega_0\sin\left[\frac{n\pi}{2(N+1)}\right] \tag{1-89}$$

where $\omega_0 = (f/m)^{1/2}$, and n and ω are treated as continuous.   Then, solving for n, it is found

$$n = \frac{2}{\pi}(N+1)\arcsin\left(\frac{\omega}{2\omega_0}\right) \tag{1-90}$$

This is differentiated with respect to ω to obtain

$$\frac{dn}{d\omega} = \frac{(N+1)}{\pi\omega_0}\frac{1}{\sqrt{1-(\omega/2\omega_0)^2}} \tag{1-91}$$

as the number of frequencies per unit range of frequency. This is plotted in Fig. 1-14 as the "discrete model," because it is based on the treatment of discrete particles connected by springs.   It can be seen that there is a much higher density of high frequencies than low frequencies.   The maximum frequency is

$$\omega_{max} = \omega(N) = 2\omega_0\sin\left[\frac{N\pi}{2(N+1)}\right] \tag{1-92}$$

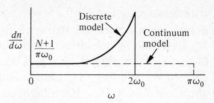

FIGURE 1-14 Distribution of modes vs frequency for the one-dimensional lattice. The discrete model involves a set of point particles coupled by springs, whereas the continuum model uses a continuous elastic medium.

which is nearly equal to $2\omega_0$ for large values of N. Note that integration of Eq. (1-91) from $\omega = 0$ to $\omega = \omega_{max}$ yields the total number of modes as

$$\int dn = \frac{N+1}{\pi\omega_0} \int_0^{\omega_{max}} \frac{d\omega}{\sqrt{1-(\omega/2\omega_0)^2}} = N+1 \cong N$$

The normal modes can be identified by substituting the appropriate values of k_n into Eq. (1-84). Neglecting the time dependence, the amplitude of particle j in normal mode n is

$$A_{jn} = \sin\left(\frac{jn\pi}{(N+1)}\right) \tag{1-93}$$

As a simple illustration, consider the case where $N = 4$. Then, the normal modes are found from Eq. (1-93) as illustrated below:

(a) $n = 1$

$$A_{11} = \sin\frac{\pi}{5} = 0.59$$

$$A_{21} = \sin\frac{2\pi}{5} = 0.95$$

$$A_{31} = \sin\frac{3\pi}{5} = 0.95$$

$$A_{41} = \sin\frac{4\pi}{5} = 0.59$$

(b) $n = 4$

$$A_{14} = \sin\frac{4\pi}{5} = 0.59$$

$$A_{24} = \sin\frac{8\pi}{5} = -0.95$$

$$A_{34} = \sin\frac{12\pi}{5} = 0.95$$

$$A_{44} = \sin\frac{16\pi}{5} = -0.59$$

with analogous results for $n = 2$ and $n = 3$. Each normal mode can be drawn by attaching an arrow of length $|A_{jn}|$ to atom j, with the sign of A_{jn} determining the direction of the motion. Thus, the normal modes have the form shown in Fig. 1-15. These motions can be seen more clearly by plotting A_{jn} perpendicular to the direction of motion, as shown in Fig. 1-16. Each higher mode involves the addition of another half wave length to the motion within fixed boundaries. The higher numbered modes have higher frequencies because they involve more severe perturbations to the bonds since adjacent atoms tend to move in opposite directions. In the lower frequencies, the atoms tend to move together as a unit without large changes in the *relative* positions.

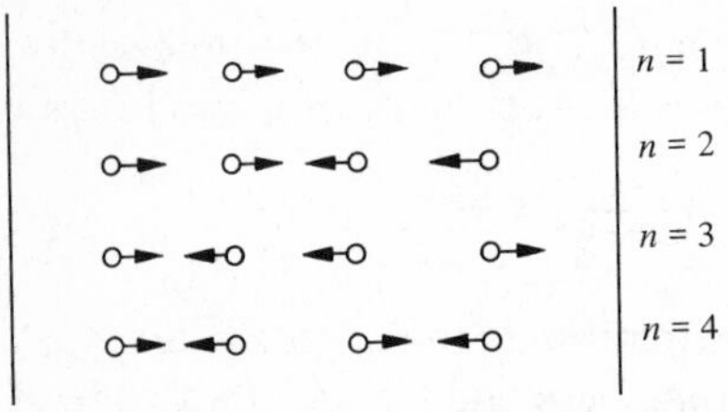

FIGURE 1-15 Normal modes of longitudinal vibration of four particles coupled by springs in a one-dimensional array.

An approximate way of treating such a one-dimensional solid is to treat the material as a continuum that is capable of elastic vibrations. The N atoms connected by springs are replaced by a hypothetical continuous elastic band of length L. The equilibrium bond distance is denoted as r, and from Figs. 1-11, 1-12, and 1-13, it follows that

$$(N+1)r = L \tag{1-94}$$

For a continuous elastic medium, one may replace Eq. (1-83) by the formula

$$m\ddot{\xi} = fr^2 \frac{\partial^2 \xi}{\partial x^2} \tag{1-95}$$

where x is the distance along the band measured from the left end. The solutions to this equation have the form

$$\xi_n(x, t) = \sin\left(\frac{n\pi x}{L}\right) \sin\left(\omega_n t + \alpha_n\right) \tag{1-96}$$

analogous to Eq. (1-84). The frequency of the nth normal mode is

$$\omega_n = \frac{\pi n}{L}\left(\frac{fr^2}{m}\right)^{1/2} = \frac{n\pi}{N+1}\left(\frac{f}{m}\right)^{1/2} \tag{1-97}$$

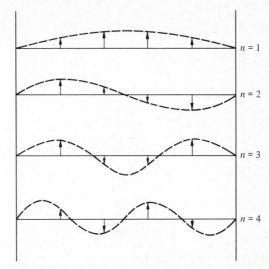

FIGURE 1-16 Normal modes of four linear coupled particles with the displacements plotted perpendicular to the actual directions of motion so as to better illustrate the wave forms.

This is clearly the same as Eq. (1-88) if $n \ll N + 1$. Therefore, the continuum model agrees with the particulate model only for the lower frequency normal modes which involve group movement of large numbers of atoms. For high frequencies, involving relative motion of adjacent atoms, the continuum picture fails badly. A continuum has an infinite number of normal modes. Therefore, in order to bring this model into greater agreement with the particle model, one must arbitrarily restrict the number of normal modes to the first N modes. The variation of frequency with n derived from Eq. (1-97) is plotted as a dashed line in Fig. 1-17, along with the correct function Eq. (1-88). The continuum model only works well for small n. The distribution of normal modes per unit frequency range is

$$\frac{dn}{d\omega} = \frac{N + 1}{\pi \omega_0} = \text{const} \qquad (1\text{-}98)$$

in the continuum approximation. This could have been obtained from Eq. (1-91) by assuming $\omega_0 \gg \omega$. The frequency distribution of the continuum model is compared with the discrete atom model in Fig. 1-14. In order to keep the total number of modes equal to N in the continuum model, put

$$\int_0^{\omega \max} \frac{dn}{d\omega} \, d\omega = N$$

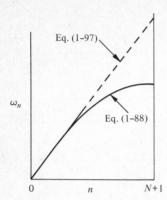

FIGURE 1-17 Angular frequency of the normal modes of a one-dimensional lattice for the discrete model (solid line) and continuum model (dashed line).

Thus,

$$\omega_{max} = \frac{\pi \omega_0 N}{N+1} \qquad (1\text{-}99)$$

in the continuum model. This is considerably larger than Eq. (1-92) for the discrete model, where $\omega_{max} \cong 2\omega_0$. The extra range of frequencies is needed because there has to be the same total number of modes, namely, N, in each case. The continuum model is not necessary to treat a one-dimensional solid. However, the continuum model greatly simplifies the treatment of two- and three-dimensional lattices. Although the continuum model gives a poor description of the high frequencies, it adequately represents low frequencies, which exert the major influence on the energy and specific heat at low temperature.

The vibrations of a three-dimensional isotropic solid are not easy to analyze. We shall not deal with this subject in this book, except to mention a few results that come from a more complete treatment.[9] In the continuum model of a three-dimensional solid, one considers a small block of material in the solid, and applies the classical laws of motion to that block. An equation analogous to Eq. (1-95) is found for the displacement in each direction (x, y, or z). These equations are somewhat more complicated than Eq. (1-95). Just as the solutions to Eq. (1-95) can be represented as wave forms in coordinate x along the one-dimensional solid, it turns out that the solutions to the three-dimensional equations

---

[9] J. C. Slater and N. H. Frank, *Mechanics* (New York: McGraw-Hill Book Co., 1947), Chapt. XII.

are also displacements which vary with position in the solid as wave forms. The displacements in the x direction can be propagated through the solid in the x, y, and z directions. The latter two are called transverse waves, and correspond to motion like that shown in Fig. 1-16. Propagation of x-displacements in the x-direction constitute longitudinal waves, as illustrated in Fig. 1-15. Each solution of the three dimensional problem can be put into the form

$$\xi(x, y, x, t) = \sin\left(\frac{n_x \pi x}{L}\right) \sin\left(\frac{n_y \pi y}{L}\right) \sin\left(\frac{n_z \pi z}{L}\right) \sin(\omega_n t + \alpha_n)$$

for a cubical solid of side L, where n_x, n_y, and n_z are integers, and ξ is the displacement of a point in the solid from its equilibrium position of x, y, z at time t. This equation is analogous to Eq. (1-96) in the one-dimensional case. The standing waves in Eq. (1-100) may be thought of as resulting from superpositions of traveling waves moving in the direction $\mathbf{k}$ where $\mathbf{k}$ has components

$$k_x = \frac{n_x \pi}{L}, \qquad k_y = \frac{n_y \pi}{L} \quad \text{and} \quad k_z = \frac{n_z \pi}{L} \qquad (1\text{-}101)$$

The displacement ξ in Eq. (1-100) can either be along the $\mathbf{k}$ direction (longitudinal wave), or in two directions perpendicular to $\mathbf{k}$ (transverse waves). The effective force constant [f in Eq. (1-95)] is larger for longitudinal displacements than for transverse displacements. For either kind of wave, the frequency is determined by the relation

$$\omega(n_x, n_y, n_z) = \frac{\pi}{L}\left(\frac{fr^2}{m}\right)^{1/2} (n_x^2 + n_y^2 + n_z^2)^{1/2} \qquad (1\text{-}102)$$

Note that this is the same as Eq. (1-97) for the one-dimensional solid with $(n_x^2 + n_y^2 + n_z^2)^{1/2}$ replacing n. The bond distance is r, as before. However, f is different in the transverse and longitudinal modes. For either kind of mode, the different frequencies may be displayed by plotting a point for each combination of the integers n_x, n_y, and n_z on cartesian axes. Consider the plot given in Fig. 1-18. Since each interior point corresponds to a mode, and each point is part of 8 cubes with 8 points per cube, the density of modes per unit volume is one mode per unit volume. For large values of n_x, n_y, and n_z the integers may be regarded as varying continuously. The number of modes with frequency between ω and $\omega + d\omega$ is the volume of the spherical shell $4\pi n^2 \, dn$ over one octant, where

$$n = (n_x^2 + n_y^2 + n_z^2)^{1/2}$$

is a radius from the origin. Thus, if Eq. (1-102) is rewritten

$$n = \gamma \omega$$

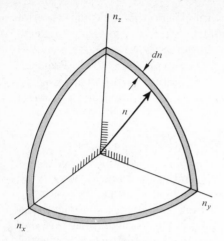

FIGURE 1-18 Distribution of modes of vibration for a three-dimensional lattice. A mode exists for each set of integer values of n_x, n_y, and n_z greater than unity. For large n_x, n_y, and n_z, these quantities may be treated as continuous, and the number of modes in a shell of thickness dn at a radial distance n from the origin may be calculated.

where $\gamma = (L/\pi)(m/fr^2)^{1/2}$, the number of modes of one type in the interval ω, $\omega + d\omega$ is

$$dg(\omega) = \tfrac{1}{2}\pi\gamma^3\omega^2\, d\omega \qquad (1\text{-}103)$$

This holds for either kind of mode. Since there are three modes per set of n_x, n_y, n_z, we may write $\gamma = \gamma_T$ for transverse waves, and $\gamma = \gamma_L$ for longitudinal waves, and the total number of modes in the interval ω, $\omega + d\omega$ is

$$dg(\omega) = \tfrac{1}{2}\pi(2\gamma_T^3 + \gamma_L^3)\omega^2\, d\omega \qquad (1\text{-}104)$$

As we have emphasized, the continuum model is merely an approximation to the discrete atom model for low frequencies. A solid containing N atoms has $3N$ different normal modes of vibration. Therefore, one must arbitrarily cut off the distribution of frequencies at an upper limit, ω_m, such that

$$\int_0^{\omega_m} dg(\omega) = 3N \qquad (1\text{-}105)$$

From Eqs. (1-104) and (1-105), it follows that

$$\omega_m^3 = \frac{6}{\pi}\frac{3N}{(2\gamma_T^3 + \gamma_L^3)} \qquad (1\text{-}106)$$

so that Eq. (1-104) becomes

$$dg(\omega) = 9N \frac{\omega^2 \, d\omega}{\omega_m^3} \tag{1-107}$$

As in the case of the one-dimensional crystal, a more rigorous discrete model will give a frequency distribution that will agree with this function only at frequencies ω considerably smaller than ω_m.

It has now been shown that in the continuum approximation for a three-dimensional isotropic crystal lattice of N atoms, the distribution function for normal mode frequencies is $dg(\omega)/d\omega = 9N\omega^2/\omega_m^3$, where ω_m is the maximum frequency and is given by Eq. (1-106). This function is only accurate for $\omega \ll \omega_m$, but it will be used for all ω.

Now consider the general question of the thermodynamic behavior of a collection of oscillators which are assumed to be very weakly coupled so that thermal equilibrium exists between them, but which may nevertheless be treated as independent because these interactions are weak. Let the oscillators have a broad distribution of frequencies governed by an arbitrary function $g(\omega)$. Thus, the number of oscillators with frequency in the range between ω and $\omega + d\omega$ is $g(\omega) \, d\omega$, and

$$\int_0^{\omega_m} \frac{dg}{d\omega} \, d\omega = 3N \tag{1-108}$$

The Einstein treatment of a collection of $3N$ oscillators, all of the same frequency ω, gave the total energy at temperature T:

$$E = \frac{3N\hbar\omega}{2} + 3NkT \left(\frac{\hbar\omega/kT}{e^{\hbar\omega/kT} - 1} \right) \tag{1-109}$$

The average energy per Einstein oscillator is, therefore,

$$\bar{\epsilon} = \frac{\hbar\omega}{2} + kT \frac{\theta(\omega)/T}{e^{\theta(\omega)/T} - 1} \tag{1-110}$$

where $\theta(\omega) = \hbar\omega/k$. For a collection of $3N/2$ oscillators of frequency ω_1 and $3N/2$ with ω_2, in thermal equilibrium, the average energy of the oscillators of type 1 will be given by Eq. (1-110) with $\omega = \omega_1$, and the average energy of type 2 oscillators will be given by Eq. (1-110) with $\omega = \omega_2$. Thus, in the general case of a continuous distribution of frequencies, $dg(\omega)/d\omega$, the average energy of an oscillator of frequency ω will be given by Eq. (1-110), and the total energy of $3N$ oscillators will be

$$\int_0^{\omega_m} \left\{ \frac{\hbar\omega}{2} + \frac{k\theta(\omega)}{(e^{\theta(\omega)/T} - 1)} \right\} \frac{dg}{d\omega} \, d\omega \tag{1-111}$$

In the Debye continuum approximation, $g(\omega) = 9N\omega^2/\omega_m^3$, and therefore,

$$E = \frac{9N}{\omega_m^3} \int_0^{\omega_m} \left\{ \frac{\hbar}{2} \omega^3 + \frac{\hbar\omega^3}{(e^{\hbar\omega/kT} - 1)} \right\} d\omega$$

Let $x = \hbar\omega/kT$, so that

$$E = \frac{9Nk^4T^4}{\hbar^3\omega_m^3} \int_0^{x_m} \left\{\frac{x^3}{2} + \frac{x^3}{e^x - 1}\right\} dx$$

$$E = \tfrac{9}{8}N\hbar\omega_m + 3NkT\left\{\frac{3}{x_m^3} \int_0^{x_m} \frac{x^3\,dx}{e^x - 1}\right\} \tag{1-112}$$

The integral in Eq. (1-112) can be evaluated analytically in the limits of high and low temperature, where x_m is $\ll 1$ and $\gg 1$.

At high T, the upper limit of integration is small, and therefore the denominator can be expanded in a power series about $x = 0$. Since $x \ll 1$,

$$e^x - 1 = x + x^2/2 + x^3/6 + \cdots$$

and the integrand is

$$\frac{x^3}{e^x - 1} = \frac{x^2}{1 + x/2 + x^2/6 + \cdots} \tag{1-113}$$

A power series $(1 + ax + bx^2 + \cdots)^{-1}$ can be converted to a new series $(1 + Ax + Bx^2 + \cdots)$ by demanding that the product of the series

$$(1 + ax + bx^2 + \cdots)(1 + Ax + Bx^2 + \cdots)$$

be equal to unity.   Hence,

$$1 + (a + A)x + (aA + b + B)x^2 + \cdots = 1 \tag{1-114}$$

Therefore, the coefficient of each power of x in Eq. (1-114) must be zero. Hence $A = -a$, and $B = a^2 - b$, and Eq. (1-113) becomes

$$x^2(1 - x/2 + x^2/12 - \cdots)$$

and Eq. (1-112) becomes

$$E = \tfrac{9}{8}N\hbar\omega_m + 3NkT\left\{\frac{3}{x_m^3} \int_0^{x_m} [x^2 - x^3/2 + x^4/12 - \cdots]\,dx\right\}$$

$$E = 3NkT\left(1 + \frac{x_m^2}{20} - \cdots\right) \tag{1-115}$$

At very high T, x_m is nearly zero, and

$$E \cong 3NkT \tag{1-116}$$

The specific heat $c_v = (\partial E/\partial T)_v$ is, therefore,

$$c_v = 3Nk\left(1 - \frac{x_m^2}{20} + \cdots\right) \tag{1-117}$$

which approaches the Einstein result $3Nk$ at very high T.

At low T, x_m is very large, and the upper limit of the integral in

Eq. (1-112) can be approximately replaced by ∞. In this case, the integral can be evaluated exactly:

$$\int_0^\infty \frac{x^3\,dx}{e^x - 1} = \frac{\pi^4}{15} \tag{1-118}$$

Therefore, at low T,

$$E \cong \tfrac{9}{8}N\hbar\omega_m + \frac{3\pi^4}{5} N\hbar\omega_m \left(\frac{T}{\theta_m}\right)^4 \tag{1-119}$$

where $\theta_m = \hbar\omega_m/k$. Thus, at low T,

$$c_v \cong \tfrac{12}{5}\pi^4 Nk \left(\frac{T}{\theta_m}\right)^3 \tag{1-120}$$

For intermediate T, the integral in Eq. (1-112) must be evaluated numerically. A comparison of the frequency spectra used in the Debye and Einstein calculations is shown in Fig. 1-19. The $g(\omega)$ in the Einstein calculation is a delta function since all oscillators have frequency ω_E. The Debye $g(\omega)$ is proportional to ω^2.

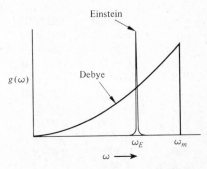

FIGURE 1-19 Distributions of angular frequencies of vibration in the Einstein and Debye models of a solid.

The specific heats predicted by the Einstein and Debye formulations are compared in Fig. 1-20. The Debye function can be fitted to experimental data more accurately than the Einstein function. However, since the Debye calculation is after all only an approximation, even it cannot be fitted exactly over wide temperature ranges. An example of this approximate nature is given by the fact that the values of θ_m, obtained by fitting experimental data to the Debye function at different T, vary slightly. This is illustrated in the following table.

The term $\tfrac{9}{8}N\hbar\omega_m$ in Eq. (1-115) is the zero point energy of the oscillators. For a collection of $3N$ oscillators with frequency spectrum

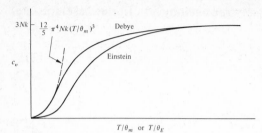

FIGURE 1-20 Specific heat of a crystalline solid according to the Einstein and Debye models

Table 1-1 Heat capacity of silver at different temperatures[a]

Temperatures ($°K$)	c_v (obs) (Cal/mole $- deg^2$)	c_v Calculated Einstein	c_v Calculated Debye	Θ Calculated from T^3 Law	Θ Calculated from Debye Function
1.35	0.000254	8.76×10^{-49}		165	
2	0.000626	1.39×10^{-32}		181	
3	0.00157	6.16×10^{-20}		200	
4	0.00303	5.92×10^{-15}		214	
5	0.00509	1.62×10^{-11}		225	
6	0.00891	3.24×10^{-9}		224	
7	0.0151	1.30×10^{-7}	0.0172	219	
8	0.0236	2.00×10^{-6}	0.0257	216	
10	0.0475	1.27×10^{-4}	0.0502	214	
12	0.0830	0.0010	0.0870	213	
14	0.1336	0.0052	0.137	212	
16	0.2020	0.0180	0.207	211	
20	0.3995	0.0945	0.394	209	
28.56	1.027	0.579	1.014		209
36.16	1.694	1.252	1.69		210
47.09	2.582	2.272	2.60		211
55.88	3.186	2.946	3.22		212
65.19	3.673	3.521	3.73		214
74.56	4.039	3.976	4.13		217
83.91	4.326	4.309	4.45		220
103.14	4.797	4.795	4.86		220
124.20	5.084	5.124	5.17		225
144.38	5.373	5.323	5.37		210
166.78	5.463	5.476	5.51		222
190.17	5.578	5.581	5.61		220
205.30	5.605	5.633	5.66		226

[a]C. Kittel, *Introduction to Solid State Physics*, copyright © 1953, by permission of John Wiley & Sons, Inc.

$dg(\omega) = (\text{const}) \, \omega^2 \, d\omega$ the zero point energy is

$$E_{T=0} = \int_0^{\omega_m} \left(\frac{\hbar\omega}{2}\right) dg(\omega) = \frac{\hbar\omega_m^4}{8} (\text{const})$$

But since

$$3N = \int_0^{\omega_m} dg(\omega) = (\text{const}) \frac{\omega_m^3}{3}$$

it follows that

$$E_{T=0} = \tfrac{9}{8} N \hbar \omega_m$$

For further reading on improvements upon the Debye model, see Hill.[10]

Problems

1. Using the table of factorials below, calculate the ratio of the probabilities of being dealt bridge hands with 5-5-3-0 and 7-5-1-0 distributions.

$$
\begin{aligned}
&F(0) = 1 && F(5) = 120 \\
&F(1) = 1 && F(6) = 720 \\
&F(2) = 2 && F(7) = 5040 \\
&F(3) = 6 && F(13) = 4.7900 \times 10^8 \\
&F(4) = 24
\end{aligned}
$$

2. List the permutations of the letters a, b, and c.
3. How many different ways are there to put 15 billiard balls into six pool-table pockets such that $n_1 = 12$, $n_2 = 3$, and all other n_i ($i = 3$ to 6) are 0?
4. Without using a slide rule or any calculating device, make quick semi-quantitative estimates of the partition function, energy, entropy, and number of particles in each energy level, of a solid crystal lattice composed of particles with energy levels as shown below at $T \cong 0$, $T \cong \epsilon/k$, and $T \cong 10\,\epsilon/k$.

$g_4 = 1$ _____ $\cdots$ _____ 4th excited level $\epsilon_4 = 100\epsilon$

$\cdots$

$\cdots$

$g_3 = 4$ _____ 3rd excited level $\epsilon_3 = 1.1\,\epsilon$

$g_2 = 1$ _____ 2nd excited level $\epsilon_2 = \epsilon$

$\cdots$

$\cdots$

$g_1 = 3$ _____ 1st excited level $\epsilon_1 = 0.2\,\epsilon$

$g_0 = 2$ _____ ground level $\epsilon_0 = 0$

---

[10]T. L. Hill, *Introduction to Statistical Mechanics* (Reading, Mass.: Addison Wesley, 1960).

5. Show that the zero point energy of an Einstein solid is the same as that of a Debye solid if ω_E is chosen as $\frac{3}{4}$ of ω_m in the Debye model.

6. What is the limiting form for the partition function of a crystal lattice of harmonic oscillators at high T?   [Hint: Expand the exponentials in Eq. (1-75).] From this, determine E and c_v at high T.

chapter 2

THE DARWIN-FOWLER METHOD:
Average Distribution for
Localized Particles

2.1 INTRODUCTION

Chapter 1 dealt with an assembly of N particles, and with calculation of the most probable distribution $n_1^*, n_2^*, n_3^*, \ldots$ of the particles
among the various quantum states, using Stirling's approximation. It
was shown that for purposes of calculating macroscopic thermodynamic
properties, it suffices to treat the most probable distribution as if it were
the only distribution possible, even though this is not strictly correct.
One objectionable aspect of this procedure is that Stirling's approximation:

$$\ln n_i! \cong n_i \ln n_i - n_i$$

is used for all the n_i, even when n_i is not necessarily large. In the Darwin-
Fowler treatment, this is avoided by the calculation of the *average* distribution instead of the most probable distribution. If there is a collection of N particles, there are many possible sets of the n_i that will conserve
N and E. Each of these sets constitutes a macrostate. The number of
particles in state i in macrostate s is defined as n_{si}. The thermodynamic
degeneracy of the sth macrostate is denoted as W_s. Then the average
distribution is defined as[1]

$$\langle n_i \rangle = \frac{\displaystyle\sum_{\substack{s=\text{all} \\ \text{macrostates}}} W_s n_{si}}{\displaystyle\sum_s W_s} \tag{2-1}$$

It is assumed that if a large number of successive measurements of n_i
are made on a system, the average of all the measured results will be
equal to $\langle n_i \rangle$ regardless of whether n_i is large or small. This should be
clearly distinguished from n_i^*, which is the set of n_i that maximizes W_s.

Before proceeding to the evaluation of Eq. (2-1), we will first re-
express W_s in terms of *energy levels* instead of states. Let n_i represent

---

[1]The sum over macrostates is implied for $\sum_i n_{si} = N$, and $\sum_i \epsilon_i n_{si} = E$.

the number of particles in the ith energy level (with degeneracy g_i). Then Eq. (2-1) may be applied directly. However, W_s refers to the number of ways to put n_1 particles in level 1, n_2 in level 2, and so forth, and is not given by Eq. (1-7). To determine W_s in terms of these n_i, we use the following argument. Consider the ith energy level as a cell with g_i compartments:

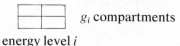
g_i compartments

energy level i

If there are n_i particles in this cell, the number of possible arrangements inside the cell is

$$\underbrace{g_i \cdot g_i \cdot g_i \cdots g_i}_{n_i \text{ terms}} = g_i^{n_i}$$

if the particles are distinguishable, since each particle can be put in any of the g_i compartments. Now consider all the energy levels as cells lined up as follows:

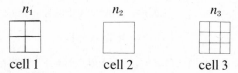

n_1 n_2 n_3

cell 1   cell 2   cell 3

The number of ways to put n_1 particles in cell 1, n_2 into cell 2, and so forth, neglecting the substructure of the cells, is

$$\frac{N!}{\prod\limits_{\substack{i=\text{all} \\ \text{levels}}} n_i!}$$

But each such arrangement is really

$$g_1^{n_1} g_2^{n_2} g_3^{n_3} \cdots$$

arrangements, because each arrangement within one cell can be combined with every arrangement in the other cells. Thus, the total number of microscopic arrangements of macrostate s is

$$W_s = \frac{N!}{\prod\limits_{\substack{i=\text{all} \\ \text{levels}}} n_i!} \prod\limits_{\substack{i=\text{all} \\ \text{levels}}} g_i^{n_i} \tag{2-2}$$

This is to be compared with

$$W_s = \frac{N!}{\prod\limits_{\substack{i=\text{all} \\ \text{states}}} n_i!} \tag{2-3}$$

which is the number of ways to put N particles into nondivisible compartments with n_1 in the first, n_2 in the second, and so forth.

It will be assumed that the thermodynamic behavior depends only on how many particles are in the various energy levels, and *not* on which states within an energy level are populated. Therefore, the average macrostate is calculated over all possible arrangements of the particles among the levels. It should be carefully noted that if Chapter 1 were repeated using Eq. (2-2) instead of (2-3), all the results of Chapter 1 would be obtained as before without change except that the n_i would refer to levels, not states.

2.2 EVALUATION OF $\sum_s W_s$ IN TERMS OF THE MULTINOMIAL THEOREM

Now proceed to the evaluation of Eq. (2-1). First calculate the denominator

$$\Omega = \sum_{\substack{s=\text{all}\\ \text{macrostates}}} W_s = \sum_s \frac{N!}{\prod_{\substack{i=\text{all}\\ \text{levels}}} n_{si}!} \prod_{\substack{i=\text{all}\\ \text{levels}}} g_i{}^{n_{si}} \tag{2-4}$$

To do this, it is necessary to make use of the *multinomial theorem*, which shall be cited without proof. First, let us remark that the *binominal theorem* for expanding $(z_0 + z_1)^N$ is

$$(z_0 + z_1)^N = z_0{}^N + N z_0{}^{N-1} z_1 + \frac{N(N-1)}{2!} z_0{}^{N-2} z_1{}^2$$

$$+ \cdots + \frac{N(N-1)\ldots N-n+1}{n!} z_0{}^{N-\alpha} z_1{}^\alpha \tag{2-5}$$

$$+ \cdots z_1{}^N$$

where z_0 and z_1 are arbitrary quantities, N is an integer, and $n+1$ is an index giving the number of the term in the sum. In more concise form,

$$(z_0 + z_1)^N = \sum_{n_1=0}^{N} \frac{N!}{n!(N-n)!} z_0{}^{N-n} z_1{}^n \tag{2-6}$$

Similarly, for a large number of variables, the multinomial theorem states that

$$(z_0 + z_1 + z_2 + \cdots + z_p)^N = \sum_{\substack{s=\text{all sets}\\ \text{of integers}\\ n_{si} \text{ such that}\\ \sum_i n_{si}=N}} \frac{N! z_0{}^{n_{s0}} z_1{}^{n_{s1}} z_2{}^{n_{s2}} \cdots z_p{}^{n_{sp}}}{n_{s0}! n_{s1}! n_{s2}! \cdots n_{sp}!} \tag{2-7}$$

In more compact form,

$$\left(\sum_{i=1}^{p} z_i\right)^N = \sum_{\substack{\text{all } s \\ \text{such that} \\ \sum_i n_{si}=N}} \frac{N! \prod_{i=1}^{p} z_i^{n_{si}}}{\prod_{i=1}^{p} n_{si}!} \tag{2-8}$$

In this equation, p and the n_{si} are integers which are not yet identified with any physical quantities. Equation (2-8) is the *multinomial theorem*. It is used to express Eq. (2-4) in more useful form. If the arbitrary substitution is made in Eq. (2-8) of replacing z_i by g_i, and letting p be the highest possible energy level, then Eq. (2-8) becomes

$$\left(\sum_{\substack{i=\text{all} \\ \text{levels}}} g_i\right)^N = \sum_{\substack{\text{all } s \\ \text{such that} \\ \sum_i n_{si}=N}} \left\{ \frac{N! \prod_{\substack{i=\text{all} \\ \text{levels}}} g_i^{n_{si}}}{\prod_{\substack{i=\text{all} \\ \text{levels}}} n_{si}!} \right\} \tag{2-9}$$

On comparison with Eq. (2-4), it is seen that Eq. (2-9) is not quite equal to Ω. The sum in Eq. (2-4) over s = macrostates is taken subject to restrictive conditions for N and E, whereas in Eq. (2-9), E does not enter. Thus

$$\Omega = \sum_{\substack{s \\ \text{such that} \\ \sum_i n_{si}=N \\ \sum_i \epsilon_i n_{si}=E}} \frac{N! \prod_i g_i^{n_{si}}}{\prod_i n_{si}!} \tag{2-10}$$

Equation (2-10) will be used shortly. First, it is necessary to digress on integration processes in the complex plane.

2.3 EVALUATION OF $\sum_s W_s$ BY COMPLEX INTEGRATION

Consider the following definition. The quantity Q is defined as

$$Q = \sum_{i=1}^{p} g_i z^{\epsilon_i} \tag{2-11}$$

where g_i and ϵ_i are mathematical constants, z is some arbitrary (complex) variable, and p is an integer. Then, if the multinomial theorem is used, with z_i arbitrarily chosen as $g_i z^{\epsilon_i}$,

$$Q^N = \left(\sum_{i=1}^{p} z_i\right)^N = \sum_{\substack{\text{all } s \\ \text{such that} \\ \sum_i n_{si}=N}} \frac{N! \prod_{i=1}^{p} g_i^{n_{si}} z^{\sum_{i=1}^{p} n_{si}\epsilon_i}}{\prod_{i=1}^{p} n_{si}!} \tag{2-12}$$

If this quantity is integrated along a contour in the complex plane, a useful result is obtained.

Only a brief elementary discussion of integration in the complex plane will be given here. A complex number z is written as

$$z = x + iy$$

where x and y are the real and imaginary parts of z. A complex number occupies a point in the complex plane where x is plotted along the abscissa and y along the ordinate. A point in the complex plane may also be described in terms of the distance r from the origin, and the phase angle θ, by

$$z = re^{i\theta}$$

as shown in Fig. 2-1. Now suppose we desire to calculate an integral of a function $f(z)$ along a contour in the complex plane.

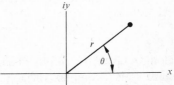

FIGURE 2-1 A point P in the complex plane located by coordinates x, y as $x + iy$, or by r and θ as $re^{i\theta}$.

$$I = \oint f(z)\, dz \qquad (2\text{-}13)$$

where the symbol $\oint$ means that dz is taken along a closed path in the plane. It is necessary to specify the contour in order to carry out the integral because dz is not uniquely defined unless a path is given. A useful path is a circle of radius r as shown in Fig. 2-2. Along this path, $z = re^{i\theta}$, with $r = \text{const}$, and θ varies from 0 to 2π. Thus,

$$dz = \frac{\partial}{\partial\theta}(re^{i\theta})\, d\theta = re^{i\theta}(i\, d\theta)$$

along a circular contour. Now, integrate the function $f(z) = z^{-1}$ along a circular contour of arbitrary radius r. It is found that

$$I_1 = \oint \frac{dz}{z} = \int_0^{2\pi} \frac{re^{i\theta}id\theta}{re^{i\theta}} = \int_0^{2\pi} id\theta = 2\pi i$$

Similarly, if $f(z) = z^n$,

$$I_2 = \oint z^n\, dz = \int_0^{2\pi} r^n e^{in\theta} re^{i\theta} id\theta = \int_0^{2\pi} r^{n+1} e^{i(n+1)\theta} id\theta$$

$$I_2 = \frac{r^{n+1}}{n+1} [e^{i(n+1)\theta}]_0^{2\pi} = 0 \qquad \text{if } n \neq -1$$

$$I_2 = 2\pi i \qquad \text{if } n = -1$$

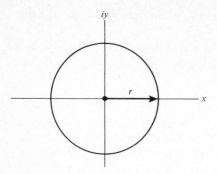

FIGURE 2-2 The circular path of radius
r in the complex plane.

Now consider the integral

$$I_3 = \oint \frac{Q^N \, dz}{z^{E+1}} \tag{2-14}$$

along a circular contour in the complex plane, where N and E are constants, z is a complex variable, and Q^N is given in Eq. (2-12) as a function
of z.   Using Eq. (2-12), it is found that

$$I_3 = \sum_{\substack{s,\dashv \\ \sum_i n_{si}=N}} \oint \frac{N! \prod_i g_i{}^{n_{si}} z^{\sum_i n_{si}\epsilon_i} \, dz}{\prod_i n_{si}! z^E z} \tag{2-15}$$

where the symbol $\dashv$ means "such that".   The only terms in this sum
which are nonzero are those for which

$$E = \sum_i n_{si}\epsilon_i$$

because $\oint dz/z^j = 2\pi i \delta_{j1}$.[2]   The terms which are nonzero, contain the
integral $\oint dz/z = 2\pi i$.   Therefore, Eq. (2-15) becomes

$$I_3 = \sum_{\substack{\text{all } s,\dashv \\ \sum_i n_{si}=N \\ \sum_i n_{si}\epsilon_i=E}} \left\{ \frac{N! \prod_i g_i{}^{n_{si}}}{\prod_i n_{si}!} \right\} (2\pi i) \tag{2-16}$$

From Eq. (2-10), it can be seen that

$$I_3 = 2\pi i \Omega \tag{2-17}$$

Then, from Eq. (2-14),

---

[2]δ_{jk} is defined as 0 if $j \neq k$, and 1 if $j = k$.

$$\Omega = \frac{1}{2\pi i} \oint \frac{Q^N \, dz}{z^{E+1}} = \sum_s W_s \qquad (2\text{-}18)$$

In this equation, Q is to be regarded as a function of the "dummy" complex variable of integration z. Equation (2-18) will be retained for the denominator of Eq. (2-1).

2.4 EVALUATION OF $\sum_s W_s n_{si}$ BY COMPLEX INTEGRATION

Next consider the numerator of Eq. (2-1) by similar methods. The term under the summation sign in the numerator of Eq. (2-1) is denoted as t_{si}, and is

$$t_{si} = n_{si} \left(\frac{N! \prod_j g_j^{n_{sj}}}{\prod_j n_{sj}!} \right) \qquad (2\text{-}19)$$

This may be expanded as

$$t_{si} = \frac{N(N-1)! [g_0^{n_{s0}} g_1^{n_{s1}} \cdots (g_i g_i^{n_{si}-1}) \cdots g_p^{n_{sp}}] \not n_{si}}{n_{s0}! n_{s1}! \cdots n_{s,i-1}! [\not n_{si}(n_{si}-1)!] n_{s,i+1} \cdots n_{sp}} \qquad (2\text{-}20)$$

The terms n_{si} in numerator and denominator are cancelled, and we define a new set of quantities as follows

$$l_0^s = n_{s0}$$
$$l_1^s = n_{s1}$$
$$\cdot$$
$$\cdot$$
$$\cdot$$
$$l_{i-1}^s = n_{s,i-1}$$
$$l_i^s = (n_{si}) - 1$$
$$l_{i+1}^s = n_{s,i+1}$$
$$\cdot$$
$$\cdot$$
$$\cdot$$

That is, all $l_k^s = n_{sk}$ except for $k = i$, in which case $l_i^s = n_{si} - 1$. Then

$$t_{si} = \frac{N g_i (N-1)! \prod_j g_j^{l_j^s}}{\prod_j l_j^s!} \qquad (2\text{-}21)$$

The numerator in Eq. (2-1) is

$$\sum_s t_{si} = N g_i \sum_{\substack{\text{all } s, + \\ \left(\sum_j n_{sj} = N \\ \sum_j n_{sj} \epsilon_j = E \right)}} \frac{(N-1)! \prod_j g_j^{l_j^s}}{\prod_j l_j^s!} \qquad (2\text{-}22)$$

Equation (2-22) can be expressed in terms of an integral in the complex plane by means of the following argument.   The integral I_4 is defined as

$$I_4 = \oint \frac{Q^{N-1} \, dz}{z^{E-\epsilon_i+1}} = \int \frac{Q^{N-1}}{z^{E-\epsilon_i}} \frac{dz}{z} \tag{2-23}$$

But according to Eqs. (2-16), (2-17), and (2-18),

$$\oint \frac{Q^N \, dz}{z^{E+1}} = 2\pi i \sum_{\substack{\text{all } s\to \\ \sum_j n_{sj}=N \\ \sum_j n_{sj}\epsilon_j=E}} \left\{ \frac{N! \, \prod_j g_j^{\, n_{sj}}}{\prod_j n_{sj}!} \right\} \tag{2-24}$$

If Eq. (2-24) is applied with $N-1$ replacing N and $E-\epsilon_i$ replacing E, the result is

$$I_4 = 2\pi i \sum_{\substack{\text{all } s\to \\ \sum_j l_j^s=N-1 \\ \sum_j l_j^s\epsilon_j=E-\epsilon_i}} \frac{(N-1)! \, \prod_j g_j^{\, l_j^s}}{\prod_j l_j^s!} \tag{2-25}$$

But if $\sum_j l_j^s = N-1$ and $\sum_j l_j^s\epsilon_j = E - \epsilon_i$, then it follows automatically that $\sum_j n_{sj} = N$ and $\sum_j n_{sj}\epsilon_j = E$. Thus, from Eqs. (2-22) and (2-25), we obtain

$$\sum_s t_{si} = \frac{Ng_i}{2\pi i} \oint \frac{Q^{N-1} \, dz}{z^{E-\epsilon_i} z} = \frac{N}{2\pi i} \oint \left(\frac{g_i z^{\epsilon_i}}{Q} \right) \frac{Q^N \, dz}{z^{E+1}} \tag{2-26}$$

2.5   METHOD OF STEEPEST DESCENTS

When Eqs. (2-26), (2-18), and (2-1) are combined, the following is obtained:

$$\langle n_i \rangle = \frac{N \oint (g_i z^{\epsilon_i}/Q) Q^N (dz/z^{E+1})}{\oint Q^N (dz/z^{E+1})} \tag{2-27}$$

where z is a complex variable of integration, and Q is a function of z given in Eq. (2-12).   The method of steepest descents is a procedure for evaluating the integral in the numerator of this expression in which it is shown that the main contribution to the integral comes from a small region of the complex plane near $z = z_m$, a constant.   Since the integrand is very small for $z \neq z_m$, it follows that $(g_i z_m^{\epsilon_i}/Q)$ may be taken out of the integral, and

$$\langle n_i \rangle \cong Ng_i z_m^{\epsilon_i}/Q \tag{2-28}$$

since the remaining integrals cancel.

When z_m and Q are evaluated, it is found that $z_m = e^{-1/kT}$ and $Q = \sum_j g_j e^{-\epsilon_j/kT}$. Therefore, $\langle n_i \rangle$ is the same as n_i^*, calculated in Chapter 1.

Now proceed to show that Eq. (2-28) results from (Eq. 2-27). A quantity $\phi(z)$ is defined such that

$$[\phi(z)]^N = Q^N/z^E \tag{2-29}$$

Then

$$\phi(z) = \frac{Q}{z^{E/N}} = \frac{g_0 z^{\epsilon_0} + g_1 z^{\epsilon_1} + \cdots + g_p z^{\epsilon_p}}{z^{(n_0\epsilon_0/N + n_1\epsilon_1/N + \cdots n_p\epsilon_p/N)}} \tag{2-30}$$

and

$$\phi(z) = \sum_{j=0}^{p} g_j z^{\epsilon_j - \bar{\epsilon}} \tag{2-31}$$

where

$$\bar{\epsilon} = \sum_{k=0}^{p} \epsilon_k n_k / N \tag{2-32}$$

is the average energy per particle.

If $\epsilon_j > \bar{\epsilon}$, the power of z will be > 0 for term j, and if $\epsilon_j < \bar{\epsilon}$, the power will be < 0. There will be some level, say q, for which $\epsilon_{q+i} > \bar{\epsilon}$, and $\epsilon_q < \bar{\epsilon}$. Therefore, Eq. (2-31) can be put in the form

$$\phi(z) = \sum_{j=0}^{q} \frac{g_i}{z^{\alpha_j}} + \sum_{j=q+1}^{p} g_j z^{\beta_j}$$

where $\alpha_j = \bar{\epsilon} - \epsilon_j$ and $\beta_j = \epsilon_j - \bar{\epsilon}$ are > 0. As $z \to 0$, $\phi(z) \to \infty$ because of the terms in the first sum, whereas when $z \to 1$, $\phi(z)$ becomes very large because of the second sum (if there are a large number of energy levels).

A plot of $\phi(z)$ along the real axis [that is, $z = x + (0)iy$] is given in Fig. 2-3. The point on the real axis at which a minimum is reached is called z_m.

At $z = z_m$, $\partial\phi/\partial x = 0$. From the definition of $\phi(z)$, it follows that if the derivative along the real axis is taken,

$$\left(\frac{\partial\phi}{\partial x}\right)_{z_m} = 0 = z_m^{-E/N}\left(\frac{\partial Q}{\partial x}\right)_{z_m} - \frac{E}{N} z_m^{-E/N-1} Q(z_m)$$

Thus

$$\left(\frac{\partial Q}{\partial x}\right)_{z_m} = \frac{E}{N} Q(z_m) z_m^{-1}$$

and hence the equality

$$\left(\frac{\partial \ln Q}{\partial x}\right)_{z_m} = \frac{E}{N} z_m^{-1} \tag{2-33}$$

holds at $z = z_m$.

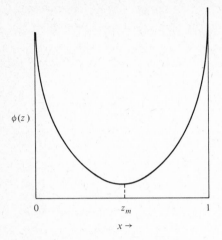

FIGURE 2-3 The function $\phi(z)$ vs x for $z = x$ along the real axis between 0 and 1. A minimum occurs at z_m.

It will now be shown that the function $[\phi(z)]^N$, although a minimum along the real axis, has a sharp maximum perpendicular to the real axis. Therefore, the form of $[\phi(z)]^N$ near the real axis is a "saddle" in the complex plane, as shown in Fig. 2-4. In this figure $\phi(z)$ is plotted in three dimensions vs x and iy, where $z = x + iy$. At any value of x between 0 and 1, $[\phi(z)]^N$ falls off sharply as $|iy|$ increases from zero. Furthermore, it can be shown that along a circular contour in the complex plane, $[\phi(z)]^N$ has a sharp maximum at the real axis where $z \cong x + (0)\,iy$. Thus, if $[\phi(z)]^N$ is integrated around such a contour, the integrand will

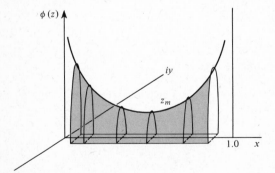

FIGURE 2-4 The function $\phi(z)$ vs x and iy in the complex plane. There is a rounded minimum vs x at z_m, and a very sharp maximum vs y at $y = 0$, leading to a saddle configuration centered over the x axis.

only be appreciable near the real axis. To show that $[\phi(z)]^N$ has a maximum at the real axis along a circular contour in the complex plane, set

$$z = r\,e^{i\theta}$$

Evidently, the real axis corresponds to $\theta = 0$, and a circular contour involves keeping $r = \text{const}$, and varying θ from 0 to 2π. If $\phi(z) = Q(z)/z^{E/N}$ has a maximum on this contour, then it can be shown that it is due to $Q(z)$ having a maximum. The denominator does not play an important role in this respect.[3] The variation of $Q(r\,e^{i\theta})$ as θ varies with $r = \text{const}$ is rather complicated because

$$Q = \sum_{j=0}^{\infty} g_j r^{\epsilon_j} e^{i\epsilon_j\theta}$$

Instead of examining the entire sum, let us temporarily consider only two arbitrary terms corresponding to $j = l$ and $j = m$. This part of the sum, to be denoted q, is

$$q = g_l r^{\epsilon_l} e^{i\epsilon_l\theta} + g_m r^{\epsilon_m} e^{i\epsilon_m\theta}$$

When it is said that a complex quantity has a maximum, it is meant that the absolute value is a maximum, regardless of the real and imaginary parts. Therefore, calculate the absolute value of q:

$$|q| = \{[g_l r^{\epsilon_l}\cos\epsilon_l\theta + g_m r^{\epsilon_m}\cos\epsilon_m\theta]^2 + [g_l r^{\epsilon_l}\sin\epsilon_l\theta + g_m r^{\epsilon_m}\sin\epsilon_m\theta]^2\}^{1/2}$$

$$|q| = \{[g_l r^{\epsilon_l}]^2 + [g_m r^{\epsilon_m}]^2 + 2g_l g_m r^{\epsilon_l\epsilon_m}(\cos\epsilon_m\theta\cos\epsilon_l\theta + \sin\epsilon_m\theta\sin\epsilon_l\theta)\}^{1/2}$$

$$|q| = \{g_l^2 r^{2\epsilon_l} + g_m^2 r^{2\epsilon_m} + 2g_l g_m r^{\epsilon_l+\epsilon_m}\cos[(\epsilon_m - \epsilon_l)\theta]\}^{1/2}$$

In general, for the entire sum,

$$|Q| = \left\{\sum_l g_l^2 r^{2\epsilon_l} + 2\sum_l\sum_{m<l} g_l g_m r^{\epsilon_l+\epsilon_m}\cos[(\epsilon_m - \epsilon_l)\theta]\right\}^{1/2}, \quad (2\text{-}34)$$

Since $|Q|$ is of the form

$$\left\{A + \sum_l\sum_m B_{lm}\cos[(\epsilon_m - \epsilon_l)\theta]\right\}^{1/2},$$

it will evidently have a maximum at $\theta = 0$ where the cosine term is unity for all (l, m) combinations. Some of the cosine terms will return to unity at values of θ such that $(\epsilon_m - \epsilon_l) = (2\pi) \times (\text{integer})$. However, different combinations of levels will reach such values at different values of θ and $|Q|$ will be less than at $\theta = 0$. Since $\phi(z)$ is a maximum at $\theta = 0$, $[\phi(z)]^N$ will be an extremely sharp maximum if N is a large number.

---

[3] Along the contour $z = r\,e^{i\theta}$ with r fixed, the absolute value of the denominator is $r^{E/N}|e^{i\theta E/N}| = r^{E/N} = \text{const}$.

(The reader is advised to compare the sharpness of the maxima in the functions $\cos x$, $[\cos x]^2$, and $[\cos x]^{10}$ at $x = 0$ in order to verify this statement.) Therefore, it has been shown that in performing contour integrals around a circular contour in the complex plane, if the integrand contains the factor $[\phi(z)]^N$, it is only appreciable in the immediate vicinity of the real axis. This is illustrated in Fig. 2-5.

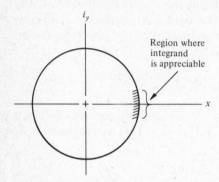

FIGURE 2-5 Region of the circular contour over which the integrand in the numerator of Eq. (2-35) is significant, due to the factor $(\phi(z))^N$.

Now return to Eq. (2-27). This may be written

$$\langle n_i \rangle = \frac{N \oint (g_i z^{\epsilon_i}/Q) [\phi(z)]^N \, dz/z}{\oint [\phi(z)]^N \, dz/z} \qquad (2\text{-}35)$$

Since $[\phi(z)]^N$ is appreciable only near the real axis, expand $[\phi(z)]^N$ about $z = z_m$. It turns out to be more convenient to expand $\ln [\phi(z)]^N$ rather than $[\phi(z)]^N$. In expanding a function of a complex variable, account has to be taken of variations in both the real and imaginary parts of z. Thus the expansion of an arbitrary function $F(z)$ about $z_0 = x_0 + iy_0$ is

$$F(z) = F(z_0) + \left(\frac{\partial F}{\partial x}\right)_{z_0} (x - x_0) + i \left(\frac{\partial F}{\partial y}\right)_{z_0} (y - y_0) \qquad (2\text{-}36)$$

$$+ \frac{1}{2}\left\{\left(\frac{\partial^2 F}{\partial x^2}\right)_{z_0} (x - x_0)^2 - \left(\frac{\partial^2 F}{\partial y^2}\right)_{z_0} (y - y_0)^2\right.$$

$$+ i \left(\frac{\partial^2 F}{\partial x \partial y}\right)_{z_0} (x - x_0)(y - y_0)\bigg\}$$

to second order.

Since $(\partial^2 F/\partial x \partial y)_{z_0} = 0$ at a saddle point,

$$\ln [\phi(z)]^N = N \ln \phi(z) = N \ln \phi(z_m) + \frac{1}{\phi(z_m)} \left(\frac{\partial \phi}{\partial x}\right)_{z_m} (x - z_m)$$

$$+ \frac{i}{\phi(z_m)} \left(\frac{\partial \phi}{\partial y}\right)_{z_m} (y - 0) \tag{2-37}$$

$$+ \frac{1}{2} \left\{ \left(\frac{\partial^2 \ln \phi}{\partial x^2}\right)_{z_m} (x - z_m)^2 - \left(\frac{\partial^2 \ln \phi}{\partial y^2}\right)_{z_m} y^2 \right\} + \cdots$$

But it has been shown that at z_m, $(\partial \phi/\partial x) = (\partial \phi/\partial y) = 0$. Furthermore, since ϕ has a very sharp maximum in the y direction and a rounded minimum in the x direction, it follows that

$$\left| \left(\frac{\partial^2 \ln \phi}{\partial y^2}\right)_{z_m} \right| \gg \left| \left(\frac{\partial^2 \ln \phi}{\partial x^2}\right)_{z_m} \right|$$

Therefore, Eq. (2-37) may be written

$$\ln [\phi(z)]^N = N \ln \phi(z_m) - \frac{1}{2} \left(\frac{\partial^2 \ln \phi}{\partial y^2}\right)_{z_m} y^2 + \cdots \tag{2-38}$$

for variations in z near $z = z_m$. It should be clear at this point that in performing the contour integral in the numerator of Eq. (2-35), it will be convenient to choose a circular contour of radius z_m in order to use Eq. (2-38) to simplify. Since $(\partial \phi/\partial y)_{z_m} = 0$, it follows that

$$\left(\frac{\partial^2 \ln \phi}{\partial y^2}\right)_{z_m} = \frac{\partial}{\partial y} \left(\frac{1}{\phi} \frac{\partial \phi}{\partial y}\right)_{z_m} = \left[\frac{1}{\phi} \left(\frac{\partial^2 \phi}{\partial y^2}\right)\right]_{z_m} \equiv \frac{1}{\phi_m} \phi''_m$$

Therefore,

$$\ln [\phi(z)]^N = N \ln \phi_m - \frac{\phi''_m y^2}{2 \phi_m} + \cdots \tag{2-39}$$

If $\phi''_m y^2/2\phi_m$ is written as $\ln [\exp (\phi''_m y^2/2\phi_m)]$, it follows that

$$[\phi(z)]^N \cong [\phi(z_m)]^N \exp \left\{-\frac{N}{2} \frac{\phi''_m}{\phi_m} y^2\right\} \tag{2-40}$$

For a circular contour of radius z_m, y is very nearly equal to $z_m \theta$ for small y, as shown in Fig. 2-6. Therefore, Eq. (2-40) may be approximated as:

$$[\phi(z)]^N \cong [\phi(z_m)]^N \exp \left\{-\frac{N}{2} \frac{\phi''_m}{\phi_m} z_m{}^2 \theta^2\right\} \tag{2-41}$$

The exponential function in Eq. (2-41) has a very sharp maximum at $\theta = 0$, and falls off rapidly for $\theta \neq 0$. Therefore, when Eq. (2-41) is substituted into Eq. (2-35), the factor $(g_i z^{\epsilon i}/Q)_{z_m}$ may be taken out of the

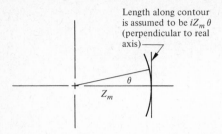

FIGURE 2-6 Illustration of how the length along the circular contour in Fig. 2-5 is approximated as $iz_m\theta$ for small regions about the x-axis.

integral sign in the numerator and the remaining integrals cancel.   Thus

$$\langle n_i \rangle \cong N[g_i z_m{}^{\epsilon_i}/Q(z_m)] \tag{2-42}$$

is a very good approximation if N is large.   Equation (2-42) may be rewritten

$$\langle n_i \rangle = \frac{N g_i z_m{}^{\epsilon_i}}{\sum\limits_i g_i z_m{}^{\epsilon_i}} \tag{2-43}$$

If we choose a temperature scale commensurate with that used in Chapter 1, we may identify the denominator Q as the partition function, and $z_m = e^{-1/kT}$.   Equation (2-43) then has the same form as the most probable distribution $n_i{}^*$, calculated in Chapter 1.

The entropy of a substance can be calculated by using Eq. (2-18), which may be written

$$\Omega = \frac{1}{2\pi i} \oint [\phi(z)]^N \frac{dz}{z} \tag{2-44}$$

To evaluate the integral, a circular contour of radius z_m is chosen, and Eq. (2-41) is applied.   Therefore,

$$\Omega \cong \frac{1}{2\pi i} [\phi(z_m)]^N \oint \exp\left\{-\frac{N}{2} \frac{\phi_m''}{\phi_m} z_m{}^2 \theta^2\right\} (id\theta) \tag{2-45}$$

in which $dz/z \cong iz_m \, d\theta/z_m = id\theta$ has been used.

It should be carefully remembered that the integrand in Eq. (2-45) is only appreciable in the immediate vicinity of the real axis where $|\theta|$ is small.   As θ increases, the integrand $\exp[-(\text{const})\theta^2]$ falls off rapidly. Therefore, the exact path of integration is not important as long as it contains the region near $\theta = 0$ shown in Fig. 2-5, where the integrand is significant.   Thus, as a purely mathematical operation, $\oint$ may be replaced by $\int_{-\infty}^{+\infty}$. Both integrals contain the small region where $\theta \cong 0$ and the integrand is appreciable. No contribution to the integral comes from large $|\theta|$. Since $\int_{-\infty}^{\infty} e^{-\alpha x^2} dx = (\pi/\alpha)^{1/2}$, it follows that

$$\Omega = \frac{[\phi(z_m)]^N}{\{2\pi N \phi''(z_m) z_m^2/\phi(z_m)\}^{1/2}} = \frac{Q^N(z_m) z_m^{-E} [\phi(z_m)]^{1/2}}{\{2\pi N \phi''(z_m) z_m^2\}^{1/2}} \quad (2\text{-}46)$$

The entropy is then

$$S = k \ln \Omega = Nk \ln Q(z_m) - kE \ln z_m$$

$$+ \frac{k}{2} \{\ln \phi(z_m) - \ln (2\pi N) - \ln [z_m^2 \phi''(z_m)]\} \quad (2\text{-}47)$$

If z_m is again chosen as $e^{-1/kT}$, we obtain $-k \ln z_m = 1/T$ and

$$Q(z_m) = \sum_{\substack{i=\text{all} \\ \text{levels}}} g_i e^{-\epsilon_i/kT} = \text{partition function} = Q$$

$$S = Nk \ln Q + E/T + (\text{small terms}) \quad (2\text{-}48)$$

Thus, the expression for S is the same as obtained in Chapter 1 if small terms are neglected.

In this chapter, we have used an alternative approach from Chapter 1 to calculate the statistical mechanical behavior of a system. We have shown that the average value $\langle n_i \rangle$ is the same as the value of n_i, n_i^*, which maximizes W. However, we have not had to resort to the use of Stirling's approximation for all the n_i in calculating $\langle n_i \rangle$. Except for small terms, S is also the same in both formulations (Chapters 1 and 2).

Problem

1. Show that the "small terms" in Eq. (2-48) are indeed small for a crystal lattice of $N = 10^{20}$ particles at $T = 1000°K$, using the Einstein model to determine Q and E, with $\theta_{\text{vib}} = 300°K$. Use Eq. (2-30) for $\phi(z)$, and set $z_m = \exp(-1/kT)$.

chapter 3

INDEPENDENT NONLOCALIZED PARTICLES

3.1 MICROSTATES AND MACROSTATES

A collection of N mass point particles are free to move about inside a container. The total energy is fixed at E. It is presumed that each particle can exist in a number of quantum states $\psi_1, \psi_2, \psi_3, \ldots$, some of which may be degenerate in energy. Each energy level is represented as a cell that is subdivided into compartments, the number of compartments being equal to the degeneracy of the level. A basic assumption is made that the thermodynamic (macroscopic) behavior of a substance depends only on the number of particles in each of the levels, and not on the distribution of particles among the compartments within the cells. Thus, the thermodynamic behavior of a substance is presumed to depend only on how many particles are in each energy level, and not on the distribution of particles among different quantum states corresponding to the same level.

Since identical particles must be treated as completely indistinguishable, the definitions of microstates and macrostates for *localized* particles are not adequate here. The particles cannot be labeled, and if it were not for degeneracy[1], the microstates and macrostates would be the same. Since the thermodynamic behavior of a substance depends on the macrostate, we define a macrostate as a specification of how many particles are in each of the energy levels, without regard to how they are distributed among the degenerate quantum states within each level. A microstate is defined as a specification of how many particles are in each of the degenerate quantum states corresponding to every energy level. As a simple illustration, let us assume the particles have two energy levels with degeneracies $g_1 = 2$ and $g_2 = 3$. Consider the case where $N = 2$ particles.

---

[1] That is, the possibility of different microscopic configurations corresponding to the rearrangement of particles among the degenerate states of an energy level.

The energy levels may be represented as

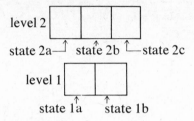

The conceivable macrostates are

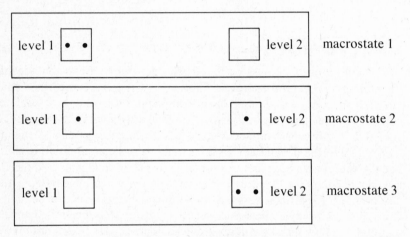

in which no reference is made to the distribution of the indistinguishable particles among the compartments within each cell. Each macrostate has several corresponding microstates. The three microstates corresponding to macrostate 1 are

Different microstates arise only from degeneracy, and *not* from permutations of distinguishable particles.   There are six microstates corresponding to macrostate 2.   One of them is

PROBLEM: Construct diagrams to represent the other five microstates of macrostate 2.

Finally, there are six microstates corresponding to macrostate 3.

As in the case of localized particles, it is assumed here that all microstates that conserve E and N are equally *a priori* probable.   The thermodynamic degeneracy W is defined as the number of microstates corresponding to a macrostate.   The goal is to determine the most probable distribution of particles n_1^* in level 1, n_2^* in level 2, and so forth. This is done by determining which set of n_i maximizes $W(n_1, n_2, \ldots)$.

Before proceeding further, a basic postulate regarding particle spins must be adopted.   It is postulated that all particles have a quantized spin angular momentum $L_s = s\hbar$, where s is the spin quantum number.   Every particle of a particular species has a particular value of s.   Only the values $s = 0, \frac{1}{2}, 1, \frac{3}{2}, 2, \frac{5}{2}, \ldots$ occur in particles observed in nature.   Those particles for which $s = 0, 1, 2, \ldots$ are called *Bosons*, and those for which $s = \frac{1}{2}, \frac{3}{2}, \frac{5}{2}, \ldots$ are called *Fermions*.   It is further postulated that the complete quantum mechanical wave function for a collection of identical particles is symmetric (antisymmetric) if the particles are Bosons (Fermions).   These postulates are required in order to obtain agreement between physical theories and experiment.   The effect of these postulates for Fermions is that no two particles can have the same set of quantum numbers.   Thus, no more than one particle is allowed in a compartment if the particles are Fermions.   For Bosons, this restriction does not apply.   In determining the most probable distribution, the calculation divides into two separate treatments, one for Fermions and one for Bosons.

3.2   BOSE-EINSTEIN STATISTICS—INTRODUCTION

This section deals with calculation of W for any arbitrary distribution of N Bosons among the energy levels with n_1 particles in level 1, n_2 in level 2, and so forth.   Then the particular set n_i^* that maximizes W will be determined.   As in the case of localized particles considered in Chapter 1, it can be shown that use of this most probable distribution leads to predictions that closely approximate the macroscopic behavior of the substance.

First, consider a single, arbitrary energy level i with degeneracy g_i. There is some arbitrary number of particles n_i in this level. Since the particles are Bosons, there are no restrictions on how many particles can go into each compartment. The goal is to determine the number of different ways n_i indistinguishable particles can be placed in g_i compartments. Let the compartments be labeled $1, 2, 3, \ldots, g_i$:

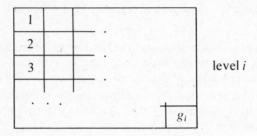

level i

The particles will *temporarily* be labeled $a, b, c, d, \ldots$ even though they are actually indistinguishable. Later on in the calculation, this improper action will be corrected. Any arbitrary distribution of the particles in the compartments can be described by a sequence such as

$$1\,a\,b\,2\,c\,3\,d\,e\,f\,4\,5\,g\,6 \ldots$$

provided the sequence *begins with a number*. It is implied in such an expression that a number specifies a compartment in the cell and the letters following that number represent the particles in the compartment. The above sequence means that particles a and b are in 1, c is in 2, d, e, and f are in 3, and so forth. Each such sequence would be that part of a microstate corresponding to level i if the particles were distinguishable.

Since there are g_i compartments, there are g_i ways to begin such a sequence. Once a sequence is begun by choosing one of the g_i numbers, the remaining $(n_i + g_i - 1)$ numbers and letters may be permuted in any arbitrary way. There are $(n_i + g_i - 1)!$ such permutations. Thus, there are a total of $g_i(n_i + g_i - 1)!$ permutations of the g_i numbers and n_i letters such that each sequence begins with a number. However, the particles are not really distinguishable, and we must undo our assumption that the particles can be labeled with letters. The effect of the temporary assumption that the particles can be labeled is to create sequences that are counted separately but which are actually the same. Consider the sequences

$$1\,a\,b\,2\,c\,3\,d\,e\,f \ldots \quad \text{and} \quad 2\,c\,1\,a\,b\,3\,d\,e\,f$$

These differ only in that certain blocks, each consisting of a number followed by the letters after it, are permuted relative to one another.

However, the letters following each number are the same in each sequence, and therefore, the two sequences are exactly the same. The number of permutations of the g_i blocks, consisting of a number followed by the letters after it, is $g_i!$. Therefore, the total number of sequences must be divided by $g_i!$ to obtain the number of sequences regardless of permutations of the compartments. Furthermore, permutation of the particles without changing the number in each of the compartments merely shifts letters, and since the particles should not have been chosen as distinguishable in the first place, none of these permutations is actually different. There are $n_i!$ permutations of the letters. Therefore, the total number of sequences should also be divided by $n_i!$. Thus, the number of different sequences for indistinguishable particles is

$$W_i = \frac{g_i(n_i + g_i - 1)!}{g_i!n_i!} = \frac{(n_i + g_i - 1)!}{(g_i - 1)!n_i!} \tag{3-1}$$

As an example of the use of this formula, consider the arbitrary case where $n_i = 2$ and $g_i = 4$. The formula for W_i gives $W_i = 5!/(3!2!) = 10$. The microstates are

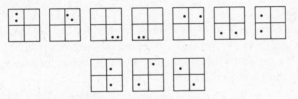

For each energy level i, with degeneracy g_i, there is a W_i. Any arrangement of the particles within the compartments of one cell can occur simultaneously with any arbitrary arrangement within the other cells. Therefore, the total number of arrangements is the product of W_i over all cells:

$$W = W_{tot} = \prod_{\substack{i=\text{all} \\ \text{levels}}} W_i = \prod_i \frac{(g_i + n_i - 1)!}{(g_i - 1)!n_i!} \tag{3-2}$$

for a set of cells with degeneracies g_i and populations n_i.

Our next task is to vary the arbitrary populations n_i and find the set n_i^* which maximizes W. If W is a maximum, then $\ln W$ is also a maximum. Taking the natural logarithm of W,

$$\ln W = \sum_i \{\ln [(g_i + n_i - 1)!] - \ln [(g_i - 1)!] - \ln n_i!\} \tag{3-3}$$

It will be assumed that n_i and g_i are large enough that Stirling's approximation can be used for $\ln n_i!$ and $\ln [(g_i - 1)!]$. Then

$$\ln W \cong \sum_i \{(g_i + n_i - 1)\ln (g_i + n_i - 1) - (g_i - 1)\ln (g_i - 1) - n_i \ln n_i\} \tag{3-4}$$

It will be assumed that $g_i + n_i$ is large enough that $g_i + n_i - 1$ can be replaced by $g_i + n_i$ in the first term. As in the case of distinguishable particles treated in Chapter 1, the n_i will be treated as if they were continuous variables, even though they are actually extremely large integers. For small variations, this is a sufficiently good approximation. Therefore, a maximum in W is sought by setting equal to zero the first derivative of $\ln W$ with respect to n_i:

$$d \ln W = 0 = \sum_i \left\{ (g_i + n_i) \frac{dn_i}{(g_i + n_i)} + \ln (g_i + n_i) \, dn_i \right.$$
$$\left. - n_i \frac{dn_i}{n_i} - \ln n_i \, dn_i \right\} \tag{3-5}$$

or

$$\sum_i [\ln (g_i + n_i) - \ln n_i] \, dn_i = 0 \tag{3-6}$$

The variations in the n_i are subject to the restrictive conditions

$$\sum_i dn_i = 0 \tag{3-7}$$

$$\sum_i \epsilon_i \, dn_i = 0 \tag{3-8}$$

The variables dn_i in Eq. (3-6) cannot be treated as independent unless Lagrange's method is applied to remove the restrictive conditions. Therefore, we multiply Eqs. (3-7) and (3-8) by the constants $-\alpha$ and $-\beta$, respectively, and add to Eq. (3-6). The result is

$$\sum_i \{\ln (g_i + n_i) - \ln n_i - \alpha - \beta \epsilon_i\} \, dn_i = 0 \tag{3-9}$$

and the dn_i may be treated as linearly independent. Thus, each term in braces in Eq. (3-9) is equal to zero, and

$$\ln \left(\frac{(g_i + n_i)}{n_i} \right) = \alpha + \beta \epsilon_i$$

$$\frac{g_i}{n_i} + 1 = e^\alpha \, e^{\beta \epsilon_i}$$

$$\frac{n_i}{g_i} = \frac{1}{e^\alpha \, e^{\beta \epsilon_i} - 1} \tag{3-10}$$

If it were not for the term -1 in the denominator of Eq. (3-10), this would be the same as Eq. (1-33) for localized particles.

There is a special case of great practical importance, where the energy levels are highly degenerate and $g_i \gg n_i$ for all levels. An expression for $\ln W$ can be derived in this limit. For such "dilute" systems, Eq. (3-4) may be rewritten, noting that $g_i \gg 1$:

$$\ln W \cong \sum_i \{(g_i + n_i) \ln (g_i + n_i) - g_i \ln g_i - n_i \ln n_i\} \tag{3-11}$$

Since $g_i \gg n_i$ by assumption, $\ln W$ for dilute systems may be written

$$\ln W \cong \sum_i \{g_i \ln (g_i + n_i) + n_i \ln g_i - g_i \ln g_i - n_i \ln n_i\} \qquad (3\text{-}12)$$

Note that n_i cannot be neglected compared to g_i in the first term because $\ln (g_i + n_i)$ is multiplied by g_i, which is large. It is not evident that the difference $g_i \ln (g_i + n_i) - g_i \ln g_i$ is small compared to other terms in Eq. (3-12). However, it is permitted to neglect n_i compared to g_i in $(g_i + n_i)$ when this factor multiplies $\ln (g_i + n_i)$. The quantity $\ln (g_i + n_i)$ may be written

$$\ln (g_i + n_i) = \ln [g_i (1 + n_i/g_i)] = \ln g_i + \ln (1 + n_i/g_i)$$

$$\cong \ln g_i + \left(\frac{n_i}{g_i} + \cdots\right) \qquad (3\text{-}13)$$

Thus,

$$\ln W \cong \sum_i \{g_i \ln g_i + n_i + n_i \ln g_i - g_i \ln g_i - n_i \ln n_i\} \qquad (3\text{-}14)$$

The terms n_i and $-n_i \ln n_i$ may be grouped to use Stirling's approximation "in reverse" as

$$n_i - n_i \ln n_i \cong -\ln n_i! \qquad (3\text{-}15)$$

Thus,

$$\ln W \cong \sum_i (n_i \ln g_i - \ln n_i!) = \sum_i [\ln g_i^{n_i} - \ln n_i!] \qquad (3\text{-}16)$$

or

$$W = \prod_{\substack{i=\text{all} \\ \text{levels}}} \frac{g_i^{n_i}}{n_i!} \qquad (3\text{-}17)$$

for dilute systems. This formula is a good approximation to Eq. (3-2) if $n_i \ll g_i$. In a subsequent section, it will be demonstrated that a gas of ordinary molecules at moderate temperature and pressure qualifies as a dilute system, to a good approximation.

The relationship between the dilute limit of the statistics of non-localized particles and the statistics of localized particles is of great significance. To see this more clearly, the statistics of localized particles are rederived in terms of energy levels, rather than in terms of quantum states as was done in Chapter 1. Consider N particles, which may be treated as if they were distinguishable because they occupy distinguishable lattice sites, to be distributed among cells representing the energy levels. Each level i is subdivided into g_i compartments. The number of different ways to distribute the particles among the energy levels such that n_1 are in level 1, n_2 in level 2, and so forth, without considering the internal structure of the cells, is

$$\frac{N!}{\prod\limits_{\substack{i=\text{all} \\ \text{levels}}} n_i!}$$

The number of different arrangements within each cell is determined as follows.   Each of the n_i particles in cell i may be put into any of the g_i compartments.   Therefore, there are $g_i^{n_i}$ different ways to put the n_i particles into cell i.   Considering all levels, the total number of ways to put N particles into the subdivided cells is

$$W = N! \prod_{\substack{i=\text{all} \\ \text{levels}}} \frac{g_i^{n_i}}{n_i!} \tag{3-18}$$

This distribution over levels is essentially the same as Eq. (1-10) over all states, when summed over $s = $ all macrostates.

$$\sum_{\substack{\text{all} \\ \text{macrostates}}} \prod_{\substack{j=\text{all} \\ \text{states}}} \frac{1}{n_j!} = \sum_{\substack{\text{all} \\ \text{macrostates}}} \prod_{\substack{i=\text{all} \\ \text{levels}}} \frac{g_i^{n_i}}{n_i!} \tag{3-19}$$

Equation (3-18) is the thermodynamic degeneracy of a system of localized particles and may be denoted W_{loc}.   Equation (3-17) is the dilute limit for nonlocalized particles, and will be denoted W_{nonloc}.   It is evident that

$$W_{\text{nonloc}} = \frac{W_{\text{loc}}}{N!} \tag{3-20}$$

The reason for this relation is as follows.   In the dilute limit of nonlocalized statistics, $g_i \gg n_i$ for all levels.   Therefore, configurations with more than one particle in a compartment are of minor importance. The different configurations of the system are obtained mostly by putting the particles, one per compartment, into different combinations of compartments.   For each arrangement with indistinguishable particles, there are $N!$ possible permutations of N distinguishable objects.   When an ensemble of nonlocalized particles is not in the dilute limit, the relation between W_{nonloc} [given in Eq. (3-2) for Bosons] and W_{loc} is not as simple, due to the preponderance of microstates with more than one particle per compartment.

The distribution function n_i for nonlocalized particles in the dilute limit can be calculated two ways.   One can use Eq. (3-18) for W, set $dW = 0$ subject to the restrictive conditions Eqs. (3-7) and (3-8), and solve for n_i^* in the usual way.   The result is identical with a simpler procedure that we present next.   In the dilute limit, since $n_i/g_i \ll 1$, Eq. (3-10) may be written

$$\frac{n_i}{g_i} = \frac{1}{e^\alpha e^{\beta \epsilon_i} - 1} \ll 1$$

Therefore,

$$e^\alpha e^{\beta \epsilon_i} \gg 1$$

and a good approximation to n_i/g_i is

$$\frac{n_i}{g_i} \cong \frac{1}{e^\alpha e^{\beta \epsilon_i}} \tag{3-21}$$

The entropy of a perfect gas of Bosons in the dilute limit is obtained as follows:

$$S = k \ln W = k \sum_{\substack{i=\text{all} \\ \text{levels}}} [n_i \ln g_i - n_i \ln n_i + n_i] \tag{3-22}$$

using Stirling's approximation for the n_i. This may be rewritten as

$$S = k \sum_i n_i [1 - \ln (n_i/g_i)] \tag{3-23}$$

From Eq. (3-21), which is the same as Eq. (1-33), it follows that

$$(n_i/g_i) = (N/Q) \, e^{-\epsilon_i/kT} \tag{3-24}$$

Therefore,

$$S = k \{N \ln Q + E/kT - (N \ln N - N)\} \tag{3-25}$$

Using Stirling's approximation "in reverse" for N, we find

$$S_{\text{nonloc}} = S_{\text{loc}} - k \ln N! \tag{3-26}$$

where

$$S_{\text{loc}} = Nk \ln Q + E/T \tag{3-27}$$

as in Eq. (1-53). Equation (3-26) is merely another manifestation of Eq. (3-20). The factors of $N!$ in Eq. (3-20), and of $k \ln N!$ in Eq. (3-26) are corrections (in the dilute case) to the formulas of Chapter 1 due to indistinguishability of the particles.

3.3 FERMI-DIRAC STATISTICS – INTRODUCTION

Fermi-Dirac statistics apply to particles known as Fermions which have a particle spin quantum number $\frac{1}{2}, \frac{3}{2}, \frac{5}{2} \ldots$. At most, one Fermion is allowed in a compartment. The calculation of W in this case is very simple. Consider cell i having g_i compartments and containing n_i particles. Since there can be, at most, one particle per compartment, it follows that $n_i \leqq g_i$. If n_i compartments of cell i are filled, $g_i - n_i$ are empty. The number of ways to distribute g_i objects into two categories, $g_i - n_i$ empty, and n_i filled, is

$$W_i = \frac{g_i!}{(g_i - n_i)! n_i!} \tag{3-28}$$

Therefore, the total number of ways to put N particles into the cells with n_1 in the first cell, n_2 in the second, and so forth, is

$$W = \prod_{\substack{i=\text{all} \\ \text{levels} \\ \text{(cells)}}} \frac{g_i!}{(g_i - n_i)! n_i!} \tag{3-29}$$

This should be compared with Eq. (3-2) obtained for Bosons. The most probable set of n_i are determined in the usual way, by setting $d \ln W = 0$ subject to Eqs. (3-7) and (3-8). Thus,

$$\ln W = \sum_i \{\ln g_i! - \ln n_i! - \ln [(g_i - n_i)!]\} \tag{3-30}$$

It will be assumed that $(g_i - n_i) \gg 1$. This implies that $g_i \gg 1$, but it is not necessary that $g_i \gg n_i$. For example, if g_i is, say, 5×10^6, and $n_i = 4 \times 10^6$, $g_i - n_i = 1 \times 10^6$, and thus $(g_i - n_i) \gg 1$ even though g_i is *not* $\gg n_i$. Then, using Stirling's approximation for n_i and g_i,

$$\ln W = \sum_i \{g_i \ln g_i - \cancel{g_i} - n_i \ln n_i + \cancel{n_i}$$

$$- (g_i - n_i) \ln (g_i - n_i) + \cancel{g_i} - \cancel{n_i}\} \tag{3-31}$$

and

$$d \ln W = \sum_i \ln \left[\frac{(g_i - n_i)}{n_i} \right] dn_i = 0 \tag{3-32}$$

for a maximum W. If Eqs. (3-7) and (3-8) are multiplied by $-\alpha$ and $-\beta$, respectively, and added to Eq. (3-32), the result is

$$\sum_i \left\{ \ln \left[\frac{(g_i - n_i)}{n_i} \right] - \alpha - \beta \epsilon_i \right\} dn_i = 0 \tag{3-33}$$

The dn_i may now be treated as independent variables, and the term in braces in Eq. (3-33) is, therefore, set equal to zero. Thus,

$$\frac{g_i}{n_i} - 1 = e^{\alpha} e^{\beta \epsilon_i}$$

and hence

$$\frac{n_i}{g_i} = \frac{1}{e^{\alpha} e^{\beta \epsilon_i} + 1} \tag{3-34}$$

If $g_i \gg n_i$, the dilute limit of Fermi-Dirac statistics is obtained, and Eq. (3-34) reduces to Eq. (3-21). In this same limit, Eq. (3-31) reduces to Eq. (3-16), and thus, the dilute limit of Fermi-Dirac statistics is the same as the dilute limit of Bose-Einstein statistics. This limit is called *corrected Boltzmann* statistics because it involves use of all the results of Chapter 1 for localized particles (Boltzmann statistics), but it is corrected for the indistinguishability of the particles.

Thus, in summary, for both types of statistics

$$n_i = \frac{g_i}{e^{\alpha} e^{\beta \epsilon_i} \pm 1} \tag{3-35}$$

where the $+$ sign holds for Fermions and the $-$ sign for Bosons. If $n_i \ll g_i$, the corrected Boltzmann limit requires that $e^{\alpha} e^{\beta \epsilon_i} \gg 1$, and the factor ± 1 in the denominator of Eq. (3-35) is negligible. In Table 3-1, the important assumptions and results are summarized.

TABLE 3-1

Localized Particles	Nonlocalized Particles	
Boltzmann Statistics	Fermi-Dirac Statistics	Bose-Einstein Statistics
$W = \dfrac{N!}{\prod\limits_{\substack{i=\text{all}\\ \text{states}}} n_i!}$	$W = \prod\limits_{\substack{i=\text{all}\\ \text{levels}}} \dfrac{g_i!}{(g_i - n_i)!n_i!}$	$W = \prod\limits_{\substack{i=\text{all}\\ \text{levels}}} \dfrac{(n_i + g_i - 1)!}{(g_i - 1)!n_i!}$
ith level: $n_i = \dfrac{g_i}{e^\alpha e^{\beta \epsilon_i}}$ if $n_i \gg 1$ $e^\alpha = Q/N$ $\beta = 1/kT$	$g_i \gtreqqless n_i$ $n_i = \dfrac{g_i}{e^\alpha e^{\beta \epsilon_i} + 1}$ if $g_i \gg 1, n_i \gg 1$ $(g_i - n_i) \gg 1$	$n_i = \dfrac{g_i}{e^\alpha e^{\beta \epsilon_i} - 1}$ if $g_i \gg 1, n_i \gg 1$

For dilute systems
$(g_i \gg n_i)$
corrected Boltzmann statistics

$$W = \prod_{\substack{i=\text{all}\\ \text{levels}}} \frac{g_i^{n_i}}{n_i!}$$

$$n_i = \frac{g_i}{e^\alpha e^{\beta \epsilon_i}}$$

3.4 THE PERFECT GAS

This chapter deals with independent nonlocalized particles which do not possess internal structure. Since the particles are independent, they do not interact with one another, and since they are nonlocalized, the substance is in the gaseous state. This is simply what is usually referred to as a perfect or ideal gas. If a gas of helium atoms at ordinary temperatures and moderately low[2] pressures is considered, the assumption of independency turns out to be a very good one because the interatomic forces are quite weak. The energy of a helium atom is composed of two parts; the translational energy of the atom through space, and the internal excitation energy of the electrons about the nucleus. The first electronic excited state lies ~ 20 eV above the ground state, and since $kT \cong 0.03$ eV at room temperature, the population (proportional to

---

[2]Say, less than ~ 1 atm.

$e^{-20/0.03}$) of the excited state is negligibly small. Therefore, it is a very good approximation to regard the helium atoms as point masses containing no internal structure or energy. Hence, the statistics presented in this chapter should be representative of helium gas under moderate conditions. Helium atoms happen to be Bosons because $s = 0$. However, the ensuing discussion will not utilize this fact. Therefore, the discussion is more general than helium, and applied to any monatomic gas regardless of the spin of the atoms. The purpose will be to show that the magnitudes of the important quantities are such that a monatomic gas at moderate pressure and temperature is in the dilute limit, and corrected Boltzmann statistics may be employed.

Assume that the gas is in a rectangular container.[3] According to quantum mechanics, the energy levels of a particle in a rectangular container are

$$\epsilon_n = \frac{h^2}{8m} \left\{ \frac{n_x^2}{L_x^2} + \frac{n_y^2}{L_y^2} + \frac{n_z^2}{L_z^2} \right\} \tag{3-36}$$

where m is the particle mass, L_j is the length of side j, n_j is the quantum number associated with direction j, and j can be x, y, or z. The energy levels have the form of an integer multiplied by a basic energy unit $\epsilon = h^2/8mL^2$. If the mass of helium atoms is used, and L is taken as 1 cm, the basic energy unit turns out to be $\sim 5 \times 10^{-31}$ erg. The spacing between energy levels is of this general magnitude. The characteristic temperature $\theta = \epsilon/k$, which determines how many of the levels are highly populated, is $\theta \cong 3 \times 10^{-15}$°K. At moderate temperatures, $T \gg \theta$, and a huge number of translational excited states are populated. Because of the extremely small spacing between translational energy levels, large numbers of states adjacent in energy may be grouped to form conceptual energy levels that have very large degeneracies and still have only very small widths in energy. This is illustrated in Fig. 3-1. It will be shown later that the conceptual energy levels may be chosen with very large degeneracies, and still the population of a conceptual level will be small compared to its degeneracy. Therefore, corrected Boltzmann statistics may be applied to the conceptual levels. Since the energy width of a conceptual level is small compared to the precision to within which an energy measurement can be made, no significant error is introduced by the conceptual levels. This will be discussed at greater length shortly. Temporarily, assume that $g_i \gg n_i$ for the conceptual levels, and use corrected Boltzmann statistics. The basis for this procedure will be reexamined later.

Corrected Boltzmann statistics will be used for an ideal gas in terms

---

[3]This is not really necessary, but it makes the illustration easier to visualize.

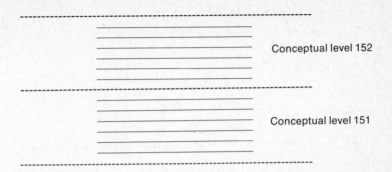

FIGURE 3-1 Conceptual levels of translation with very high degeneracy.

of conceptualized energy levels with large degeneracy. The algebra is somewhat simplified if we choose $L_x = L_y = L_z = L$. The partition function for an atom is

$$Q = \sum_{n_x=1}^{\infty} \sum_{n_y=1}^{\infty} \sum_{n_z=1}^{\infty} \exp\{-[n_x^2 + n_y^2 + n_z^2]\theta/T\} \tag{3-37}$$

where, as before,

$$\theta = \epsilon/k = h^2/8mkL^2 \tag{3-38}$$

The triple sum in Eq. (3-37) may be rewritten

$$Q = \sum_{n_x=1}^{\infty} e^{-n_x^2\theta/T} \sum_{n_y=1}^{\infty} e^{-n_y^2\theta/T} \sum_{n_z=1}^{\infty} e^{-n_z^2\theta/T} \tag{3-39}$$

Since n_x, n_y and n_z are *dummy variables*, each of the sums in Eq. (3-39) is equal. Thus,

$$Q = Q_{1D}^3 \tag{3-40}$$

where

$$Q_{1D} = \sum_{n=1}^{\infty} e^{-n^2\theta/T} \tag{3-41}$$

is the partition function for a hypothetical one-dimensional gas. In evaluating the sum in Eq. (3-41), it should be noted that $\theta \cong 3 \times 10^{-15}$°K, and $\theta/T \cong 10^{-17}$. For values of n such that $n^2 \ll 10^{17}$, all terms in the sum are essentially unity. It is only for n^2 comparable or larger than 10^{17} that the terms in the sum become significantly reduced below unity. This is illustrated in Fig. 3-2. To evaluate the sum in Eq. (3-41), each term in the sum is multiplied by $\Delta n = 1$, as illustrated schematically in Fig. 3-3. Figure 3-3 is, of course, highly distorted due to compression along the horizontal axis, in order to fit the diagram into a reasonable size. In actuality, $e^{-n^2\theta/T}$ does not fall significantly below unity until n is greater

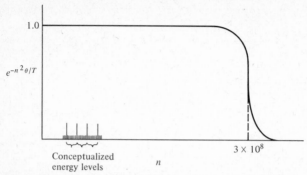

Figure 3-2 Plot of exp $(-n^2\theta/T)$ vs n for a particle in a box
with adjacent levels grouped into conceptual levels. Since
$\theta/T \cong 10^{-17}$, the exponential function is nearly unity for
$n < 10^8$.

than 10^8. Thus,

$$Q_{1D} = \sum_{n=1}^{\infty} e^{-n^2\theta/T}\, \Delta n \tag{3-42}$$

The area of a typical rectangle shown in Fig. 3-3 is $e^{-n^2\theta/T}\, \Delta n$. If $e^{-n^2\theta/T}$
varies sufficiently slowly with n, the sum can be replaced by an integral.
Thus,

$$Q_{1D} \cong \int_0^\infty e^{-n^2\theta/T}\, dn = \frac{1}{2}\sqrt{\frac{\pi T}{\theta}} = \left(\frac{2\pi mkT}{h^2}\right)^{1/2} L \tag{3-43}$$

and

$$Q = Q_{1D}^3 = \left(\frac{2\pi mkT}{h^2}\right)^{3/2} V \tag{3-44}$$

where $V = L^3$ is the volume of the container.

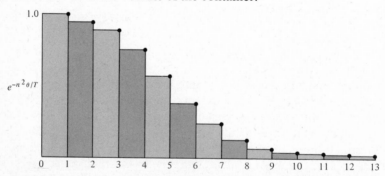

FIGURE 3-3 Crude schematic representation of the integration of the sum
for the partition function of a free particle, as in Eq. (3-42). In actuality, there
are many more levels and the variation of exp $(-n^2\theta/T)$ with n is much more
gradual.

The entropy of an ideal gas according to corrected Boltzmann statistics is

$$S = k \ln W = Nk \ln Q + NkT \, d \ln Q/dT - k \ln N! \qquad (3\text{-}45)$$

where $NkT \, d \ln Q/dT = E/T$. From Eq. (3-44),

$$\ln Q = \ln V + \tfrac{3}{2} \ln T + \tfrac{3}{2} \ln (2\pi mk/h^2) \qquad (3\text{-}46)$$

$$\frac{d \ln Q}{dT} = \frac{3}{2} \cdot \frac{1}{T} \qquad (3\text{-}47)$$

Therefore,

$$S = k \left\{ -\ln N! + N \ln V + \frac{3}{2} N \ln T + \frac{3N}{2} \ln \left(\frac{2\pi mk}{h^2}\right) + \frac{3N}{2} \right\} \qquad (3\text{-}48)$$

If we deal with one mole, N is Avogadro's number, $Nk = R$, and $3R/2 = c_v$ (the specific heat per mole at constant volume). Using Stirling's approximation for $N!$, we obtain, finally,

$$S = R \ln V + c_v \ln T + \frac{3R}{2} \ln \left(\frac{2\pi mk}{h^2}\right) + \frac{5R}{2} - R \ln N \qquad (3\text{-}49)$$

Equation (3-49) gives the absolute entropy of an ideal gas, and is often called the Sackur-Tetrode Equation. In a physical process where one mole of a perfect gas is taken from initial conditions V_1, T_1 to final conditions V_2, T_2, the change in entropy is

$$\Delta S = R \ln (V_2/V_1) + c_v \ln (T_2/T_1) \qquad (3\text{-}50)$$

This is the same as the result obtained from purely thermodynamic arguments, provided that k in Eqs. (1-42), (1-53), and (3-45) is chosen equal to the Boltzmann constant.

This section concludes by reexamining the basis for the use of corrected Boltzmann statistics for a gas, rather than Fermi-Dirac or Bose-Einstein statistics. First, calculate Q for a gas with $m = 20$ amu, $T = 300°K$ and $V = 1$ cm³. The sum in Eq. (3-41) can be approximated by a sum of terms equal to unity for $n \lesssim 4 \times 10^8$, and equal to zero for $n \gtrsim 4 \times 10^8$:

$$Q_{1D} \cong \sum_{n=1}^{\sim 4 \times 10^8} (1) + \sum_{\sim 4 \times 10^8}^{\infty} (0) \cong 4 \times 10^8$$

and $Q = (Q_{1D})^3 \cong 10^{26}$. At a pressure of 1 atm, $N \cong 6 \times 10^{23}$ molecules/22,400 cm³ $\cong 3 \times 10^{19}$ molecules/cm³. Therefore,

$$\frac{Q}{N} \cong \frac{10^{26}}{3 \times 10^{19}} \cong 3 \times 10^6$$

The population of any arbitrary quantum state j is

$$n_j = \frac{N}{Q} e^{-j^2\theta/T}$$

If j is less than $\sim 10^8$, $n_j \cong N/Q \cong 3 \times 10^{-7}$. For higher values of j, n_j is even less. Thus, the probability of a quantum state being populated is $\sim 3 \times 10^{-7}$ or less. For a collection of, say, 10^{12} adjacent quantum states comprising arbitrary conceptual energy level l, g_l would be 10^{12} and n_l would be $\sim 3 \times 10^{-7} \times 10^{12} = 3 \times 10^5$. Thus, $n_l \gg 1$, $g_l \gg 1$, and $g_l \gg n_l$, which are the conditions necessary for the dilute limit of Fermi-Dirac or Bose-Einstein statistics. The energy width of this "band" of states can be calculated as follows. In order to obtain 10^{12} different states, one must include 10^4 different values of n_x, 10^4 of n_y, and 10^4 of n_z. Since the energy of a state is equal to $(n_x^2 + n_y^2 + n_z^2)\epsilon$, the range of energies is of magnitude $3 \times 10^8 \epsilon$. But, as shown previously, $\epsilon \cong 5 \times 10^{-31}$ erg. Therefore, the width of a conceptual energy "level" is $\sim 10^{-23}$ erg $= 10^{-11}$ eV. No scientific experiments involving measurements of energy to this kind of accuracy are of any interest. Therefore, for any practical purposes, the use of conceptual energy levels does not lead to significant error.

Problem

1. Calculate the partition function of O_2 gas at $T = 300°K$, $1000°K$ and $3000°K$, treating the O_2 molecules as if they were mass points with mass 3B amu.

chapter 4

THE CLASSICAL LIMIT

4.1 CLASSICAL STATISTICAL MECHANICS

All particles obey quantum mechanics. A particle constrained to move in one dimension in a potential well, such as that shown in Fig. 4-1, has energy levels $\epsilon_1, \epsilon_2, \epsilon_3, \ldots$. Classical mechanics is an approximation which treats the energy levels as a continuum, and this becomes better for higher quantum numbers. If kT is very large compared to the spacing between energy levels, then a large number of levels will be substantially populated. In this limit, the upper quantum states play a dominant role, and classical mechanics should be a good approximation. It was seen in the previous chapter, that in this limit, the sum in the expression for the partition function can be replaced by an integral. Rather than setting up the description of a particle in terms of quantum mechanics, and taking the classical limit by replacing the sum by an integral, the goal here will be to determine the result of beginning with classical mechanics.

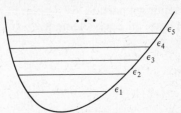

FIGURE 4-1 Quantum energy levels of a particle in a potential well.

Consider the classical mechanical description of a particle moving in the potential shown in Fig. 4-2. If the particle has energy E (which, according to classical mechanics, can be any value and is not related to the quantum energy levels), then it oscillates back and forth between a and b. At a and b, its momentum p is zero, whereas at point c, p is a

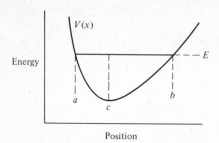

FIGURE 4-2 Potential energy vs position for a particle in a potential well. The classical turning points occur at *a* and *b* and the maximum kinetic energy occurs at *c*.

maximum. In general, for any value of x, conservation of energy requires that

$$E = \frac{p^2(x)}{2m} + V(x) \qquad (4\text{-}1)$$

and for fixed E, a solution of Eq. (4-1) for p yields a result with the general form shown in Fig. 4-3. The curve is symmetrical about the $p = 0$ axis because the negative of any value of p is also a solution of Eq. (4-1) at any x. A plot of p vs x such as given in Fig. 4-3 is called a locus in *phase space*. According to classical mechanics, the state of a point mass particle can be specified by giving its position x and its momentum p. Therefore, the microscopic configuration of a collection of particles is specified by giving (x, p) for all particles. This is best done by representing a particle by a point in phase space, and specifying the number of points in each element $dpdx$ of phase space. Therefore, the classical

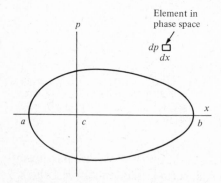

FIGURE 4-3 Phase-space plot of the classical motion of the particle described in Fig. 4-2.

mechanical description of an ensemble of particles involves determining the density of points in phase space.

In actuality, the particles obey quantum mechanics. If the energy levels of a particle are as specified in Fig. 4-1, these levels in phase space may be plotted by giving $p(x)$ for classical energies corresponding to the quantum levels, as shown in Fig. 4-4. According to quantum statistical mechanics, the number of particles in level i, n_i, is given by

$$\frac{n_i}{N} = \frac{e^{-\epsilon_i/kT}}{\sum_i e^{-\epsilon_i/kT}} \tag{4-2}$$

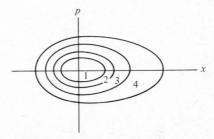

FIGURE 4-4 Phse-space plots of the classical motions corresponding to classical energies chosen equal to the quantum energy levels of a particle in a potential well.

Note that g_i factors have not been included because the degeneracies of levels of a particle constrained to move in a one-dimensional potential well are unity. If kT is very large compared to the spacing between levels, then a region of phase space may be taken, which is small compared to the total area over which levels are substantially populated, but large enough that many levels are included, as shown in Fig. 4-5. Even

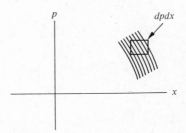

FIGURE 4-5 An element of phase space including many closely packed energy levels.

though many levels are included, the approximation may nevertheless be made that the energy is a constant, E, over this region, because the variation in the ϵ_i of the levels is small compared to kT. For a group of, say, j adjacent states with nearly the same energy E, the total population is

$$\frac{\sum\limits_{i=l}^{l+j} n_i}{N} \cong \frac{j\,e^{-E/kT}}{\sum\limits_{\text{all }i} e^{-\epsilon_i/kT}} \tag{4-3}$$

where l is some large quantum number which defines E. The area of phase space occupied by these j states as shown in Fig. 4-6, is denoted as H. This area can be determined from quantum mechanics from the WKB (Wentzel-Kramers-Brillouin) approximation.[1] According to this result, the energy of a state with high quantum number n is determined by the relation

$$2\int_a^b p\,dx = (n+\tfrac{1}{2})h \tag{4-4}$$

where p is the momentum $\{p = [2m(E-V(x))]^{1/2}\}$, h is Planck's constant, and a and b are the turning points corresponding to energy E. But the left side of Eq. (4-4) is the area in phase space enclosed by the energy level, as shown in Fig. 4-7. For two adjacent states with quantum numbers $n = l$ and $n = l+1$, the annular area in phase space between the two states is $(l+1+\tfrac{1}{2})h - (l+\tfrac{1}{2})h = h$. Thus, the addition of a quantum state produces an additional area h in phase space. When j states are added, the additional area required in phase space is $H = jh$.

At this point, an assumption is made which is analogous to the assumption in quantum statistical mechanics that all states of equal energy

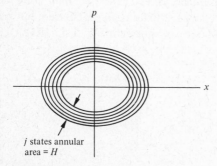

j states annular area $= H$

FIGURE 4-6 The annular area of j quantum states in phase space (equal to H).

[1]D. Rapp, *Quantum Mechanics* (New York: Holt, Rinehart and Winston, 1971).

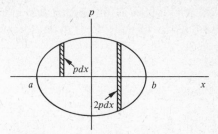

FIGURE 4-7 Elements of area in phase space.

are equally *a priori* probable. It is assumed that the number of particles in any two elements of phase space corresponding to the same energy are proportional to the areas of the elements in phase space. Therefore, the ratio of the number of particles dN in a small region of phase space $dpdx$, as shown in Fig. 4-8, to the total population of the j states is the ratio of areas $dpdx/jh$, and hence,

$$dN = \frac{dp\,dx}{jh} \sum_{i=1}^{j} n_i \qquad (4-5)$$

When Eq. (4-5) is combined with Eq. (4-3), the following is obtained:

$$dN = \frac{N\,e^{-E/kT}\,dp\,dx/h}{\sum\limits_{\substack{i=\text{all} \\ \text{states}}} e^{-\epsilon/kT}} \qquad (4-6)$$

The denominator will be evaluated in the following way. Consider the j states in the sum which fit into area $H = jh$. The contribution to the partition function from these states is

$$\sum_{i=1}^{j} e^{-\epsilon_i/kT} \cong j\,e^{-E/kT} = \frac{H}{h}\,e^{-E/kT} = \frac{1}{h} \iint\limits_{\substack{\text{area} \\ H}} e^{-E/kT}\,dp\,dx \qquad (4-7)$$

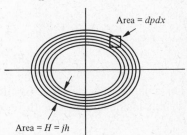

FIGURE 4-8 An element of area in phase space, *dpdx*, compared with the total area, *H*, of *j* states ($H=jh$).

in which it is assumed that $E \cong$ const over area H. Similarly, by summing over all such areas in phase space,

$$Q = \sum_{\substack{i=\text{all} \\ \text{states}}} e^{-\epsilon_i/kT} \cong \frac{1}{h} \underset{\substack{\text{all} \\ \text{phase space}}}{\int\int} e^{-E/kT} \, dp \, dx \tag{4-8}$$

Thus,

$$dN = \frac{N \, e^{-E/kT} \, dp \, dx/h}{\int_{-\infty}^{\infty} \int_{-\infty}^{\infty} e^{-E/kT} \, dp \, dx/h} \tag{4-9}$$

The classical approximation to the partition function is given in Eq. (4-8). This is the general procedure for replacing the sum by an integral.

It can be shown that for a particle in r dimensions characterized by coordinates $q_1, q_2, \ldots, q_r$ and momenta $p_1, p_2, \ldots, p_r$,

$$dN(p_1, p_2, \ldots, p_r; q_1, q_2, \ldots, q_r) =$$
$$\frac{N \, e^{-E/kT} \, dp_1 \, dp_2 \ldots dp_r \, dq_1 \, dq_2 \ldots dq_r/h^r}{\underset{2r}{\int \cdots \int} e^{-E/kT} \, dp_1 \, dp_2 \ldots dp_r \, dq_1 \ldots dp_r/h^r} \tag{4-10}$$

The distribution law in classical statistical mechanics [Eqs. (4-9) or (4-10)] is very analogous to Eq. (4-2) for quantum statistics. For any specific particle, E must be evaluated as a function of p and x, and the classical distribution function must be evaluated. Several examples are given in the next few paragraphs.

4.2   APPLICATIONS OF CLASSICAL STATISTICAL MECHANICS

4.2.1   Particle in a Three-Dimensional Container

In order to apply classical statistical mechanics, the total energy of a particle is expressed in terms of x, y, z, p_x, p_y, and p_z. The potential energy of a particle in a container is zero inside the container, and ∞ outside the container. Thus,

$$E = \frac{p_x^2}{2m} + \frac{p_y^2}{2m} + \frac{p_z^2}{2m} + \infty \quad \text{(outside)}$$

$$E = \frac{p_x^2}{2m} + \frac{p_y^2}{2m} + \frac{p_z^2}{2m} \quad \text{(inside)}$$

To evaluate the classical partition function, the integral is divided into regions inside and outside the container:

$$Q_{\text{class}} = \frac{1}{h^3} \int_{-\infty}^{\infty} \int_{-\infty}^{\infty} \int_{-\infty}^{\infty} \left\{ \underset{\text{outside}}{\int\int\int} e^{-\infty} + \underset{\text{inside}}{\int\int\int} e^{0} \right\} e^{-p_x^2/2mkT}$$
$$\times e^{-p_y^2/2mkT} \, e^{-p_z^2/2mkT} \, dx \, dy \, dz \, dp_x \, dp_y \, dp_z$$

The integral over the region of $dx\,dy\,dz$ outside the container yields zero, and the integral inside the container is equal to the volume V. The integrals over dp_x, dp_y, and dp_z may be separated and are equal. Thus,

$$Q_{\text{class}} = \frac{1}{h^3} V \left(\int_{-\infty}^{\infty} e^{-p^2/2mkT}\, dp \right)^3 = \left(\frac{2\pi mkT}{h^2} \right)^{3/2} V \qquad (4\text{-}11)$$

which is in agreement with the result obtained in Chapter 3.

4.2.2 The One-Dimensional Harmonic Oscillator

Consider a particle in one dimension bound by the potential $V(x) = \frac{1}{2} f x^2$. This is a harmonic oscillator with frequency $\nu = (2\pi)^{-1} (f/m)^{1/2}$. According to quantum mechanics, the energy levels of an oscillator are $\epsilon_n = (n + \frac{1}{2}) h\nu$ where $n = 0, 1, 2, \ldots$. As was shown in Chapter 1, the quantum-mechanical partition function is

$$Q_{\text{qu}} = \sum_{n=0}^{\infty} e^{-(n+(1/2))h\nu/kT} = \frac{e^{-\theta/2T}}{1 - e^{-\theta/T}} \qquad (4\text{-}12)$$

where $\theta = h\nu/k$. To calculate Q_{class}, write

$$E = p^2/2m + \tfrac{1}{2} f x^2 \qquad (4\text{-}13)$$

Then,

$$Q_{\text{class}} = \int_{-\infty}^{\infty} \int_{-\infty}^{\infty} e^{-p^2/2mkT}\, e^{-fx^2/2kT}\, dp\, dx/h \qquad (4\text{-}14)$$

The integrals can be separated, and

$$Q_{\text{class}} = \left(\frac{2\pi mkT}{h^2} \right)^{1/2} \left(\frac{2\pi kT}{f} \right)^{1/2} = \frac{kT}{h\nu} = \frac{T}{\theta} \qquad (4\text{-}15)$$

The relationship between Q_{class} and Q_{qu} can be seen by evaluating Q_{qu} in the limit that $kT \gg h\nu$, the spacing between levels. Thus, $\theta \gg T$, and the numerator of Q_{qu} is $\cong 1$. The denominator may be expanded in a power series

$$1 - e^{-\theta/T} = 1 - \left[1 - \frac{\theta}{T} + \frac{\theta^2}{2T^2} - \cdots \right] = \frac{\theta}{T} - \frac{\theta^2}{2T^2} + \cdots \qquad (4\text{-}16)$$

Since $\theta \ll T$, the denominator is closely approximated by θ/T, and therefore, as $\theta/T \to 0$, $Q_{\text{qu}} \to T/\theta$. Hence, when kT is large compared to the separation between levels, $Q_{\text{qu}} \to Q_{\text{class}}$.

4.2.3 Molecular Velocity Distribution in a Gas

The distribution of N noninteracting particles with p_x between p_x and $p_x + dp_x$, p_y between p_y and $p_y = dp_y$, p_z between p_z and $p_z + dp_z$, x

between x and $x + dx$, y between y and $y + dy$, and z between z and $z + dz$ is

$$\frac{dN(p_x, p_y, p_z, x, y, z)}{N} = e^{-(p_x{}^2 + p_y{}^2 + p_z{}^2)/2mkT}\frac{dp_x\, dp_y\, dp_z}{h^3}$$

$$\times \left(\frac{e^{-V(x,y,z)/kT}\, dx\, dy\, dz}{\left(\dfrac{2\pi mkT}{h^2}\right)^{3/2} U} \right) \qquad (4\text{-}17)$$

where V is a function equal to zero inside a container of volume U, and ∞ at the walls. The distribution of particles with momenta between p_x and $p_x + dp_x$, p_y and $p_y + dp_y$, and p_z and $p_z + dp_z$, integrated over the entire container, is

$$\frac{dN(p_x, p_y, p_z)}{N} = e^{-(p_x{}^2 + p_y{}^2 + p_z{}^2)/2mkT}\,(2\pi mkT)^{-3/2}\, dp_x\, dp_y\, dp_z \qquad (4\text{-}18)$$

The momenta can be written in terms of velocities by dividing by mass, for example, $p_x = mv_x$. Therefore, the distribution of molecular velocities is

$$\frac{dN}{N}(v_x, v_y, v_z) = e^{-m(v_x{}^2 + v_y{}^2 + v_z{}^2)/2kT}\,(m/2\pi kT)^{3/2}\, dv_x\, dv_y\, dv_z \qquad (4\text{-}19)$$

The velocity of a particle can be specified by giving the cartesian components v_x, v_y, and v_z, of the velocity vector **v**. The velocity vector may also be given in terms of spherical polar coordinates v, θ_v, ϕ_v, locating the velocity vector of length v in space, as shown in Fig. 4-9. The speed is $v = (v_x{}^2 + v_y{}^2 + v_z{}^2)^{1/2}$, and the volume element is $dv_x\, dv_y\, dv_z = v^2\, dv \sin\theta_v\, d\theta_v\, d\phi_v$. When the substitutions are made in Eq. (4-19), and the result integrated over all directions of the velocity vector (θ_v from 0 to π, ϕ_v from 0 to 2π) the resulting Maxwellian distribution of velocity

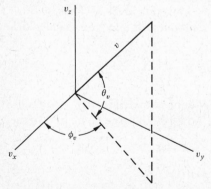

FIGURE 4-9 Spherical polar coordinates used to define the velocity vector.

vector lengths is

$$\frac{dN(v)}{N} = e^{-mv^2/2kT} 4\pi \, (m/2\pi kT)^{3/2} v^2 \, dv \qquad (4\text{-}20)$$

This distribution is shown in Fig. 4-10 for several temperatures.

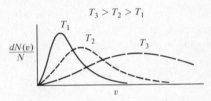

FIGURE 4-10 Distributions of molecular velocities for various temperatures, where $T_3 > T_2 > T_1$.

4.2.4 Alignment of Polar Gases in an Electric Field

Consider an ideal gas of diatomic molecules which have permanent dipole moments. The permanent dipole moment is produced by the fact that each atom of the diatomic molecule has a different net effective nuclear charge attracting the bonding electrons of the molecule. For example, in the molecule HCl, the electrons spend much more time on average near the Cl nucleus because it has a higher nuclear charge than H. There is a net separation of charge in a direction to make the H positive and the Cl negative. The permanent dipole moment may be schematically represented as a separation of charge with $+q$ on the H and $-q$ on the Cl, where q is a fraction of an electronic charge. If the bond distance is r_e, and the z-axis is defined along the bond, the dipole moment along the bond axis is

$$D = \sum_{\substack{i=\text{all} \\ \text{charges}}} q_i z_i = q r_e \qquad (4\text{-}21)$$

as may be seen by putting the origin at one nucleus.

Suppose this dipole is placed in a uniform electric field that has magnitude ξ in the $+z$ direction, as shown in Fig. 4-11. The force on an

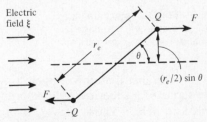

FIGURE 4-11 Forces on a dipole in a uniform electric field.

electric charge $\pm q$ in a uniform electric field is $\pm q\xi$. The net effect of the field on a dipole is the torque

$$\tau = -2\left[q\xi\left(\frac{r_e}{2}\sin\theta\right)\right] \tag{4-22}$$

where $((a/2 \sin\theta)$ is the "lever arm" and the minus sign is taken because the torque acts to decrease θ. The potential energy associated with the rotation of the molecule in the field is

$$V = -\int_{\theta_0}^{\theta} \tau\, d\theta \tag{4-23}$$

which is the work required to turn the dipole from some arbitrary angle θ_0 where $V = 0$, to angle θ. The minus sign is put in Eq. (4-23) because work must be done on the dipole to increase θ. If Eq. (4-22) is substituted into Eq. (4-23), the result is

$$V = \int_{\theta_0}^{\theta} q\xi r_e \sin\theta\, d\theta = [-q\xi r_e \cos\theta]_{\theta_0}^{\theta} = q\xi r_e(\cos\theta_0 - \cos\theta)$$

Let the zero of potential energy be chosen at $\theta_0 = \pi/2$ so that

$$V = -(qr_e)\xi\cos\theta = -D\xi\cos\theta \tag{4-24}$$

A plot of $V(\theta)$ is given in Fig. 4-12. The most stable position ($V = -D\xi$) corresponds to $\theta = 0$. The least stable orientation is for $\theta = \pi$.

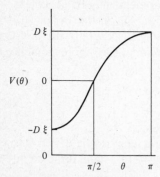

FIGURE 4-12 Potential energy of a dipole vs the angle it makes with the external electric field.

The distribution of dipole orientations in space may be described in terms of a spherical polar coordinate system with polar axis along the electric field. The angles θ, ϕ are the coordinates, as shown in Fig. 4-13. Since there is no variation in the electric field with ϕ, the distribution of dipoles in ϕ must be uniform. Therefore, the distribution of

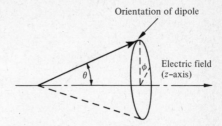

FIGURE 4-13 Cone of angles describing the orientation of a dipole with respect to the electric field.

dipoles in θ is

$$\frac{dN(\theta)}{N} = \frac{e^{-V(\theta)/kT} \sin \theta \, d\theta}{\int_0^\pi e^{-V(\theta)/kT} \sin \theta \, d\theta} \tag{4-25}$$

Note that the factor $\sin \theta \, d\theta$ is the volume element in coordinate θ. The denominator in Eq. (4-25) is the partition function for dimension θ. Thus

$$Q = \int_0^\pi e^{+D\xi \cos \theta/kT} \sin \theta \, d\theta \tag{4-26}$$

If $\chi = \cos \theta$ and $y = D\xi/kT$ are defined, it can be shown that

$$Q = 2(\sinh y)/y \tag{4-27}$$

Thus,

$$\frac{dN(\theta)}{N} = \frac{e^{y \cos \theta} \sin \theta \, d\theta}{2 \sinh y/y} \tag{4-28}$$

where $dN(\theta)$ is the number of dipoles oriented between θ and $\theta + d\theta$, and N is the total number of dipoles. At fixed T, the denominator of Eq. (4-28) is constant. The distribution function depends on the two factors $e^{y \cos \theta}$ and $\sin \theta \, d\theta$. The factor $e^{y \cos \theta}$ is the effect of the external field which tends to line the particles up at $\theta = 0$. The net result of the tendency of the random thermal energy to oppose this alignment, is the distribution function $e^{y \cos \theta} \sin \theta d\theta$ peaked at $\theta = 0$. At low T, y is larger, and the distribution in θ becomes narrower. The factor $\sin \theta$ is due to the fact that for each possible orientation angle θ, there is actually a cone of possible orientations, depending on angle ϕ, as shown in Fig. 4-13. The circumference of a section of the cone is proportional to $\sin \theta$. Therefore, orientations close to $\theta = \pi/2$ are inherently more probable than orientations near $\theta = 0$ or $\theta = \pi$ due to this effect. The separate factors $\sin \theta$ and $e^{y \cos \theta}$, and the product of the two, are shown in Fig. 4-14.

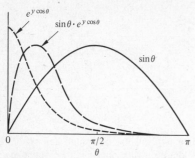

FIGURE 4-14 Functions of importance in determining the dipole moment along the field; exp (y cos θ), sin θ, and sin θ exp (y cos θ).

The average dipole moment per molecule along the external electric field is denoted as $\langle D \rangle$. The average dipole moment perpendicular to the field is zero because the distribution perpendicular to the field is symmetrical about the field direction. To obtain $\langle D \rangle$, we sum (D cos θ) $dN(\theta)$ over all possible orientations:

$$\langle D \rangle = \int_0^\pi (D \cos \theta) \, dN(\theta) \Big/ \int_0^\pi dN(\theta) \tag{4-29}$$

The function $dN(\theta) = $ (const) $e^{y\chi} d\chi$, where $\chi = \cos \theta$. Therefore,

$$\langle D \rangle = D \frac{\displaystyle\int_1^{-1} e^{y\chi} \chi \, d\chi}{\displaystyle\int_1^{-1} e^{y\chi} \, d\chi} \tag{4-30}$$

$$\langle D \rangle = D \left\{ \coth y - \frac{1}{y} \right\} = D \cdot L(y) \tag{4-31}$$

and $L(y)$ is the term in braces in Eq. (4-31).
The function $L(y)$ is plotted in Fig. 4-15. As $T \to \infty$, y becomes very

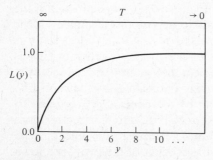

FIGURE 4-15 The function $L(y)$.

small, and

$$\coth y \cong \frac{1}{y} + \frac{y}{3} - \frac{y^3}{45} + \cdots \qquad (4\text{-}32)$$

Thus, as $T \to \infty$, $L(y) \to y/3$, and $\langle D \rangle \to yD/3 = D^2\xi/3KT$. For very large T, $\langle D \rangle$ is very small, and the random thermal energy negates the aligning effect of the electric field. At low T, y is large, and $L(y) \cong 1$. Thus, at low T, $\langle D \rangle \cong D$, and essentially all the dipoles are completely aligned. At intermediate T, $\langle D \rangle$ varies between 0 and D.

So far, only the effect of a *permanent* dipole moment has been considered. Molecules and atoms in an external electric field can also acquire an *induced* dipole moment due to polarization (separation of charge) as illustrated in Fig. 4-16. It is assumed that the induced dipole moment along the field, D_{ind} is proportional to the field and independent of the orientation of the dipole. Thus

$$D_{ind} = \alpha\xi \qquad (4\text{-}33)$$

where α is the *polarizability* of the molecule. The average dipole moment per molecule, along the field, is given by

$$\langle D_{tot} \rangle = \alpha\xi + \langle D_{perm} \rangle \qquad (4\text{-}34)$$

where $\langle D_{perm} \rangle$ is the contribution due to the permanent dipole moment. Note that $\alpha\xi$ acts regardless of T if α is independent of orientation. For all ordinary molecules at moderate T, it turns out that $y \ll 1$, so

$$\langle D_{perm} \rangle \cong Dy/3 = D^2\xi/3kT \qquad (4\text{-}35)$$

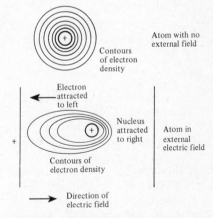

FIGURE 4-16 Schematic plot of electron density contours for an atom in an s state, with and without an external electric field.

Thus,

$$\langle D_{tot} \rangle \cong D^2 \xi/3kT + \alpha\xi = (D^2/3kT + \alpha)\xi \qquad (4\text{-}36)$$

The polarization per unit volume is $(N/V)\langle D_{tot} \rangle$. By definition, this is equal to $(K-1)\xi$, where K is the dielectric constant. Thus, it is predicted that

$$K - 1 = (D^2/3kT + \alpha)N/V \qquad (4\text{-}37)$$

If K is measured as a function of T, and $(V/N)(K-1)$ is plotted vs $1/T$, the result should be a straight line with intercept α and slope $D^2/3k$. Thus, α and D can be obtained from a measurement of K vs T. This is illustrated schematically in Fig. 4-17.

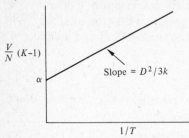

FIGURE 4-17 Schematic plot for determining the permanent dipole moment, D, and polarizability, α, of a substance from the experimental determination of the temperature dependence of the dielectric constant K.

Problems

1. Show that for an ensemble of N particles, each of which have classical energy that can be expressed as a sum of square terms in coordinates and momenta:

$$E_{\text{class}} = \sum_{i=1}^{\alpha} a_i q_i^2 + \sum_{j=1}^{\beta} b_j p_j^2$$

that contributions of each square term to the energy and specific heat of the ensemble are $NkT/2$ and $Nk/2$, respectively.

2. From the results of problem 1, determine the contributions of the rotations and vibrations of a diatomic molecule to the specific heat at high T.

3. Consider an ensemble of gaseous noninteracting particles in a gravitational field subject to the force $F = -mg$.
   (a) Determine the potential field from which the gravitational force can be derived.
   (b) How does the velocity distribution compare with the velocity distribution of a perfect gas not subject to a gravitational field at the same temperature?
   (c) What is the spatial distribution of particles? (Let the total number of particles in an ∞ vertical column of unit cross sectional area be N.) Assume the atmosphere is isothermal.

(d) Discuss the results of (c) in terms of the opposing forces of random thermal motion and gravitation at equilibrium at various temperatures.

4. (a) Set up the expression for the classical partition function of a diatomic molecule using cartesian coordinates x_1, y_1, z_1 and x_2, y_2, z_2 of the two atoms relative to a set of axes fixed in space. Show that one can immediately integrate over the six conjugate momenta, and end up with the result

$$Q = \left(\frac{2\pi m_1 kT}{h^2}\right)^{3/2} \left(\frac{2\pi m_2 kT}{h^2}\right)^{3/2} \int \cdots \int_6 e^{-U/kT} \, dx_1 \, dy_1 \, dz_1 \, dx_2 \, dy_2 \, dz_2$$

where m_i is the mass of atom i, and U is the potential energy. One may assume that $U(x_1, \ldots, z_2)$ depends only on the distance between atoms 1 and 2, and not on the position of the center of mass of the molecule. Therefore, it is convenient to transform to new coordinates corresponding to the three cartesian coordinates of the center of mass, and the three cartesian coordinates of particle 2 relative to particle 1. These coordinates will be denoted as x_m, y_m, z_m, and x_{21}, y_{21}, z_{21}, respectively.

(b) Find the volume element $dx_m \, dy_m \, dz_m \, dx_{21} \, dy_{21} \, dz_{21}$ in terms of $dx_1 \, dy_1 \, dz_1 \, dx_2 \, dy_2 \, dz_2$. See Appendix VI for the Jacobian procedure.

(c) Integrate over $dx_m \, dy_m \, dz_m$ and show that

$$Q = \left(\frac{2\pi m_1 kT}{h^2}\right)^{3/2} \left(\frac{2\pi m_2 kT}{h^2}\right)^{3/2} V \int \int \int e^{-U/kT} \, dx_{21} \, dy_{21} \, dz_{21}$$

where V is the volume.

(d) Since U depends only on the distance between atoms 1 and 2, and not on their orientation in space, transform from x_{21}, y_{21}, z_{21} to spherical polar coordinates r, θ, ϕ of atom 2 relative to atom 1 as origin. Show that

$$dx_{21} \, dy_{21} \, dz_{21} = r^2 \, dr \sin \theta \, d\theta \, d\phi$$

(e) Since U depends only on r, and not on θ or ϕ, integrate over $d\theta \, d\phi$ to obtain

$$Q = \left(\frac{2\pi m_1 kT}{h^2}\right)^{3/2} \left(\frac{2\pi m_2 kT}{h^2}\right)^{3/2} V \cdot 4\pi \int_0^\infty e^{-U(r)/kT} r^2 \, dr$$

(f) Assume $U(r) = \frac{1}{2} f(r - r_{eq})^2$, and assume that $e^{-U/kT}$ falls off rapidly as r deviates from r_{eq}, so that an approximate integration can be obtained by taking r^2 out of the integral sign as r_{eq}^2. Show that the remaining integral can be performed to yield

$$Q = \left(\frac{2\pi m_1 kT}{h^2}\right)^{3/2} \left(\frac{2\pi m_2 kT}{h^2}\right)^{3/2} 4\pi V r_{eq}^2 \cdot \left(\frac{2\pi kT}{f}\right)^{1/2}$$

(g) Show that this can be rearranged as

$$Q = V \left(\frac{2\pi (m_1 + m_2) kT}{h^2}\right)^{3/2} \left(\frac{2 I kT}{\hbar^2}\right) \left(\frac{kT}{\hbar \omega}\right)$$

where $I = \mu r_{eq}^2$, $\omega^2 = f/\mu$, $\mu = m_1 m_2 / (m_1 + m_2)$. The first term corresponds to translations of the molecule, the second to rotation, and the third to vibration.

chapter 5

PARTITION FUNCTIONS AND THERMODYNAMIC FUNCTIONS OF AN IDEAL GAS

5.1 THERMODYNAMIC PROPERTIES AND PARTITION FUNCTIONS

Sections 1.5, 1.6, 1.8, and 3.4 have already indicated that the thermodynamic properties of a perfect gas can be evaluated from the temperature dependence of the partition function. Chapter 1 dealt with independent *localized* systems. In Sec. 1.5, the various thermodynamic functions are expressed in terms of the partition function for a molecule. For localized systems, therefore,

$$S = Nk \ln Q + E/T \tag{5-1}$$

$$E = NkT^2 \left(\frac{\partial \ln Q}{\partial T}\right)_V \tag{5-2}$$

$$A = E - TS = -NkT \ln Q \tag{5-3}$$

where Q is the partition function. Chapter 3 dealt with independent *nonlocalized* systems, and it was found that in the dilute limit the thermodynamic degeneracy is $1/N!$ of what it would be for *localized* systems. Therefore, for nonlocalized systems in the dilute limit (all actual gases at moderate conditions)

$$S = Nk \ln Q + E/T - k \underbrace{\ln N!}_{\searrow (N \ln N - N)} \tag{5-4}$$

$$E = NkT^2 \left(\frac{\partial \ln Q}{\partial T}\right)_V \tag{5-5}$$

$$A = -NkT \ln Q + kT \underbrace{\ln N!}_{\searrow (N \ln N - N)} \tag{5-6}$$

The partition function of a molecule is

$$Q = \sum_{\substack{i=\text{all} \\ \text{levels}}} g_i e^{-\epsilon_i/kT} \tag{5-7}$$

5.2   GASES OF ATOMS

Gases composed of mass-points were considered in Sec. 3.4, where it is shown that the energy levels of translation in a three-dimensional rectangular box lead to the result

$$Q = \left\{ \sum_{n=1}^{\infty} e^{-n^2\theta_{tr}/T} \right\}^3 \tag{5-8}$$

where $\theta_{tr} = \pi^2\hbar^2/2mV^{2/3}k$. Since $T \gg \theta_{tr}$, the sum is replaced by an integral, and one obtains

$$Q_{tr} = \left(\frac{2\pi mkT}{h^2} \right)^{3/2} V \tag{5-9}$$

where V is the volume of the container. In Sec. 4.2.1, it was shown that in the classical limit, which should be a good approximation for $T \gg \theta_{tr}$, Eq. (5-9) is obtained for any arbitrarily shaped container of volume V. When Eq. (5-9) is substituted into Eqs. (5-4), (5-5), and (5-6), the *thermodynamic functions of a gas of mass-points* [see Eq. (3-49)] per mole is obtained:

$$S_0 = R \left\{ \tfrac{3}{2}\ln T + \tfrac{3}{2}\ln \left(\frac{2\pi mk}{h^2} \right) + \tfrac{5}{2} - \ln N + \ln V \right\} \tag{5-10}$$

$$E_0 = \tfrac{3}{2}RT \tag{5-11}$$

$$A_0 = -RT \left\{ \tfrac{3}{2}\ln \left(\frac{2\pi mk}{h^2} \right) + \tfrac{3}{2}\ln T + \ln V - \ln N + 1 \right\} \tag{5-12}$$

An atom can have electronic as well as translational energy. The electronic energy depends on the potential between the nucleus and the electrons. The equations of motion (whether in quantum or classical mechanics) of an atom can always be separated into equations for the center of mass motion and the relative internal motion since the potential energy is independent of the position of the center of mass. One may therefore rigorously write the energy levels of an atom as a sum of electronic and translational terms:

$$\epsilon = \epsilon_{tr} + \epsilon_{el} \tag{5-13}$$

Since the electronic state is independent of the translational state, the summation in Eq. (5-7) breaks up into a product of two sums

$$Q = \underbrace{\sum_{i=\text{all}} g_{tr_i} e^{-\epsilon_{tr_i}/kT}}_{\substack{\text{translational} \\ \text{levels}}} \underbrace{\sum_{j=\text{all}} g_{el_j} e^{-\epsilon_{el_j}/kT}}_{\substack{\text{electronic} \\ \text{levels}}} \tag{5-14}$$

$$Q = Q_{tr}Q_{el}$$

In general, when the energy ϵ of each level can be expressed as a sum of independent energy terms $\epsilon = \epsilon_1 + \epsilon_2 + \ldots$, it follows that the partition function can be written as a product of terms $Q = Q_1 Q_2 \ldots$.

The translational levels of an atom are the same as for a mass point in a container of volume V, and one obtains Eq. (5-9) for Q_{tr}. The electronic levels vary from substance to substance. Each electronic energy level of an atom has electron orbital and electron spin angular momenta. The orbital angular momentum is $\sqrt{L(L+1)}\,\hbar$, where L is the orbital angular momentum quantum number, and the spin angular momentum is $\sqrt{S(S+1)}\,\hbar$, where S is the spin quantum number (often referred to loosely as "the spin"). L can take on values $0, 1, 2, \ldots$, and S can be $0, \frac{1}{2}, 1, \frac{3}{2}, \ldots$. The total electronic angular momentum in an atom is $\sqrt{J(J+1)}\,\hbar$, where J is the quantum number for total angular momentum, and can take on values $0, \frac{1}{2}, 1, \ldots$. For any atomic energy level, there are $2J+1$ different states with the same energy, corresponding to different orientations of the total angular momentum vector. Each total angular momentum state can be regarded as formed from a vector addition of an L vector and an S vector. The possible values of J range from $|L-S|$ to $L+S$. The total number of states that can be formed from an L, S combination is

$$\sum_{J=|L-S|}^{J=L+S} (2J+1) = (2L+1)(2S+1)$$

The spin-orbit interaction between L and S results in each of the states with different J having slightly different energies. The spectroscopic notation used to describe the state of an atom is

$$\underset{\substack{\text{spin} \\ \text{degeneracy}}}{\underbrace{2S+1}} \mathscr{L}_J \quad \begin{array}{l} \text{a letter code for } L, \text{whereby} \\ S, P, D, F, G, \ldots \text{denote} \\ L = 0, 1, 2, 3, 4, \ldots, \text{respectively} \end{array}$$

$$J = \text{total angular momentum}$$

For example, the ground level of oxygen has $L = 1$ and $S = 1$, so it is a ^3P level. Oxygen atoms with $L = 1$ and $S = 1$ can have $J = 0$, 1, or 2. Each of these sublevels has slightly different energy. Thus, the energies of the $^3P_2$, $^3P_1$ and $^3P_0$ levels of the oxygen atom are 0.00, 0.02, and 0.03 eV, respectively. They form a set[1] of, respectively, $5 + 3 + 1 = 9$ states with nearly the same energy. For atoms of higher atomic number, the splitting of energy with J increases markedly. Usually, the next set of levels lies considerably higher in energy. For example, in oxygen, the next levels above the ^3P levels are ^1D, and are about 1.75 eV above the ^3P levels. If the spacing between electronic levels is large compared to

---

[1]The degeneracy of a level with total electronic angular momentum J is $2J+1$.

kT, a good approximation is to replace the electronic partition function by g_0 for the ground level, treating the states with varying J as if they were degenerate.   Thus, in oxygen, one would have $Q_{el} \cong 5+3+1 = 9$ as a rough approximation.   A more accurate value would be

$$Q_{el} = \sum_{\substack{i=\text{all} \\ \text{electronic} \\ \text{levels}}} (2J+1)_i \, e^{-\epsilon_{el_i}/kt} \tag{5-15}$$

The thermodynamic properties of a monatomic gas per mole are given by

$$S_1 = S_0 + R \ln Q_{el} + E_{el}/T \tag{5-16}$$

$$E_1 = E_0 + E_{el} \tag{5-17}$$

$$E_{el} = RT^2 \left(\frac{\partial \ln Q_{el}}{\partial T}\right)_V \tag{5-18}$$

$$A_1 = A_0 - RT \ln Q_{el} \tag{5-19}$$

5.3   DIATOMIC MOLECULES

The motion of a molecule can be analyzed in terms of the motion of the center of mass, and the relative internal motion of the particles of which it is composed.   Since the potential energy of the molecule depends only on the relative positions of the nuclei and electrons, and not on the position of the center of mass, the translational motion of the center of mass can be rigorously separated from the internal motion, and the energy levels have the form

$$\epsilon = \epsilon_{int} + \epsilon_{tr} \tag{5-20}$$

where ϵ_{int} refers to relative internal motion.   The energy levels of the motion of the center of mass are the same as for a point mass with the same mass as the molecule.   Therefore, the partition function of a diatomic molecule is

$$Q_2 = Q_{int}Q_{tr} \tag{5-21}$$

where Q_{tr} is given in Eq. (5-9), and Q_{int} is Eq. (5-7) for the internal energies and degeneracies.

The internal motion can be analyzed in terms of the *Born-Oppenheimer approximation*, where it is assumed that the light electrons move rapidly compared to the ponderous nuclei.   It therefore follows that the electronic states and energy levels can be evaluated with the nuclei treated as if they were stationary.   A single electronic state will have an electronic energy $\epsilon_{el}(R)$ which is a function of the internuclear distance R, the energy at any R being calculated with the nuclei treated as if they were at rest. Since the electrons rapidly adjust to the slower motion of the nuclei, the electronic energy $\epsilon_{el}(R)$ acts as a potential for nuclear motion in any particular electronic state. As a result of this procedure, the internal energy levels of a diatomic molecule are the totality of all levels associated with nuclear motion in each of the possible electronic states. This is illustrated in Fig. 5-1 for the O_2 molecule. The energy units eV are electron-volts ($1 \text{ eV} = 1.6 \times 10^{-12}$ erg). At infinite separation of the nuclei, the energy of O_2 is simply the sum of energies of two O atoms. The ground electronic state of O has the spectroscopic[2] notation ^3P, and the first excited state is ^1D. The ^1D state lies about 1.5 eV higher than the ^3P state. When two oxygen atoms approach one another, a number of molecular electronic states can be formed. The electronic energies $\epsilon_{el}(R)$ of a few of the states[3] are illustrated in Fig. 5-1. The so-called X, a, b, and A states go to the ground asymptotic state, whereas the B state goes to $O(^3P) + O(^1D)$. The probability that any particular molecular state will be formed when two oxygen atoms in particular electronic states combine is proportional to the electronic degeneracy of the molecular electronic state. This quantity will be denoted g_{el_i} for the ith molecular electronic state. The degeneracy of an electronic level is determined by the total electronic angular momentum. The spectro-

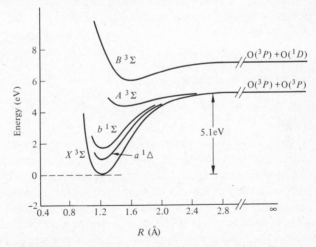

FIGURE 5-1 Electronic energies vs internuclear distance for several electronic states of molecular oxygen (O_2).

[2]This neglects reference to the total electronic angular momentum.
[3]The spin-orbit interaction is not considered here and the small splitting between states with different J is neglected.

scopic notation for diatomic molecules is

$$
\underset{\substack{\text{spin}\\ \text{degeneracy}}}{\underbrace{2S+1}} \Lambda_J
\begin{cases}
\text{orbital angular momentum quantum number}\\[4pt]
\Sigma, \pi, \Delta, \ldots \text{ denote}\\[4pt]
0, 1, 2, \ldots, \text{respectively}
\end{cases}
$$

total angular momentum
quantum number

For each value of S, there are $2S+1$ values of m_S, the component of S along the bond axis. The possible values of J along the bond are $\Lambda + m_S$, and each such level has a degeneracy of 2 if $\Lambda \neq 0$.[4] For example, the ground level of the NO molecule is $^2\pi$, and therefore, there are two sublevels: $^2\pi_{1/2}$ and $^2\pi_{3/2}$. Each of these sublevels is doubly degenerate. The spin-orbit interaction results in these levels having slightly different energy (~ 0.01 eV). If this slight difference is neglected, the ground level of NO may be regarded as a four-fold degenerate $^2\pi$ level.

A diatomic molecule in any arbitrary electronic state has a number of possible energy levels due to motion of the nuclei. The motion of the nuclei can be analyzed by noting from Fig. 5-1 that in a particular electronic state, there is a most stable internuclear distance (bottom of the potential well). If the nuclei have energy, it can be in the form of oscillations of the internuclear distance about equilibrium

$$\leftarrow\!\circ \qquad \circ\!\rightarrow$$

and/or rotation of the molecule.

A commonly used approximation is the assumption of separation of vibrational and rotational motion. It is assumed that the vibrational amplitude is small enough that the vibrations do not change the moment of inertia of the molecule, and that the rotational speed is low enough that centrifugal forces do not produce anharmonicity in the vibrational motion. This is usually good within a few percent or better for most diatomic molecules. With this approximation, the energy levels of a diatomic molecule due to nuclear motion may be taken as a sum of energies of a rigid rotator and a one-dimensional oscillator moving in the potential $\epsilon_{el}(R)$.

---

[4]If $\Lambda = 0$, the degeneracy is unity. The degeneracy is 2 for $\Lambda \neq 0$ because the projection of the orbital angular momentum vector on the bond axis can be $\pm \Lambda$.

$$\epsilon_{int} = \epsilon_{el} + \epsilon_{nuc} \quad \text{(Born-Oppenheimer)}$$

$$\epsilon_{nuc} = \epsilon_{rot} + \epsilon_{vib} \text{ (Vibrational-rotational separation)} \quad (5\text{-}22)$$

The energy levels and degeneracies[5] of a linear rigid rotator are

$$\epsilon_l = \frac{\hbar^2}{2I} l(l+1)$$
$$g_l = 2l+1 \quad (5\text{-}23)$$

where I is the moment of inertia $[I = \mu R_{eq}^2$ where $\mu =$ reduced mass of the two atoms $= m_A m_B/(m_A + m_B)$, R_{eq} is the distance at which the minimum in $\epsilon_{el}(R)$ occurs, and l is the rotational quantum number which can take on values $0, 1, 2, \ldots$. The rotational partition function Q_{rot} is

$$Q_{rot} = \sum_{l=0}^{\infty} (2l+1) \, e^{-l(l+1)\theta_{rot}/T} \quad (5\text{-}24)$$

where

$$\theta_{rot} = \hbar^2/2Ik \quad (5\text{-}25)$$

The vibrational energy levels can usually be closely approximated by the harmonic oscillator approximation in which $\epsilon_{el}(R)$ is expanded in a power series about R_{eq}. It is found that if $\epsilon_{el}(R_{eq})$ is chosen as the zero of energy,

$$\epsilon_{el}(R) = \tfrac{1}{2}f(R - R_{eq})^2 + \cdots \quad (5\text{-}26)$$

where $f = (\partial^2 \epsilon_{el}/\partial R^2)_{R=R_{eq}}$ is the force constant, and the fact that $(\partial \epsilon_{el}/\partial R)_{R_{eq}} = 0$ has been used. In the harmonic approximation, we fit a parabola to the potential curve at the bottom of the well as shown in Fig. 5-2. It gives good approximations to the lower energy levels but fails badly for higher levels where the energy levels become closely packed as shown in Fig. 5-3. With this approximation, the vibrational

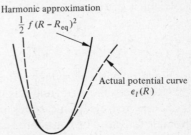

Harmonic approximation
$\tfrac{1}{2}f(R-R_{eq})^2$

Actual potential curve
$\epsilon_l(R)$

FIGURE 5-2 Harmonic approximation $[\tfrac{1}{2}f(R-R_{eq})^2]$ to the true potential curve $\epsilon_{el}(R)$.

[5]D. Rapp, *Quantum Mechanics* (New York: Holt, Rinehart and Winston, 1971).

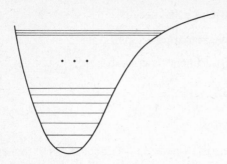

FIGURE 5-3 Vibrational levels of a molecule including anharmonicity.

motion is the same as for a harmonic oscillator, and the vibrational energy levels are[6]

$$\epsilon_{vib} = (n + \tfrac{1}{2})h\nu \tag{5-27}$$

where n is the vibrational quantum number $(0, 1, 2, 3, \ldots)$ and ν is the vibrational frequency $(\nu = (2\pi)^{-1}(f/\mu)^{1/2})$. The vibrational partition function is

$$Q_{vib} = \sum_{n=0}^{\infty} e^{-(n+(1/2))\theta_{vib}/T} \tag{5-28}$$

where

$$\theta_{vib} = h\nu/k \tag{5-29}$$

The sum in Eq. (5-28) was evaluated in Sec. 1.8.   The result is

$$Q_{vib} = \frac{e^{-\theta_{vib}/2T}}{(1 - e^{-\theta_{vib}/T})} \tag{5-30}$$

According to Eqs. (5-21) and (5-22), the partition function of a diatomic molecule is therefore:

$$Q_2 = Q_{tr} \sum_{\substack{i=\text{all} \\ \text{electronic} \\ \text{states}}} g_{el_i} e^{-\epsilon_{el i}/kT} \{Q_{rot}Q_{vib}\}_{\text{electronic state } i} \tag{5-31}$$

It should be particularly noted that Q_{rot} and Q_{vib} will vary from electronic state to electronic state, and it is *not* correct to write $Q_2 = Q_{tr}Q_{el}Q_{rot}Q_{vib}$. If the excited electronic states lie high enough in energy that only the ground state need be considered in the sum in Eq. (5-31),

$$Q_2 \cong g_{el_0}Q_{tr}Q_{rot}Q_{vib} \tag{5-32}$$

An example should make this material easier to understand.   Con-

---

[6]D. Rapp, *Quantum Mechanics* (New York: Holt, Rinehart and Winston, 1971).

sider the O_2 molecule as illustrated in Fig. 5-1. The vibrational frequencies and equilibrium bond distances of the nuclei in the first few electronic states are tabulated below.

TABLE 5-1 Properties of lower electronic states of O_2

State	ν/c (cm^{-1})	R_{eq} $(\mathring{A})$	θ_{vib} $(°K)$	θ_{rot} $(°K)$
$X\,^3\Sigma$	1580	1.207	2270	2.07
$a\,^1\Delta$	1509	1.216	2170	2.04
$b\,^1\Sigma$	1433	1.227	2060	2.01
$A\,^3\Sigma$	819	(1.42)	(1180)	(1.50)

As discussed in Chapter 1, the partition function (based on the lowest level as a zero of energy) is a measure of how many states of a particle are substantially populated.   In terms of calculations, Eq. (5-9) is

$$Q_{tr} = 1.87 \times 10^{20}(mT)^{3/2}V \qquad (5\text{-}33)$$

with m in atomic mass units, T in $°K$ and V in cm^3.   For O_2, $m = 32$, and if V is chosen equal to 1 cm^3, the values listed in Table 5-2 are obtained.

TABLE 5-2 Partition functions of O_2 per cm^3

$T(°K)$	Q_{tr}	Electronic State	Q_{rot}	$Q_{vib}{}^a$	$g_{el}\,e^{-\epsilon_{el}/kT}$
300	1.76×10^{26}	X	72.4	1.00	3
		a	73.5	1.00	7×10^{-17}
		b	74.6	1.00	3×10^{-28}
1000	1.07×10^{27}	X	241	1.12	3
		a	245	1.13	2.3×10^{-5}
		b	249	1.16	6×10^{-9}
3000	5.57×10^{27}	X	724	1.88	3
		a	735	1.94	0.045
		b	746	2.01	0.0018

[a]Based on the lowest vibrational state as the zero of energy.

Evidently, a tremendous number ($\sim 10^{26}$) of translational states are populated.   The characteristic temperature for translations, θ_{tr}, is of the order of $10^{-16°}K$.   The translational contribution to Q is independent of the electronic state.   The rotational partition function does not vary greatly over the X, a, and b states.   The rotational partition function indicates that about 70 rotational states are appreciably populated at room temperature, whereas over 700 states[7] are populated at 3000°K. The molecules are essentially all in the lowest vibrational state at room temperature, but ~ 2 vibrational states are appreciably populated at 3000°K. In general, most molecules are in the ground electronic state. At room temperature, only about $\sim 10^{-17}$ of the molecules are in excited electronic states. At 3000°K, about 2% are in excited electronic states. At 3000°K, the total partition function for O_2 would be calculated from Eq. (5-31) as

$$Q = 1.96 \times 10^{27} \{8.7 \times 724 \times 1.88 + 0.021 \times 735 \times 1.94 + 0.0021 \times 746 \times$$
$$\times 2.01 + \cdots\}$$

where the factors 1, 0.021, and 0.0021 are $\exp(-\epsilon_{el_i}/kT)$ for $i = 1, 2$, and 3.

It is instructive to calculate the spacing between energy levels. The spacing between electronic levels was shown in Fig. 5-1. For a particular electronic state, the vibrational and rotational levels are as shown in Figs. 5-4 and 5-5. The vibrational levels are spaced about 0.2 eV apart, and the rotational level spacings for the lower levels are of general magnitude 2×10^{-3} eV. For high rotational quantum numbers ($l \gtrsim 100$) the rotational spacing is so large that a state with vibrational quantum number, say, $n = 1$ can lie higher in energy than a state with $n = 2$, if the rotational quantum number is much lower in the upper vibrational state.

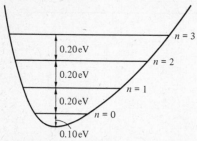

FIGURE 5-4 Vibrational energy levels of the ground electronic state of O_2 in the harmonic approximation.

The rotational partition function [Eq. (5-24)] can be evaluated in closed form if it is assumed that a large number of rotational states are substantially populated. Equation (5-24) can be multiplied by $\Delta l = 1$. [Note: see the discussion of Eq. (3-42) for an analogous situation.] If a plot of $(2l + 1) \exp[-l(l + 1)\theta_{rot}/T]$ vs l is made, the result has the general form shown in Fig. 5-6. The exact shape will depend on the ratio θ_{rot}/T. The summation in Eq. (5-24) is the sum of areas of rec-

7This corresponds to about 27 levels.

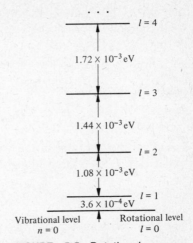

FIGURE 5-5 Rotational energy levels in the ground vibrational state of the ground electronic state of O_2.

tangles in Fig. 5-6. If $T \gg \theta_{rot}$, there are many such rectangles, and the vertical height varies slowly from rectangle to rectangle. Therefore, the sum may be approximated by an integral.

$$Q_{rot} \cong \int_0^\infty (2l+1) \exp\left[-l(l+1)\theta_{rot}/T\right] dl \qquad (5-34)$$

This integral is easily evaluated by letting $y = l(l+1)$ so that $dy = (2l+1)\, dl$. Thus one obtains $Q_{rot} = T/\theta_{rot}$. However, one must also include the *symmetry number* (see Chapter 6), so that

$$Q_{rot} = T/\sigma\theta_{rot} \qquad (5-35)$$

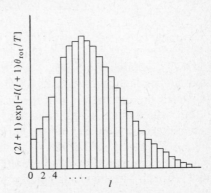

FIGURE 5-6 Plot of terms in the sum for the rotational partition function vs rotational quantum number l.

where σ is the symmetry number (2 for symmetrical[8] and 1 for unsymmetrical[9] diatomic molecules). In summary, for a diatomic molecule,

$$Q = Q_{tr} \sum_{\substack{i=\text{all} \\ \text{electronic} \\ \text{levels}}} g_{\text{el}_i}\, e^{-\epsilon_{\text{el}_i}/kT} \{Q_{rot}Q_{vib}\}_{\text{electronic level } i} \tag{5-36}$$

where

$$Q_{vib} = \frac{e^{-\theta_{vib}/2T}}{1 - e^{-\theta_{vib}/T}} \tag{5-37}$$

based on the "bottom of the well" as the zero of energy, and for most molecules, T/θ_{rot} is large enough that

$$Q_{rot} = \frac{T}{\sigma\theta_{rot}} \tag{5-38}$$

The thermodynamic properties can be written in simplified form if only the ground electronic state needs to be included in Eq. (5-36). In this case, which is fairly common,

$$S_2 = S_1 + Nk \ln Q_{rot} + Nk \ln Q_{vib} + E_{rot}/T + E_{vib}/T + Nk \ln g_{\text{elo}} \tag{5-39}$$

$$E_2 = E_1 + E_{rot} + E_{vib} \tag{5-40}$$

$$E_{rot} = NkT \tag{5-41}$$

$$E_{vib} = \frac{Nh\nu}{2} + \frac{Nh\nu\, e^{-\theta_{vib}/T}}{1 - e^{-\theta_{vib}/T}} \tag{5-42}$$

$$A_2 = A_1 - NkT \ln Q_{rot} - NkT \ln Q_{vib} - NkT \ln (g_{\text{elo}}) \tag{5-43}$$

Equation (5-41) is obtained from the relation

$$E_{rot} = \frac{1}{Q_{rot}} \int_0^\infty (2l+1)\, \exp\left[-l(l+1)\theta_{rot}/T\right]$$

$$\times \left\{\frac{\hbar^2}{2I} l(l+1)\right\} dl = NkT \tag{5-44}$$

and E_{vib} is taken from Sec. 1.8. It is instructive to calculate the specific heat of molecular oxygen as a function of temperature. We shall carry through the discussion as if the gas did not liquefy at very low temperature, which makes the argument somewhat artificial at low T. The characteristic temperatures for various modes of energy are $\theta_{tr} \cong 10^{-16}$°K, $\theta_{rot} \cong 2$°K, $\theta_{vib} \cong 2270$°K, $\theta_{el} \cong 8000$°K. At temperatures below 2°K, but large compared to θ_{tr}, the rotational, vibrational, and electronic states are "frozen" in the ground states and the only nonground state energy in

---

[8] For example, O_2.
[9] For example, HCl.

the gas is translational. The energy of translation is given by Eq. (5-11), and therefore, $c_v = (\partial E/\partial T)_V = 3Nk/2$, or $3R/2$ per mole. When T is raised through the range $1-10°K$, the upper rotational levels become populated, and the rotations contribute to the energy of the gas. For $T \gg \theta_{rot}$, E_{rot} is given by Eq. (5-41), and $c_v = 3R/2 + R = 5R/2$. When T is raised to several thousand degrees, the vibrational states become broadly populated and the energy of vibration becomes equal to Eq. (5-42) with $T \gg \theta_{vib}$. As shown in Sec. 1.8, this reduces to $E_{vib} \cong RT$ per mole, and therefore, $c_v = 5R/2 + R = 7R/2$. At still higher T, c_v increases due to population of excited electronic levels. This is illustrated in Fig. 5-7. The other thermodynamic functions also go through similar series of plateaus as the various forms of energy levels become populated.

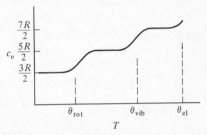

FIGURE 5-7 Specific heat vs temperature for a gas of diatomic molecules.

In molecular hydrogen, the spacing between rotational levels is much greater than in O_2, corresponding to $\theta_{rot} = 87°K$. Thus, Fig. 5-7 would actually represent H_2, from cryogenic temperatures to several thousand degrees.

At temperatures high enough that a large number of vibrational states are populated, the use of the harmonic oscillator model [Eq. (5-26) and Fig. 5-2] can lead to considerable error. A more refined treatment of the dynamics of motion of a diatomic molecule, including approximate treatment of the effects of anharmonicity in the potential curve, centrifugal stretching effects, and varying moment of inertia due to vibrations, leads to the result that the sum of rotational and vibrational energies of a diatomic molecule is

$$\epsilon_{rot, vib} = \left[(n+\tfrac{1}{2})\, h\nu + \frac{\hbar^2}{2I} l(l+1) \right]$$

$$-4 \left(\frac{\hbar^2}{2I}\right)^3 \left(\frac{1}{h\nu}\right)^2 l^2(l+1)^2 - h\nu\tilde{\chi}(n+\tfrac{1}{2})^2 - \tilde{\alpha}(n+\tfrac{1}{2})l(l+1) \quad (5\text{-}45)$$

The term in brackets is the energy in the approximation of a rigid rotator-harmonic oscillator. The term in $l^2(l+1)^2$ is a correction for variation in

I as the molecule vibrates. The term in $(n+\frac{1}{2})^2$ is due to the anharmonicity of the potential curve, and the last term is due to centrifugal stretching effects. The constants $\tilde{\chi}$ and $\tilde{\alpha}$ are determined empirically. Usually, spectroscopists give values for $\alpha_e = \tilde{\alpha}/hc$, $\chi_e = \tilde{x}/hc$, $B_e = \hbar^2/2Ihc$, and $\omega_e = \nu/c$.

5.4 POLYATOMIC MOLECULES

5.4.1 Separation of Translational and Electronic Energy

To determine the partition function of a polyatomic molecule, it is necessary to determine the energy levels and degeneracies of the molecule. The procedure is similar to that used for diatomics. The potential energy of the molecule is independent of the position of the center of mass, and therefore the motion of the center of mass can be separated out. The motion of the molecule acts like a superposition of motions of a free mass point at the center of mass with the mass of the entire molecule, plus the motion of the particles relative to the center of mass treated as fixed. This results in Eqs. (5-20) and (5-21), as in the case of diatomic molecules.

The Born–Oppenheimer separation is also made, since the nuclei are very heavy compared to the electrons. The electronic energy is calculated for fixed nuclear positions, and this energy $\epsilon_{el}(R_1, R_2, \ldots)$ acts as a potential for nuclear motion in a particular electronic state. The electronic energy depends on the coordinates $R_1, R_2, \ldots R_{3n-3}$ of the nuclei with respect to the center of mass. A plot of $\epsilon_{el}(R_1, R_2, \ldots)$ yields a potential surface corresponding to a single electronic state. The form of this surface will vary from molecule to molecule, but it is presumed that there is a stable bowl formed at a point corresponding to the equilibrium configuration of the molecule. For example, consider the simplified case of a CO_2 molecule for *linear configurations only*. The distances from the C to each of the O atoms are denoted as R_1 and R_2, and the potential surface is shown in Fig. 5-8. The lines represent contours of constant $\epsilon_{el}(R_1, R_2)$. A bowl occurs with a minimum at $R_1 = R_2 = R_e$ corresponding to the most stable configuration of CO_2. Sections 1–2 and 3–4 of the surface give the CO potential curve since the other O atom is very far away. At point 5, the system consists of $C + O + O$, all very far from one another. Lines AA' on the surface have the same energy. For nonlinear configurations, it is assumed that the entire surface moves up since the linear configuration is most stable.

5.4.2 Rotation of Polyatomic Molecules

Since the translational motion of the center of mass has already been separated out, the dynamics of the nuclear motion of a polyatomic mole-

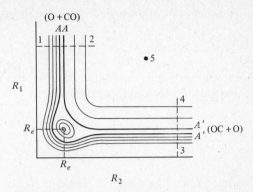

FIGURE 5-8 Schematic energy surface for the triatomic molecule CO_2 in linear configurations. The entire surface changes shape and magnitude for nonlinear configurations.

cule must be treated in terms of $3N - 3$ coordinates relative to the center of mass, where N is the number of atoms in the molecule. If it is assumed that the nuclear motion consists of superimposed small vibrations about equilibrium and rigid rotations of the molecular frame (with internuclear distances corresponding to equilibrium), the vibrational-rotational separation is made in the same way as for diatomics. In general, this approximation is fairly good for smaller molecules but becomes successively worse for larger molecules, especially if some of the main bond linkages are weak. Here the vibrational-rotational separation is used in all discussions. With this approximation, Eq. (5-36) is obtained *even for polyatomic molecules.*

Q_{rot} is only calculated for a polyatomic molecule in the classical approximation. This will be a good approximation for most molecules at room temperature and above because kT is much greater than the spacing between rotational levels. In Chapter 6, the formula for the quantum mechanical energy levels of a symmetrical rigid rotor are given. However, Q_{rot} is calculated here in the purely classical limit. The rotations of a polyatomic molecule are treated as the rotations of a rigid body. According to classical mechanics, three coordinates are required to specify the orientation of a rigid body in space. The angular momentum and angular velocity of the body are vectors with three components each. The relation between components of these vectors is

$$p_i = \sum_j I_{ij}\omega_j \tag{5-46}$$

where p_i is the ith component of the angular momentum, ω_j is the jth component of angular velocity, and I_{ij} is the (i,j)th component of the moment of inertia matrix.   The kinetic energy of rotation is

$$T = \tfrac{1}{2} \sum_{i=1}^{3} \sum_{j=1}^{3} I_{ij}\omega_i\omega_j \tag{5-47}$$

In one particular coordinate system, the I_{ij} matrix will become diagonal, and Eqs. (5-46) and (5-47) simplify to

$$p_i = I_i\omega_i \tag{5-48}$$

$$T = \tfrac{1}{2} \sum_{i=1}^{3} I_i\omega_i^2 = \tfrac{1}{2} \sum_{i=1}^{3} p_i^2/I_i \tag{5-49}$$

where I_i are the diagonal elements, I_{ii}, of the I_{ij} matrix.   The coordinate axes which diagonalize I_{ij} are called the principal axes of inertia, and the moments of inertia for rotation about these axes, I_i, are the principal moments of inertia.   For a symmetrical molecule, the principal axes are along lines of symmetry.   In the general case, the identification of the principal axes can be achieved by standard methods.[10]

To calculate the classical rotational partition function, we require the total energy of the rigid rotating body expressed in terms of a set of coordinates and conjugate momenta,[11] and the volume element in phase space.   There are three coordinates required to specify the orientation of a rigid body in space.   To evaluate the energy, it is most convenient to use Eq. (5-49).   Since there are no forces associated with the rotational motion, the total energy is equal to the kinetic energy.   Using the coordinate system in which I_{ij} is diagonal, gives, from Eqs. (5-49),

$$T = \tfrac{1}{2} \sum_{i=1}^{3} p_i^2/I_i \tag{5-50}$$

Next, it is required to obtain the volume element in phase space corresponding to the three coordinates, and conjugate momenta corresponding to the principal axes.   To obtain these, consider the diagram shown in Fig. 5-9.   The origin of coordinates is at the center of mass.   An arbitrary set of cartesian axes fixed in space is **x**, **y**, and **z**.   For any arbitrary orientation of the molecule in space, let axes **1**, **2**, and **3** be the principal axes of rotation.   A set of coordinates is required which locates axes **1**, **2**, and **3** relative to **x**, **y**, and **z**.   It is convenient to choose these coordinates as the *Eulerian angles* θ, ϕ, and ψ, as shown in Fig. 5-9.   Angle θ is the angle between axes **3** and **z**.   Axes **1** and **2** are located by drawing lines ξ and η such that ξ lies along the intersection of the **x-y** and **1-2**

[10]J. C. Slater and H. H. Frank, *Mechanics* (New York: McGraw-Hill Book Co., 1947), pp. 100–102.
[11]The momentum *conjugate* to generalized coordinate q_i is $p_i = -\partial E/\partial q_i$ where E is the total energy in classical mechanics.

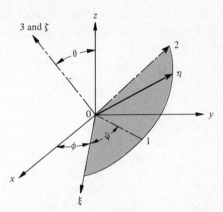

FIGURE 5-9 Coordinates of importance in studying rotation of molecules. Angles θ, ϕ, and ψ are the *Eulerian angles*.

planes, and η is in the **1-2** plane, perpendicular to both axes ξ and **3**. Angle ϕ is the angle between **x** and ξ, and lies in the **x-y** plane. The angle between ξ and **1** is ψ in the **1-2** plane, as shown in Fig. 5-9. The momenta p_1, p_2, and p_3 in Eq. (5-50) are conjugate to coordinates representing rotation about each of the principal axes. These coordinates may be determined as follows. A rotation about axis ξ corresponds to changes in angle θ with ϕ and ψ fixed, because ξ is perpendicular to **z** and ζ. Thus, if θ were to vary with time, the component of angular velocity about ξ would be

$$\omega_\xi = \dot{\theta} \qquad (5\text{-}51)$$

Similarly, since ϕ lies in the **x-y** plane, the body traces out a cone around the **z** axis when ϕ varies with time. If $\dot{\phi}$ is treated as a vector along **z**, it can be resolved into components $\dot{\phi} \sin\theta$ and $\dot{\phi} \cos\theta$, about the η and ζ axes, respectively. When ψ changes with time, it corresponds to rotation about the ζ axis. Thus, summing the contributions of $\dot{\phi}$ and $\dot{\psi}$,

$$\omega_\eta = \dot{\phi} \sin\theta$$
$$\omega_\zeta = \dot{\psi} + \dot{\phi} \cos\theta \qquad (5\text{-}52)$$

Now it is desired to obtain ω_1, ω_2, and ω_3. Since the **3** and ζ axes are identical, it is clear that $\omega_3 = \omega_\zeta$. By treating ω_ξ and ω_η as vectors along the ξ and η axes, respectively, these vectors can be resolved into components along the **1** and **2** axes. When this is done, it is found that

$$\omega_1 = \omega_\xi \cos\psi + \omega_\eta \sin\psi$$
$$\omega_2 = -\omega_\xi \sin\psi + \omega_\eta \cos\psi \qquad (5\text{-}53)$$
$$\omega_3 = \omega_\zeta$$

When Eqs. (5-53) are substituted into Eqs. (5-51) and (5-52), ω_1, ω_2, and ω_3 are determined in terms of $\dot{\theta}$, $\dot{\phi}$, $\dot{\psi}$, θ, ϕ, and ψ.

$$\omega_1 = (\cos \psi)\, \dot{\theta} + (\sin \theta \sin \psi)\, \dot{\phi}$$

$$\omega_2 = -(\sin \psi)\, \dot{\theta} + (\sin \theta \cos \psi)\, \dot{\phi} \qquad (5\text{-}54)$$

$$\omega_3 = (\cos \theta)\, \dot{\phi} + \dot{\psi}$$

For a small variation in time dt, the changes in θ, ϕ, and ψ are $d\theta = \dot{\theta}\, dt$, $d\phi = \dot{\phi}\, dt$ and $d\psi = \dot{\psi}\, dt$. The changes in the angles about axes **1**, **2**, and **3** are $d\chi_1 = \omega_1\, dt$, $d\chi_2 = \omega_2\, dt$ and $d\chi_3 = \omega_3\, dt$. Thus, when we multiply Eq. (5-54) through by dt, we obtain the same equation with $d\chi_1$, $d\chi_2$, $d\chi_3$, $d\theta$, $d\phi$, and $d\psi$ replacing ω_1, ω_2, ω_3, $\dot{\theta}$, $\dot{\phi}$, and $\dot{\psi}$, respectively. It can be shown[12] that in any transformation of coordinates from χ_1, χ_2, χ_3, to θ, ϕ, ψ, the relation between volume elements is

$$d\chi_1\, d\chi_2\, d\chi_3 = |J|\, d\theta\, d\phi\, d\psi \qquad (5\text{-}55)$$

where $|J|$ is the determinant[13] with (i,j) element $\partial \chi_i / \partial \mu_j$, and $\mu_1 = \theta$, $\mu_2 = \phi$, and $\mu_3 = \psi$. Thus, $|J|$ needs to be evaluated. From Eqs. (5-54) (rewritten after multiplying through by dt), the following is obtained:

$$|J| = \begin{vmatrix} \cos \psi & \sin \theta \sin \psi & 0 \\ -\sin \psi & \sin \theta \cos \psi & 0 \\ 0 & \cos \theta & 1 \end{vmatrix} = \sin \theta \qquad (5\text{-}56)$$

Now the classical expression for the rotational partition function can be written, using Eq. (4-10), as

$$Q_{\text{rot}} = \frac{1}{h^3} \underbrace{\int \cdots \int}_{6} e^{-E(p_1, p_2, p_3)/kT}\, dp_1\, dp_2\, dp_3\, d\chi_1\, d\chi_2\, d\chi_3 \qquad (5\text{-}57)$$

The problem with trying to evaluate Eq. (5-57) directly is that it is difficult to evaluate the limits on coordinates χ_1, χ_2, and χ_3. By using Eq. (5-55), $d\chi_1\, d\chi_2\, d\chi_3$ transforms to $\sin \theta\, d\theta\, d\phi\, d\psi$. Since the limits on θ, ϕ, and ψ are $0 \to \pi$, $0 \to 2\pi$, and $0 \to 2\pi$, respectively, the following may be written:

$$Q_{\text{rot}} = \frac{1}{h^3} \int_{p_1=-\infty}^{\infty} e^{-p_1{}^2/2I_1 kT}\, dp_1 \int_{p_2=-\infty}^{\infty} e^{-p_2{}^2/2I_2 kT}\, dp_2 \int_{p_3=-\infty}^{\infty} e^{-p_3{}^2/2I_3 kT}\, dp_3$$

$$\cdot \int_{0}^{\pi} \sin \theta\, d\theta \int_{0}^{2\pi} d\phi \int_{0}^{2\pi} d\psi \qquad (5\text{-}58)$$

---

[12]See Appendix VI.
[13]This determinant is called the *Jacobian* for the transformation.

Each of the momentum integrals is of the same form, the result of integration being[14]

$$Q_{rot} = \frac{1}{h^3} (2\pi I_1 kT)^{1/2} (2\pi I_2 kT)^{1/2} (2\pi I_3 kT)^{1/2} \cdot 2 \cdot 2\pi \cdot 2\pi \quad (5\text{-}59)$$

$$Q_{rot} = \sqrt{\pi} \left(\frac{2I_1 kT}{\hbar^2}\right)^{1/2} \left(\frac{2I_2 kT}{\hbar^2}\right)^{1/2} \left(\frac{2I_3 kT}{\hbar^2}\right)^{1/2} \quad (5\text{-}60)$$

It will be shown in Chapter 6 that this should be divided by the symmetry number σ to take account of states which are excluded in quantum mechanics if the molecule possesses symmetry.

For the special case of a linear polyatomic molecule, the rotations are the same as for a diatomic molecule, and Q_{rot} is given by Eq. (5-35).

5.4.3 Vibration of Polyatomic Molecules

The vibration of polyatomic molecules can be analyzed in terms of the harmonic oscillator model. Referring to Fig. 5-8, one expands $\epsilon_{el}(R_1, R_2, \ldots)$ about the equilibrium configuration:

$$\epsilon_{el}(R_1, R_2, \ldots) = \epsilon_{el}(R_1^e, R_2^e \ldots) + \sum_i \left(\frac{\partial \epsilon_{el}}{\partial R_i}\right)_{eq} (R_i - R_i^e)$$
$$+ \tfrac{1}{2} \sum_i \sum_j F_{ij}(R_i - R_i^e)(R_j - R_j^e) + \cdots \quad (5.61)$$

where $F_{ij} = (\partial^2 \epsilon_{el} / \partial R_i \partial R_j)_{eq}$. The electronic energy of the bottom of the bowl may be defined as the zero of energy, so $\epsilon_{el}(R_1^e, R_2^e, \ldots) = 0$. Furthermore, the mathematical property of the bottom of a bowl-shaped surface is that ϵ_{el} is a minimum in all direction, and therefore $(\partial \epsilon_{el} / \partial R_i)_{eq} = 0$. If higher order terms in Eq. (5-61) are neglected, the result is

$$\epsilon_{el}(R_1, R_2, \ldots) = \tfrac{1}{2} \sum_i \sum_j F_{ij} S_i S_j + \cdots \quad (5\text{-}62)$$

where $S_i = R_i - R_i^e$. This is entirely analogous to Eq. (5-26). Since 3 coordinates were used for the center of mass, and 3 more for rotation (only two for linear molecules), a total of $3N - 6$ ($3N - 5$ for linear molecules) internal coordinates S_i are required to give the vibrational state of a molecule. The force constant matrix F_{ij} can usually be estimated by semi-empirical techniques. It can be shown that whether classical or quantum mechanics is used the analysis of vibrational motion is greatly

---

[14]It should be noted that it is not necessary to actually calculate the principal moments of inertia. In any arbitrary coordinate system, $I_1 I_2 I_3 = \det \mathbf{I}$ where $\mathbf{I}$ is the moment of inertia matrix in the coordinate system.

simplified by using *normal coordinates.* In general the kinetic energy of vibration of a polyatomic molecule can be shown[15] to have the form

$$T = \tfrac{1}{2} \sum_i \sum_j (G^{-1})_{ij} \dot{S}_i \dot{S}_j \tag{5-63}$$

where the $(G^{-1})_{ij}$ are effective mass terms which depend on the masses of the atoms and the structure of the molecule. *Normal coordinates* are defined as the coordinate system in which F_{ij} and $(G^{-1})_{ij}$ become diagonal. In normal coordinates,

$$\epsilon_{el} = \tfrac{1}{2} \sum_i F_{ii} S_i^2$$

$$T = \tfrac{1}{2} \sum_i (G^{-1})_{ii} \dot{S}_i^2 \tag{5-64}$$

In normal coordinates, both the classical and quantum mechanical equations of motion separate into $3N - 6$ uncoupled equations, one for each *normal mode.* Each normal mode has a frequency given by

$$\nu_i = \frac{1}{2\pi} \sqrt{\frac{F_{ii}}{(G^{-1})_{ii}}} \tag{5-65}$$

where F_{ii} and $(G^{-1})_{ii}$ are from Eqs. (5-64). The quantum mechanical energy levels are then simply a sum of harmonic oscillator levels over all normal modes:

$$\epsilon_{vib} = \sum_{i=1}^{3N-6} (n_i + \tfrac{1}{2}) h \nu_i \tag{5-66}$$

Therefore, the vibrational partition function is

$$Q_{vib} = \prod_{i=1}^{3N-6} \frac{e^{-h\nu_i/2kT}}{1 - e^{-h\nu_i/kT}} \tag{5-67}$$

Each normal mode corresponds to a particular kind of motion of the molecule. For example, in CO_2, there are

(1) ←○ ○ ○→ symmetric stretch

(2) ←○ ○→ ←○ asymmetric stretch

(3) $\uparrow$○ ○$\downarrow$ $\uparrow$○ x-bend⎫
⎬(degenerate)
(4) ⊙ ⊗ ⊙ y-bend⎭

---

[15]E. B. Wilson, J. C. Decius and P. C. Cross, *Molecular Vibrations* (New York: McGraw-Hill Book Co., 1955).

To estimate the vibrational frequencies of a molecule, the $F - G$ matrix method[16] can be applied. It will be assumed that the vibrational frequencies are available for any molecule involved in statistical mechanical calculations.

5.4.4 Internal Rotation of Polyatomic Molecules

In some molecules, the possibility exists for motion involving internal rotation of one group of atoms relative to another. For example, in the ethylene molecule C_2H_4 shown below

there is the conceptual possibility that the two CH_2 groups, while retaining their triangular shapes, could rotate about the C—C bond relative to another as shown below in an end view:

This should be distinguished from overall rigid rotation of the molecule, because overall rotation would involve preservation of the relative orientation of the CH_2 groups as both CH_2 groups rotated together in space. As it turns out, internal rotation does *not* in fact occur in the ethylene molecule because the molecule is considerably more stable in the planar configuration than when distorted. Torsional vibrations, in which θ varies by a few degrees about $\theta = 0$ do occur, but this is one of the normal modes of vibration. This is one extreme case, and for such a molecule, no special treatment is required. At the other extreme, is a molecule for which the electronic energy depends only very slightly on the relative orientation of two groups. In $CH_3—NO_2$, for example, there is very little resistance to rotation of the NO_2 group relative to the CH_3. Thus, the electronic energy does not vary much with θ in the diagram

End view

[16]E. B. Wilson, J. C. Decius, and P. C. Cross, *Molecular Vibrations* (New York: McGraw-Hill Book Co., 1955).

The partition function of a molecule capable of such free internal rotation is slightly more complicated than for normal molecules. One of the normal modes of vibration which would correspond to a torsional vibration if there were a value of θ corresponding to a stable potential well, will be replaced by a free rotation in coordinate θ. Most molecules, which are conceptually capable of internal rotation, exist in the torsional vibration limit if the bond connecting the groups of atoms is a multiple bond. Many molecules with single bonds between the groups have moderate electronic-potential barriers that resist internal rotation. These molecules have a mode of "hindered rotation" that replaces what would otherwise be a torsional vibration in one extreme or a free rotation in the other extreme.

Consider a hypothetical molecule $B_2A\text{---}CD_2$ as shown below

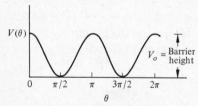

End view

where the B—A—B and D—C—D angles remain essentially constant (except for small vibrations) but the BAB and DCD groups can rotate relative to each other about the AC bond. It is conceivable that either the *opposed* or *staggered* orientations could be most stable. It will be assumed, for the sake of argument, that the staggered position is the more stable, and the potential as a function of θ has the form shown[17] in Fig. 5-10. If the potential function is assumed to be of the form

$$V(\theta) = \tfrac{1}{2}V_0(1 - \cos n\theta) \qquad (5\text{-}68)$$

where, in this case, $n = 2$, but in the general case n is the number of equivalent orientations, and the two rotating groups are symmetrical with respect to the axis of rotation, the quantum mechanical energy levels can

$V(\theta)$

$V_o = \text{Barrier height}$

0    $\pi/2$ π $3\pi/2$ 2π

θ

FIGURE 5-10 Potential energy vs angle for groups capable of relative rotation within a molecule.

[17]K. S. Pitzer, *Quantum Chemistry* (Englewood Cliffs, N.J.: Prentice-Hall, 1953), Appendix 18.

be obtained. The effective moment of inertia can be shown to be $I_{ir} = I_1 I_2 / (I_1 + I_2)$, where I_1 and I_2 are the moments of inertia of the two rotating groups about the main axis. The energy level spectrum depends on the height of the barrier and the moments of inertia. For large barrier heights and large moments of inertia, the energy levels are closely spaced compared to the barrier height shown in Fig. 5-10. A molecule in a configuration with $0 < \theta < \pi$ then has little chance to move into the region $\pi < \theta < 2\pi$, and the potential can be approximated as being harmonic in θ, as shown in Fig. 5-11. The energy levels are closely packed, and the harmonic approximation works well for the lowest levels. If kT is small compared to V_0, this model can be used for the energy levels, which correspond to the normal mode of torsional vibration. This corresponds to the limiting case of a very high barrier to internal rotation. In this case, the partition function is calculated in the usual way without specific treatment of internal rotation.

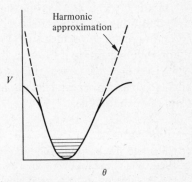

FIGURE 5-11 Harmonic approximation for large barriers to internal rotation.

The other extreme case is where the energy levels are large compared to the barrier height. In this case, the internal rotations are essentially free. The energy levels for an almost free rotor[18] with a potential that is periodic in θ with a period of $2\pi/n$ are

$$\epsilon_l = \frac{\hbar^2}{2 I_{ir}} n^2 l^2 + \frac{V_0}{2} \qquad l = 0, 1, \ldots \qquad (5\text{-}69)$$

Each level above the first is doubly degenerate. If kT is $\gg V_0$, it will be a good approximation to use Eq. (5-69) to evaluate Q_{ir} which replaces the torsional vibrational partition function. It then follows that the partition

---

[18]N. Davidson, *Statistical Mechanics* (New York: McGraw-Hill Book Co., 1962), pp. 194–202.

function for internal rotation is

$$Q_{ir} = \left\{ 1 + \sum_{l=1}^{\infty} (2l) \, e^{-\theta_{ir}l^2/T} \right\} e^{-V_0/2kT} \tag{5-70}$$

where

$$\theta_{ir} = \frac{\hbar^2 n^2}{2 I_{ir} k} \tag{5-71}$$

If kT is much greater than the spacing between levels, the sum can be multiplied by $\Delta l = 1$, and the summation sign can be replaced by an integral.   Neglecting the 1, the following is obtained:

$$Q_{ir} = e^{-V_0/2kT} \left(\frac{2 I_{ir} kT}{\hbar^2} \right)^{1/2} \frac{1}{n} \tag{5-72}$$

in the free rotator limit.   The factor n is a form of *symmetry number*, which will be discussed further in Sec. 6.6.   It can be seen that except for the factor $e^{-V_0/2kT}$, Q_{ir} has the same form as for an ordinary rotation.   If the conditions are proper for a molecule ($kT \gg V_0$ and $T \gg \theta_{ir}$), Eq. (5-72) should be used to replace the vibrational partition that would correspond to torsional vibration if V_0 were large compared to kT.

In the general case where kT is neither much greater than nor much smaller than V_0, energy levels and partition functions can be obtained from published tables.[19]

Problems

1. Calculate the partition function for the oxygen atom at 300°K and at 3000°K according to the following approximations:
   (a) Assume only the ground electronic state (^3P) is populated, and neglect the differences in energy between the $^3P_2$, $^3P_1$, and $^3P_0$ substates.
   (b) Assume only the ground electronic state is populated, but use the exact energies of the $^3P_2$, $^3P_1$, and $^3P_0$ substates.
   (c) Repeat (b), but also include the ^1D and ^1S states.
   (d) What can you conclude regarding the nature of approximations (a) and (b) at various T?
2. Calculate the contribution to the total energy of Li vapor made by the electronic and translational degrees of freedom at 300°K and 3000°K, assuming that Li vapor exists as Li atoms, and including only the lowest 3 electronic states.
3. Construct plots like the one in Fig. 5-6 for O_2 at 300°K, and for H_2 at 300°K and at 3000°K.   At what temperature would you guess the classical approximation for the rotations of H_2 becomes adequate?
4. What fraction of all the N_2 molecules in N_2 gas at 3000°K are in the first excited vibrational state of the A electronic state?

---

[19]K. S. Pitzer and W. D. Gwinn, *J. Chem. Phys.* **10**, 428 (1942).

5. What is the fraction of diatomic molecules in rotational states with rotational quantum number greater than l_m at temperature T if the rotational temperature is θ_{rot}? Hint: Take

$$\int_{l_m}^{\infty} dn(l) \Big/ \int_{0}^{\infty} dn(l)$$

6. (a) Calculate the most probable value of the rotational energy per diatomic molecule at temperature T for θ_{rot}. From this, calculate the average rotational period, treating the rotations as classical.

    (b) For the N_2 molecule in the ground electronic state, calculate numerical values for the rotational and vibrational periods at $T = 1000°K$. How many vibrations occur per rotation?

7. Calculate the electronic, translational, rotational, and vibrational energies of CO at 300°K, 1000°K, 3000°K, and 5000°K. Express E_{cm}, E_{rot}, and E_{vib} in units of kT at each temperature. Discuss in terms of problem 1 in Chapter 4.

8. Calculate the partition functions of CO_2 and H_2O at 1000°K using only the ground (singlet) electronic states. The symmetry number of each molecule is 2.

chapter 6

SYMMETRY NUMBERS

6.1 NUCLEAR CONTRIBUTION TO THE PARTITION FUNCTION

Every nucleus has nuclear energy levels and degeneracies. There-fore, to be rigorous, the energy levels of a molecule should include nuclear as well as translational, rotational, vibrational, and electronic energy levels.   If the total energy can be approximated as a sum of such contributions, then the partition function is

$$Q = Q_{el}Q_{tr}Q_{rot}Q_{vib}Q_{nuc} \tag{6-1}$$

where

$$Q_{nuc} = \sum_{\substack{i=\text{all} \\ \text{nuclear} \\ \text{levels}}} g_i \, e^{-\epsilon_i kT} \tag{6-2}$$

However, the separation between nuclear levels is usually of the order of several million eV, and $kT \cong 0.03$ eV at room temperature.   Therefore, to within an error of one part in $\exp[-10^6/0.03]$, Q_{nuc} may be approxi-mated by simply g_0, the degeneracy of the ground nuclear state.   Thus, it may be assumed that

$$Q_{nuc} = g_0 \tag{6-3}$$

In any application of statistical mechanics to a nonnuclear physical process

$$\text{initial state} \quad \rightarrow \quad \text{final state} \tag{6-4}$$

the nuclear states are unchanged in the process.   Therefore, $Q_{nuc} = g_0$ for both the initial and final states.   In computing any thermodynamic change for process (6-4), the ratio of partition functions Q_f/Q_i needs to be calculated, where f and i refer to final and initial states.   Therefore, g_0 will cancel in taking this ratio.   Hence, one need not usually bother to include g_0 in Q because it will cancel in Q_f/Q_i.

However, in those cases where a molecule possesses nuclear symmetry, a consideration of the nuclear states and energy levels cannot be neglected in taking ratios of partition functions.   This is due to certain quantum mechanical restrictions on the overall symmetry of a molecule containing symmetrically equivalent atoms.   If the energy of a molecule can be approximated as a sum of energies due to translational, rotational, vibrational, electronic, and nuclear contributions, the energy and wave function of a molecule can be represented as

$$\epsilon = \epsilon_{tr} + \epsilon_{rot} + \epsilon_{vib} + \epsilon_{el} + \epsilon_{nuc} \tag{6-5}$$

$$\Psi = \psi_{tr}\psi_{rot}\psi_{vib}\psi_{el}\psi_{nuc} \tag{6-6}$$

If a molecule does not possess any elements of symmetry (for example, HCl) then there are no restrictions on the nuclear states, and there is exact cancellation between nuclear partition functions for initial and final states in any process.   However, if the molecule possesses elements of symmetry (for example, say, H_2) certain states, which might be expected to occur in nature, in fact are absent.   Since Q is essentially a sum over all states, the partition function will be reduced.   In a process such as

$$H_2 + Cl_2 \rightarrow 2HCl$$

the partition functions of H_2 and Cl_2 are reduced compared to the partition function of HCl.   Therefore, because of symmetry, the ratio

$$\frac{Q_{HCl}Q_{HCl}}{Q_{H_2}Q_{Cl_2}}$$

has a larger value, other factors being equal, than if the reagents on the left side did not possess symmetry.

6.2   EFFECT OF SYMMETRY OPERATIONS ON MOLECULAR WAVE FUNCTIONS

6.2.1 Basic Postulates

Consider three basic postulates from quantum mechanics:

(a) All nuclei have a property called spin.   The spin is a quantity of angular momentum projected along some arbitrary axis, and has magnitude $i\hbar$, where i is the spin quantum number (often loosely referred to as simply, "the spin").   The degeneracy of the energy level associated with spin i is $2i + 1$.

(b) There are two kinds of particles, called Bosons and Fermions. Bosons may have spin $0, 1, 2, \ldots$. Fermions are particles with spin $\frac{1}{2}, \frac{3}{2}, \frac{5}{2}, \ldots$.   A particular kind of particle always has the same spin. For example, the deuteron has spin 1 and is a Boson.

(c) In any molecule containing two or more symmetrically placed Bosons (Fermions), the effect of a symmetry operation causing interchange of any two such *equivalent* particles will be to make the total wave function Ψ go to $+\Psi$ $(-\Psi)$. Thus, Ψ is said to be *symmetric* (*antisymmetric*) with respect to interchange of equivalent Bosons (Fermions).

First consider the two electrons in H_2. Interchange of the two electrons, as shown in Fig. 6-1 must cause $\Psi \rightarrow -\Psi$ since electrons are Fermions. But ψ_{rot}, ψ_{tr}, ψ_{vib} and ψ_{nuc} are all independent of the electron interchange process (that is, these functions are symmetric). Since the overall Ψ must be antisymmetric, it follows that ψ_{el} must be antisymmetric. It turns out that ψ_{el} may be written as

$$\psi_{el} = (\psi_{el})_{orb}(\psi_{el})_{spin} \tag{6-7}$$

where $(\psi_{el})_{orb}$ refers to the orbital motion of the electrons about the nuclei and $(\psi_{el})_{spin}$ refers to the electron spin function. If $(\psi_{el})_{orb}$ is antisymmetric, then $(\psi_{el})_{spin}$ must be symmetric, and vice versa. For example, $(\psi_{el})_{orb}$ for the ground electronic state and the first excited electronic state of H_2 are shown in Fig. 6-2. In the ground state $(\psi_{el})_{orb}$ is symmetric and, therefore, $(\psi_{el})_{spin}$ must be antisymmetric. The antisymmetric spin function has the spins oppositely paired (↑ ↓) leading to a total electron spin $i = 0$.

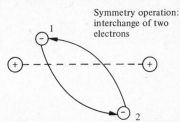

Symmetry operation: interchange of two electrons

FIGURE 6-1 Illustration of the symmetry operation of interchange of two electrons.

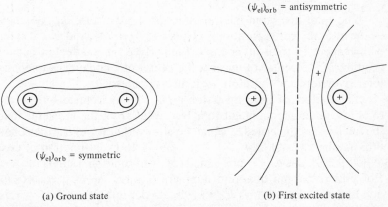

$(\psi_{el})_{orb}$ = antisymmetric

$(\psi_{el})_{orb}$ = symmetric

(a) Ground state

(b) First excited state

FIGURE 6-2 The contours of constant electron density in the ground state (a) and first excited state (b) of H_2.

Therefore, the spin degeneracy is $2i + 1 = 1$, and the ground state is a "singlet." In the first excited state $(\psi_{el})_{orb}$ is antisymmetric and, therefore, $(\psi_{el})_{spin}$ is symmetric (↑ ↑) leading to $i = 1$ and $2i + 1 = 3$. Thus the excited state is a "triplet."

6.2.2 The Total Inversion Operator

The total inversion operator for a symmetrical diatomic molecule is the operation of inverting all particles in the molecule through the midpoint between the nuclei on the bond axis, as shown in Fig. 6-3. Because this operation leaves all electron–electron and electron–nucleus distances unchanged, $|\psi_{el}|^2$ is unchanged. However, ψ_{el} can go to either $\pm \psi_{el}$ under the total inversion operator, depending on the symmetry of ψ_{el}. Electronic states that go to $\pm \psi_{el}$ under the total inversion operator are designated as $+$ or $-$ states and are said to be symmetric or antisymmetric with respect to total inversion.

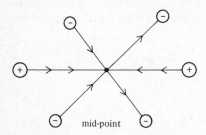

FIGURE 6-3 The symmetry operation of inversion of all particles through the origin.

The total inversion has no effect on ψ_{vib} because ψ_{vib} depends only on the internuclear separation. Thus $\psi_{vib} \to +\psi_{vib}$, and ψ_{vib} is symmetric with respect to total inversion. Similarly ψ_{tr} is unaffected because the position of the center of mass at the center of symmetry is unchanged by total inversion, and therefore ψ_{tr} is also symmetric.

The effect of total inversion on ψ_{rot} is determined by the fact that total inversion is equivalent to a 180° rotation of the molecule insofar as the nuclei are concerned. It can be shown that for even rotational quantum numbers $(l = 0, 2, 4 \ldots)$ $\psi_{rot}(\cos \theta)$ is an even function of $\cos \theta$. If $\theta \to \theta + 180°$, $\cos \theta \to -\cos \theta$, and for even rotational states $\psi_{rot} \to +\psi_{rot}$. For odd rotational quantum numbers $(l = 1, 3, 5, \ldots)$, $\psi_{rot}(\cos \theta)$ is an odd function of $\cos \theta$, and $\psi_{rot} \to -\psi_{rot}$ under total inversion. Thus, even (odd) rotational states are symmetric (antisymmetric) with respect to total inversion.

6.2.3 Inversion of Only the Electrons

The operation of inversion of only the electrons is shown in Fig. 6-4. This operation leaves all electron-nucleus and electron-electron distances unchanged. Therefore, as before, $\psi_{el} \rightarrow \pm \psi_{el}$, depending on the symmetry of the electronic state. States that go to $+ \psi_{el}$ under this operation are called *gerade* (g) and states that go to $- \psi_{el}$ are called *ungerade* (u).

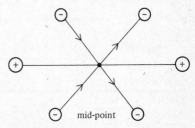

FIGURE 6-4 The symmetry operation of inversion of only the electrons through the origin.

Since the positions of the nuclei are unchanged by this operation, ψ_{tr}, ψ_{rot} and ψ_{vib} are symmetric with respect to inversion of the electrons.

6.2.4 Inversion of Only the Nuclei

The operations of interchanging the nuclei, keeping the electrons fixed, is illustrated in Fig. 6-5. This operation may be achieved by successive operations of (1) total inversion and (2) inversion of only the electrons. The first operation will invert all particles, and the second operation will bring the electrons back to their original positions, leaving only the nuclei inverted.

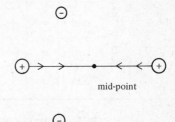

FIGURE 6-5 The symmetry operation of inversion of only the nuclei.

In order to determine the effect of inversion of the nuclei on the various wave functions, all that is necessary is to determine the effect of successive operations of total inversion and inversion of only the electrons.

Since the effect of these individual operations on the various wave functions has already been determined, the composite effect of successive application of the operations can now be determined. The results are given in Table 6-1.

TABLE 6-1

	ψ_{el}		ψ_{vib}	ψ_{rot}		ψ_{tr}
Symmetry of wave function	(+)	(−)		Even	Odd	
Step (1): total inversion	+	−	+	+	−	+
Symmetry of wave function	(g)	(u)				
Step (2): inversion of electrons	+	−	+	+	+	+

The results of Table 6-1 can be combined to yield the effect of interchange of the nuclei on $\psi_{el}\psi_{vib}\psi_{rot}\psi_{tr}$, as in Table 6-2.

TABLE 6-2

Symmetry of Wave Functions				*Effect of Nuclear Inversion on $\psi_{el}\psi_{vib}\psi_{rot}\psi_{tr}$*
vib	*tr*	*rot*	*electronic*	
does not matter	does not matter	even	g	+	+
		odd	g	+	−
		even	g	−	−
		odd	g	−	+
		even	u	+	−
		odd	u	+	+
		even	u	−	+
		odd	u	−	−

The effect of interchange of the nuclei on ψ_{nuc} depends on the symmetry of ψ_{nuc}. The following facts from quantum mechanics are stated without proof:

(a) If each nucleus of a homonuclear diatomic molecule has spin i, the total nuclear spin of the molecule, I, can have values $2i$, $2i-1$, $2i-2, \ldots 0$, depending on which nuclear state is formed.

(b) The degeneracy of these states are $2I+1$.

(c) The symmetries of these states with respect to inversion of the nuclei alternate with I so that the state with $I = 2i$ is symmetric, $2i - 1$ is antisymmetric, $2i - 2$ is symmetric, and so forth.

For any particular electronic-rotational state, the only possible nuclear states are those for which the symmetry of the overall Ψ is + for Bosons and − for Fermions. For example, consider the particular case of the first line in Table 6-2. For nuclei that are Bosons, since the symmetry of Ψ must be + with respect to interchange of the symmetrically equivalent nuclei, and since $\psi_{el}\psi_{vib}\psi_{rot}\psi_{tr}$ has + symmetry, it follows that only nuclear states with + symmetry are allowed. Antisymmetric nuclear states would only be allowed if the nuclei were Fermions. For each line in Table 6-2, a table may be constructed showing which nuclear states can be combined with various electronic-rotational states in order to make the symmetry of the overall wave function + (−) with respect to interchange of the nuclei for Bosons (Fermions). The result is given in Table 6-3. The symmetries of ψ_{nuc} have been chosen to make the symmetry of Ψ come out right.

As a first example, let us consider Hg_2^{200} which has ground electronic state with symmetry $(g, +)$. The nuclear spin of Hg^{200} is $i = 0$, so the only possible nuclear state of Hg_2^{200} has $I = 0$, which has ψ_{nuc} symmetric. If rows 1 and 2 in Table 6-3 are examined it will be found that only row 1 can exist because ψ_{nuc} cannot be antisymmetric. Therefore, only even rotational quantum states can exist for such a molecule, and the rotational states that would exist for a heteronuclear molecule with $I = 1, 3, 5, \ldots$ are forbidden by symmetry in Hg_2^{200}. The rotational levels are shown in Fig. 6-6.

. . .

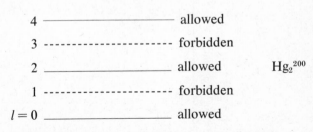

4 ———————————————	allowed	
3 ----------------------	forbidden	
2 ———————————————	allowed	Hg_2^{200}
1 ----------------------	forbidden	
$l = 0$ ———————————	allowed	

FIGURE 6-6 Allowed and forbidden rotational levels in the molecule Hg_2^{200}.

As another example, consider O_2^{16}, for which the ground electronic state is $(g, -)$. Since $i = 0$ for O^{16}, the only possible nuclear state of O_2^{16} has $I = 0$, and is symmetric. Therefore, only row 4 applies to O_2^{16}, and only odd rotational states occur, as shown in Fig. 6-7.

TABLE 6-3 Symmetry of functions

Row	ψ_{el} +, −	ψ_{el} g, u	ψ_{rot} even (e) odd (o)	$\psi_{el}\psi_{vib}\psi_{tr}\psi_{rot}$ Symmetric (S) Antisymmetric (A)	ψ_{nuc} Bosons	ψ_{nuc} Fermions	Ψ Bosons	Ψ Fermions
1	+	g	e	S	S	A	S	A
2	+	g	o	A	A	S	S	A
3	−	g	e	A	A	S	S	A
4	−	g	o	S	S	A	S	A
5	+	u	e	A	A	S	S	A
6	+	u	o	S	S	A	S	A
7	−	u	e	S	S	A	S	A
8	−	u	o	A	A	S	S	A

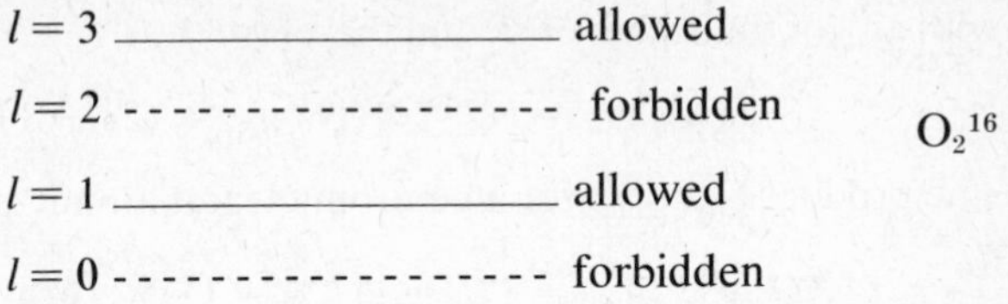

$l = 3$ _____ allowed

$l = 2$ ------------------ forbidden $\quad O_2{}^{16}$

$l = 1$ _____ allowed

$l = 0$ ----------------- forbidden

FIGURE 6-7 Allowed and forbidden rotational levels in O^{16}.

Next, consider $H_2{}^1$ for which the ground electronic state is $(g,+)$. The nuclear spin of H^1 is $i = \frac{1}{2}$, so there are two possibilities for I, namely, 1 and 0. The nuclear state with $I = 1$ is symmetric, and this corresponds to row 2 of Table 6-3, whereas the nuclear state for $I = 0$ is antisymmetric and corresponds to row 1 of Table 6-3. The possible rotational states are illustrated in Fig. 6-8.

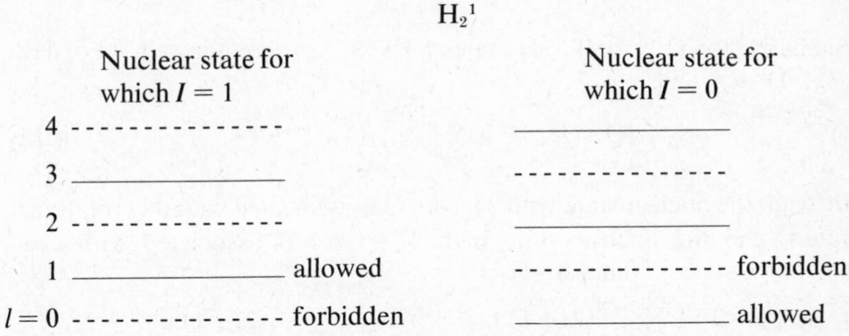

$$H_2{}^1$$

Nuclear state for which $I = 1$		Nuclear state for which $I = 0$	
4 ---------------		_____	
3 _____		---------------	
2 ---------------		_____	
1 _____	allowed	--------------	forbidden
$l = 0$ --------------	forbidden	_____	allowed

FIGURE 6-8 Allowed and forbidden rotational levels in H_2^1.

6.3 EFFECT OF SYMMETRY ON THE ROTATIONAL PARTITION FUNCTION

The nuclear partition function of an atom is

$$Q_{nuc} = \sum_{\substack{\text{all nuclear} \\ \text{levels}}} g_i \, e^{-\epsilon/kT} \tag{6-8}$$

Since the spacing between nuclear levels is generally extremely large compared to kT, a very accurate approximation is obtained by writing

$$Q_{nuc} \cong g_0 = 2i + 1 \tag{6-9}$$

where g_0 is the degeneracy of the ground level, and i is the nuclear spin. For a heteronuclear diatomic molecule consisting of nucleus A and nucleus B, any of the degenerate nuclear states of A can be associated with any of the states of B, and therefore,

$$(Q_{nuc})_{AB} = (Q_{nuc})_A (Q_{nuc})_B = (2i_A + 1)(2i_B + 1) \tag{6-10}$$

The product $Q_{rot}Q_{nuc}$ for a heteronuclear diatomic molecule is

$$(Q_{rot}Q_{nuc})_{AB} = (2i_A + 1)(2i_B + 1) \sum_{l=0}^{\infty} (2l+1) \, e^{-l(l+1)\theta_R/T} \tag{6-11}$$

For a homonuclear diatomic molecule, however,

$$(Q_{rot}Q_{nuc}) \neq (2i+1)^2 \sum_{l=0}^{\infty} (2l+1) \, e^{-l(l+1)\theta_R/T} \tag{6-12}$$

because certain rotational-nuclear state combinations are forbidden. Thus in $Hg_2{}^{200}$, $i = 0$, $2i+1 = 1$, and only $l =$ even is allowed. Then for $Hg_2{}^{200}$,

$$Q_{rot}Q_{nuc} = (2i+1)^2 \sum_{l=0,2,4,\ldots} (2l+1) \, e^{-l(l+1)\theta_R/T} \tag{6-13}$$

Similarly, for $O_2{}^{16}$, only odd states $l = 1, 3, 5 \ldots$ are allowed. For $H_2{}^1$, $(2i+1)^2 = 4$, and

$$Q_{rot}Q_{nuc} \neq 4 \sum_{l=0}^{\infty} (2l+1) \, e^{-l(l+1)\theta_R/T} \tag{6-14}$$

Instead, the nuclear state with $2I+1 = 3$ is associated with odd rotational states, and the nuclear state with $2I+1 = 1$ is associated with even rotational states. Thus for $H_2{}^1$:

$$Q_{rot}Q_{nuc} = 3 \sum_{l=1,3,5,\ldots} (2l+1) \, e^{-l(l+1)\theta_R/T} + \sum_{l=0,2,4,\ldots} (2l+1) \, e^{-l(l+1)\theta_R/T}$$

$$\tag{6-15}$$

In most cases of interest, $T \gg \theta_R$, and the sums may be replaced by integrals. Now it is necessary to show that the integral over the even states is equal to the integral over the odd states. The sum over odd rotational states may be written

$$Q_{odd} = \sum_{l=1,3,5,\ldots} (2l+1) \, e^{-l(l+1)\theta_R/T} = \sum_{n=0,1,2\ldots} (4n+3) \, e^{-(2n+1)(2n+2)\theta_R/T}$$

$$\tag{6-16}$$

where $l = 2n+1 = odd$. If the sum is multiplied by $\Delta n = 1$, and the sum replaced by an integral,

$$Q_{odd} \cong \int_0^{\infty} e^{-y\theta_R/T} \, dy/2 = \frac{1}{2}\frac{T}{\theta_R} \tag{6-17}$$

where $y = 4n^2 + 6n + 2$ and $dy = 2(4n+3) \, dn$.

Similarly.

$$Q_{even} = \sum_{l=0,2,4,...} (2l+1) \, e^{-l(l+1)\theta_R/T}$$

$$= \sum_{n=0,1,2...} (2n+1) \, e^{-2n(2n+1)\theta_R/T} \qquad (6.18)$$

where $l = 2n =$ even. Then, again multiplying by Δn and changing to an integral, the following is obtained:

$$Q_{even} \cong \frac{1}{2} \int_0^\infty e^{-y\theta_R/T} \, dy = \frac{1}{2} \frac{T}{\theta_R} \qquad (6-19)$$

where $y = 4n^2 + 2n$ and $dy = 2(4n+1) \, dn$.

If the sums can be replaced by integrals, Q_{rot} for a symmetric diatomic molecule is $\frac{1}{2}$ of that for an asymmetric molecule. The usual formula for Q_{rot} for *heteronuclear* molecules is

$$Q_{rot} = \sum_{l=0}^\infty (2l+1) \, e^{-l(l+1)\theta_R/T} \cong T/\theta_R \qquad (6-20)$$

Therefore, for heteronuclear diatomics,

$$(Q_{rot}Q_{nuc})_{AB} = (2i_A+1)(2i_B+1)T/\theta_R = GT/\theta_R \qquad (6-21)$$

where $G = (2i_A+1)(2i_B+1)$. For a *symmetrical* diatomic molecule,

$$(Q_{rot}Q_{nuc})_{AA} = (2i_A+1)^2 T/2\theta_R = GT/2\theta_R \qquad (6-22)$$

Equations (6-21) and (6-22) can be combined to produce the general formula

$$(Q_{rot}Q_{nuc}) = G(T/\sigma\theta_R) \qquad (6-23)$$

where σ is defined as the *symmetry number*, and is equal to 1 for heteronuclear diatomics, and 2 for homonuclear diatomics.

In any process

$$\text{initial state} \quad \rightarrow \quad \text{final state}$$

where one needs to take the ratio of partition functions of final and initial states, Q_f/Q_i, it will be found that the G factors will cancel. For example, in the process

$$A + B \rightarrow AB$$

Q_f/Q_i will contain the nuclear contribution

$$\frac{(Q_{nuc})_{AB}}{(Q_{nuc})_A (Q_{nuc})_B} = \frac{G}{(2i_A+1)(2i_B+1)} = 1$$

whether A is the same as B or not. However, the rotational partition function of AB will contain a factor σ is the denominator which is 2 for homonuclear, and 1 for heteronuclear molecules. Thus, the nuclear contribution to the partition functions of atoms and diatomic molecules can be neglected provided that Q_{rot} is written as

$$Q_{rot} = \frac{T}{\sigma \theta_R} \tag{6-24}$$

This is due to the fact that for symmetrical diatomics, half of the rotational quantum states that normally exist for heteronuclear diatomics are absent. As we saw in the examples, no even rotational states of $O_2{}^{16}$ exist, no odd states of $Hg_2{}^{200}$ exist, and in $H_2{}^1$, $\frac{1}{4}$ of the odd states and $\frac{3}{4}$ of the even states normally associated with a heteronuclear molecule do not exist.

It should be emphasized that the symmetry number is a purely quantum effect which does not occur in classical mechanics. When the sums that occur in the rotational partition function are replaced by integrals, the classical limit is taken. However, it is modified by the quantum symmetry effects as represented by σ.

If T is not $\gg \theta_R$, then the sums in Eqs. (6-11), (6-13), (6-14), (6-15), (6-16), and (6-18) cannot be replaced by integrals. Therefore, for low T, Eq. (6-24) is not valid, and the rotational partition function must be evaluated from the sums over states. At low temperatures, where the sum in Q_{rot} cannot be replaced by an integral, Q_{rot} for a homonuclear molecule is not exactly $\frac{1}{2}$ of Q_{rot} for a heteronuclear molecule with the same θ_R. For $Hg_2{}^{200}$,

$$Q_{rot}Q_{nuc} = 1 \cdot \sum_{l=0,2,4,\ldots} (2l+1) \, e^{-l(l+1)\theta_R/T} \tag{6-25}$$

For $H_2{}^1$,

$$Q_{rot}Q_{nuc} = 3 \sum_{l=1,3,5\ldots} (2l+1) \, e^{-l(l+1)\theta_R/T}$$

$$+ 1 \sum_{l=0,2,4\ldots} (2l+1) \, e^{-l(l+1)\theta_R/T} \tag{6-26}$$

In the general case of a homonuclear diatomic with each nucleus having spin i, there will be symmetric molecular nuclear states with spin $2i$, $2i-2$, $2i-4, \ldots$ and antisymmetric molecular nuclear states with spin $2i-1, 2i-3, \ldots$, the lowest molecular spin being 0. The total number of symmetric nuclear states will be

$$N_S = [2(2i)+1] + [2(2i-2)+1] + \cdots \tag{6-27}$$

whereas the total number of antisymmetric states is

$$N_A = [2(2i-1)+1] + [2(2i-3)+1] + \cdots \tag{6-28}$$

If i is one of the numbers $0, 1, 2, \ldots$, then

$$Q_{rot} = N_S Q_{even} + N_A Q_{odd} \tag{6-29}$$

where Q_{even} and Q_{odd} are given in Eqs. (6-16), and (6-18).   If i has one of the values $\frac{1}{2}, \frac{3}{2}, \frac{5}{2} \ldots$, then

$$Q_{rot} = N_S Q_{odd} + N_A Q_{even} \qquad (6\text{-}30)$$

PROBLEM: Prepare a table of $2i$, N_S, and N_A vs i for $i = 0, \frac{1}{2}, 1, \frac{3}{2}, 2$, and $\frac{5}{2}$. Find the values of N_S/N_A for each case.   At high T, where $T \gg \theta_R$, what is the ratio of equilibrium populations of symmetrical nuclear states to antisymmetrical nuclear states?   What is the equilibrium ratio at very low T?

6.4   ORTHO AND PARA HYDROGEN

It can be shown that the total nuclear spin I of a molecule is a property which is extremely difficult to change by molecular collision. Therefore, unless a special catalyst is present, molecules with symmetric and antisymmetric nuclear states will preserve this character regardless of any changes induced in the rotational, vibrational, electronic and translational states.   Suppose there is a container of hydrogen gas at equilibrium at some temperature T.   Hydrogen molecules in symmetric nuclear states are called *ortho* (noted with subscript o), and those in antisymmetrical nuclear states are called *para* (noted with subscript p).   At temperature T, the equilibrium ratio of numbers of ortho H_2 to para H_2 is

$$\frac{[H_2]_o}{[H_2]_p} = \frac{(3)\,Q_{odd}}{(1)\,Q_{even}}$$

The word *equilibrium* ratio should be emphasized because it is implied here that some catalyst is present to allow nuclear states to exchange and reach equilibrium.   At high T, $Q_{odd} \cong Q_{even}$, and therefore, the limiting equilibrium ratio for $T \gg \theta_R$ is

$$\frac{[H_2]_o}{[H_2]_p} = 3 \qquad \text{for } T \gg \theta_R$$

At very low T where $T \ll \theta_R$, the equilibrium distribution would have essentially all the H_2 in the ground rotational state, which happens to have para nuclear symmetry.   Thus, for $T \ll \theta_R$, $[H_2]_o/[H_2]_p \cong 0$.   At intermediate T, the equilibrium mixture is shown in Fig. 6-9.   The actual mixture in an experiment will only be given by this provided a catalyst is present to allow interconversion of ortho and para H_2 as the temperature is changed.   A catalyst usually works by dissociating the H_2 adsorbed on its surface.   When the atoms recombine on the surface they may have the nuclear spins paired in any way, and thus all nuclear states are continually interchanging on such a surface.   Of course, enough time must be provided to allow H_2 to diffuse to the surface, undergo nuclear transitions, and leave the surface.

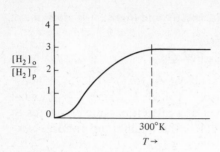

FIGURE 6-9 Equilibrium ratio of ortho-H_2 to para-H_2 vs T.

Various nonequilibrium mixtures of ortho and para H_2 can be prepared by taking H_2 to an arbitrary temperature T_0 in the presence of a catalyst, then removing the catalyst, and then changing T. The ratio of ortho to para H_2 will correspond to T_0, not T. For example, if H_2 is brought to cryogenic temperatures in the presence of a catalyst it will be nearly all para. Then if the catalyst is removed and the H_2 is brought up to room temperature, it will be nearly pure para and will not be in thermodynamic equilibrium.

In D_2, $i = 1$ and the possible nuclear states are $I = 2, 1, 0$. Since the ground electronic state is $(g, +)$ and D is a Boson, symmetric (ortho) nuclear states $(I = 2, 0)$ are associated with even rotational states and the odd nuclear state $(I = 1)$ is associated with odd rotational states. The factor $(2i + 1)^2 = 9$. The degeneracies associated with the ortho nuclear states add up to $5 + 1 = 6$, and the degeneracy of the para nuclear state is 3. At high T, the ratio $o - D_2/p - D_2 = 2$, while at low T all the D_2 becomes ortho.

In general for a $(g, +)$ ground electronic state, the ratio of ortho $- X_2$ to para $- X_2$ for any homonuclear molecule X_2 at high T is

$$\frac{[o - X_2]}{[p - X_2]} = \frac{[2(2i) + 1 + 2(2i - 2) + 1 + \cdots]}{[2(2i - 1) + 1 + 2(2i - 3) + 1 + \cdots]} = \frac{i + 1}{i}$$

6.5 POLYATOMIC SYMMETRIC TOP MOLECULES

In Sec. 6.2, it was shown that for symmetric diatomic molecules, the symmetry requirements on the overall wave function result in some of the rotational states being absent where such rotational states would exist for an unsymmetrical molecule. If kT is much greater than spacing between rotational levels, the net effect was to eliminate half the states which would exist for unsymmetrical diatomic molecules. Therefore, it was found that

$$(Q_{\text{rot}})_{\text{symmetric molecule}} = \tfrac{1}{2}(Q_{\text{rot}})_{\text{unsymmetrical molecule}} \qquad (6\text{-}31)$$

Analogous results are found in polyatomic molecules.

In Chapter 5, the classical rotational partition function of a polyatomic molecule is shown to be

$$(Q_{\text{rot}})_{\text{cl}} = \sqrt{\pi} \left(\frac{2I_A kT}{\hbar^2}\right)^{1/2} \left(\frac{2I_B kT}{\hbar^2}\right)^{1/2} \left(\frac{2I_C kT}{\hbar^2}\right)^{1/2} \tag{6-32}$$

where I_A, I_B, and I_C are moments of inertia for rotation about the principal axes of rotation. This expression contains no reference to the symmetry of a molecule, and will only be the correct high-temperature limit of the rotational partition function of a polyatomic molecule if there are no *symmetrically equivalent nuclei* in the molecule. If two or more symmetrically equivalent Bosons (Fermions) do occur in the molecule, the overall wave function must be symmetric (antisymmetric) with respect to symmetry operations which interchange these Bosons (Fermions). The symmetry of the nuclear spin functions results in some of the rotational states being excluded. If one could solve the Schroedinger equation for the rotational energy levels and wave functions of a general polyatomic molecule, evaluate the partition function as a sum over levels, and take the limit as kT becomes large compared to the spacing between levels, it would be found that

$$(Q_{\text{rot}})_{\text{qm}} = \frac{(Q_{\text{rot}})_{\text{cl}}}{\sigma} \tag{6-33}$$

where σ is the *symmetry number*[1] which results from the symmetry requirements. Unfortunately, one cannot solve exactly the general problem of the rigid rotating molecule in quantum mechanics. The quantum mechanical problem of a symmetric top (a rigid rotator for which $I_A = I_B \neq I_C$) can be solved, however, and this can be utilized to derive Eq. (6-33). The procedure to be followed is the same as for diatomic molecules. One must elucidate the rotational quantum states and determine their symmetry. Then, one must determine which combinations of nuclear spin states and rotational states produce overall wave functions with the proper overall symmetry. Unfortunately, this is very complicated, and is essentially impossible to do in general. We shall merely illustrate the procedure for a simple case.

The orientation of a symmetric top molecule can be described in terms of the angles shown in Fig. 6-10, where angles θ and ϕ determine the orientation of the main axis of symmetry, and angle χ specifies the rotational orientation about this axis. For example, in the molecule CH_3Cl, the main symmetry axis is along the C—Cl bond, and angle χ

---

[1] The symmetry number is the number of indistinguishable positions into which the molecule can be turned by simple rigid rotations.

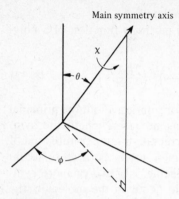

Main symmetry axis

FIGURE 6-10 Angles used to determine the orientation of the main symmetry axis of a symmetric top molecule.

specifies the orientation of the CH_3 group about this axis. It can be shown[2] that the rotational wave function for such a molecule can be put in the form

$$\psi_{\mathrm{rot}}(\theta, \phi, \chi) = \Theta_J(\theta)\, e^{im\phi}\, e^{ik\chi} \qquad (6.34)$$

where $\Theta_J(\theta)$ is a function of θ specified by quantum number J, and m and K are also quantum numbers. In order that ψ_{rot} be finite, single-valued and continuous, the only possible values of J, m, and K that can be chosen are $J = 0, 1, 2, 3, \ldots, \infty$, and for each value of J, K and m can vary from

$$K = 0, \pm 1, \pm 2, \ldots, \pm J \qquad (6\text{-}35)$$
$$m = 0, \pm 1, \pm 2, \ldots, \pm J$$

The energy of any particular state does not depend on m, and is given by

$$E_{J,K} = \frac{\hbar^2}{2}\left\{\frac{J(J+1)}{I_A} + K^2\left(\frac{1}{I_C} - \frac{1}{I_A}\right)\right\} \qquad (6\text{-}36)$$

if $I_A > I_C$. The degeneracy of such an energy level is the number of values of m, namely $2J+1$. It should be remembered that I_A is the moment of inertia which is equal to I_B by symmetry. *If there were no symmetry restrictions on the allowed wave functions*, the quantum-mechanical rotational partition function would be

$$Q_{\mathrm{rot}} = \sum_{J=0}^{\infty} \sum_{K=-J}^{J} (2J+1)\exp\left\{-\frac{\hbar^2}{2kT}\left[\frac{J(J+1)}{I_A} + K^2\left(\frac{1}{I_C} - \frac{1}{I_A}\right)\right]\right\} \qquad (6\text{-}37)$$

---

[2] L. Pauling and E. B. Wilson, Jr., *Introduction to Quantum Mechanics* (New York: McGraw-Hill Book Co., 1935).

The energy levels can be schematically illustrated

...   .

$$J = 2 \underline{\hspace{4cm}} \left.\begin{array}{l} \underline{\hspace{1cm}} K = 2 \\ \underline{\hspace{1cm}} K = 1 \\ \underline{\hspace{1cm}} K = 0 \\ \underline{\hspace{1cm}} K = -1 \\ \underline{\hspace{1cm}} K = -2 \end{array}\right\} m = (-2, -1, 0, 1, 2)$$

$$J = 1 \underline{\hspace{4cm}} \left.\begin{array}{l} \underline{\hspace{1cm}} K = 1 \\ \underline{\hspace{1cm}} K = 0 \\ \underline{\hspace{1cm}} K = -1 \end{array}\right\} m = (-1, 0, 1)$$

$$J = 0 \underline{\hspace{2cm}} \qquad\qquad K = 0, m = 0$$

Instead of summing over K from $-J$ to J, one could sum over all K from $-\infty$ to ∞, but sum J only from $|K|$ to ∞. Thus, if there were no symmetry restrictions on the rotational states, the following would be obtained:

$$Q_{\text{rot}} = \sum_{K=-\infty}^{\infty} \sum_{J=|K|}^{\infty} (2J+1) \exp\{-A[J(J+1)] + K^2(A-C)\} \qquad (6\text{-}38)$$

where $A = (\hbar^2/2I_A kT)$ and $C = (\hbar^2/2I_C kT)$.

Equation (6-38) can be rewritten

$$Q_{\text{rot}} = \sum_{K=-\infty}^{\infty} \exp[K^2(A-C)] \sum_{J=|K|}^{\infty} (2J+1) \exp[-AJ(J+1)] \qquad (6\text{-}39)$$

If $kT \gg \hbar^2/I_A$ and $kT \gg \hbar^2/I_C$, the sums can be replaced by integrals after multiplying through by $\Delta J \Delta K = 1$. Thus,

$$Q_{\text{rot}} \cong \int_{-\infty}^{\infty} \exp[(A-C)K^2] \left\{ \int_{|K|}^{\infty} (2J+1) \right.$$

$$\left. \times \exp[-AJ(J+1)] \, dJ \right\} dK \qquad (6\text{-}40)$$

The integral over dJ can be evaluated by substituting $y = AJ(J+1)$ and $dy = A(2J+1)$. The result is $A^{-1} e^{-y_{\min}}$ where $y_{\min} = A|K|(|K|+1)$. If kT is indeed large compared to the energy level spacings, it can be assumed that $|K| \gg 1$ to evaluate the integral over dK. Although this is a poor approximation for small $|K|$, the region of small $|K|$ does not contribute heavily to the integral, and no serious errors are produced. Therefore, $y_{\min}$ is approximated as

$$y_{\min} \cong A|K|^2$$

and

$$Q_{\text{rot}} \cong \int_{-\infty}^{\infty} e^{-C|K|^2} \, dKA^{-1} = \sqrt{\frac{\pi}{C}} \cdot \frac{1}{A}$$

$$Q_{\text{rot}} \cong \sqrt{\pi} \left(\frac{2I_A kT}{\hbar^2}\right)\left(\frac{2I_C kT}{\hbar^2}\right)^{1/2} \tag{6-41}$$

which is the same as Eq. (6-32) if $I_A = I_B$.

However, not all values of K are possible because of symmetry restrictions. The proper procedure at this point would be to determine the set of symmetry operations[3] that produce a molecular configuration indistinguishable from the initial system before rotation. Then the effect of these rotations on all the possible electronic, nuclear, and rotational wave functions should be determined. For a particular electronic state, it would be necessary to determine the effect of the symmetry operations only on ψ_{nuc} and ψ_{rot} for all possible nuclear and rotational states. This was carried out in previous sections for diatomic molecules. However, the number of possibilities for polyatomic molecules is manifold, and a complete discussion will not be attempted. Consider the special case of a hypothetical XY_3Z molecule illustrated in Fig. 6-11. This molecule is a symmetric top because $I_A = I_B \neq I_C$. Rotations of $2\pi/3$ and $4\pi/3$ about the X-Z bond result in indistinguishable configurations. These sym-

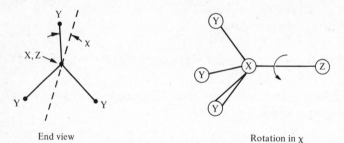

End view   Rotation in χ

Principal axes:

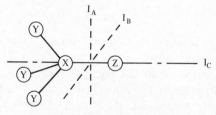

FIGURE 6-11   Axes used to define rotations of the Y_3XZ molecule.

---

[3]Rigid rotations about the principal axes.

metry operations correspond to changes in angle χ of $2\pi/3$ and $4\pi/3$, with angles θ and ϕ (see Fig. 6-10) fixed. For any particular molecule, it would be necessary to evaluate the effect of these operations on ψ_{el}, ψ_{nuc}, and ψ_{rot}. Here, only a special case is considered, for which ψ_{el} is symmetric and for which the spins of the Y nuclei are 0. The total nuclear spin for the three Y nuclei is 0, and therefore there is only a single nuclear spin state for the three Y atoms, and this is symmetric with respect to interchange of the nuclei. For this case, therefore, the effect of the symmetry operations on $\psi_{el}\psi_{nuc}$ is to make this product go to $+\psi_{el}\psi_{nuc}$. The effect of the symmetry operations on ψ_{rot} must be determined next. According to Eq. (6-34), the effect of the symmetry operations $\chi \rightarrow \chi + 2\pi/3$ and $\chi \rightarrow \chi + 4\pi/3$ will be to make $\psi_{rot} \rightarrow e^{i2\pi K/3}\psi_{rot}$ and $e^{i4\pi K/3}\psi_{rot}$. Since the Y nuclei have $i = 0$, they are Bosons, and the overall wave function is symmetric with respect to interchange. Therefore, only values of K for which $e^{i2\pi K/3}$ and $e^{i4\pi K/3}$ are equal to unity can be allowed by symmetry. It is clear that K can be $0, 3, 6, 9, \ldots$, and the values $1, 2, 4, 5, 7, 8, \ldots$ must be excluded. Thus, only $\frac{1}{3}$ of the states included in Eq. (6-39) can actually exist. When the sum is replaced by an integral, the integral will be $\frac{1}{3}$ of that calculated in Eq. (6-41). Thus, when symmetry effects are included, Eq. (6-33) is obtained, with $\sigma = 3$, in this case.

Another example is a flat planar symmetrical molecule XY_3, with the structure

The symmetry number of this molecule is 6 because one can make 6 equivalent structures by rigid rotation about an axis perpendicular to the paper passing through X, and about each of the X-Y axes. If the Y atoms were labelled 1, 2, and 3, these would be

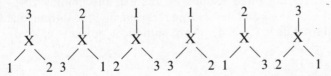

It is difficult to enumerate the specific states which are excluded by symmetry.

It must be admitted that the situation is far from satisfactory. It has merely been stated that Eq. (6-33) is correct, and illustrated how the symmetry number comes about for one minor case. A more general argument is as follows. In general, the quantum-mechanical description of a system must involve treatment of symmetrically equivalent nuclei as completely indistinguishable. On the other hand, the classical description will treat these particles as if they had labels. In the case of translational motion of an emsemble of N particles in a gas, it was found (Chapter 3) that in the limit of dilute systems at high temperature, the quantum-mechanical partition function of the gas goes to

$$(Q_{tr})_{qm} = (Q_{tr})_{cl}/N! \tag{6-42}$$

This can be rationalized on the basis that the classical description involves a phase space where the momenta and coordinates of each of the particles are included regardless of symmetry. For a gas composed of identical atoms, there will be many different regions of classical space which correspond to simply permuting the particles without changing their relative positions or moments. The classical limit of the quantum mechanical description will not contain these many regions of phase space, but only a single region corresponding to one permutation. Since there are $N!$ permutations possible for every configuration of the N particles, Eq. (6-42) follows immediately. In the same way, when we deal with the rotations of a polyatomic molecule, the classical description treats all symmetrically equivalent configurations, which can be obtained by rigid rotation, as if they were distinguishable. The result is Eq. (6-32). In the quantum-mechanical description, these configurations would be treated as a single configuration, and in effect, many regions of phase space available to the classical description would be excluded. The number of symmetrically equivalent configurations that can be reached by rigid rotation is the symmetry number, and therefore, the high temperature limit of $(Q_{rot})_{qm}$ is given by Eq. (6-33).

6.6 SYMMETRY NUMBERS AND INTERNAL ROTATION

For polyatomic molecules conceptually capable of internal rotation, the determination of the appropriate symmetry number is more complicated. There are three regions of interest; the limiting cases of small torsional vibrations and nearly free rotation, and the intermediate case of hindered rotation. Consider the hypothetical molecule

$$
\begin{array}{ccc}
B & & D \\
\diagdown & & \diagup \\
 & A\!-\!C & \\
\diagup & & \diagdown \\
B & & D
\end{array}
$$

The strict interpretation of the symmetry number as the number of equivalent configurations that can be obtained by rigid rotations of the molecule would lead to $\sigma = 2$ since the structures

$$
\begin{array}{ccc}
\begin{array}{cc}
B_1 & \quad D_1 \\
\diagdown & \diagup \\
A\!-\!C \\
\diagup & \diagdown \\
B_2 & \quad D_2
\end{array}
& \text{and} &
\begin{array}{cc}
B_2 & \quad D_2 \\
\diagdown & \diagup \\
A\!-\!C \\
\diagup & \diagdown \\
B_1 & \quad D_1
\end{array}
\end{array}
$$

are interchangeable by rigid rotation. Suppose the barrier to internal rotation is very high compared to kT so the molecule undergoes small torsional vibrations of the BAB and DCD groups out of the plane. This is in fact the case for

$$
\begin{array}{c}
\diagdown \qquad \diagup \\
C\!=\!C \\
\diagup \qquad \diagdown
\end{array}
$$

compounds. If the planar configuration is most stable, then there are two symmetrically equivalent structures (shown above) in which the molecule can exist, with very small probability of tunneling through the barrier to interchange. It might be argued that tunneling can be neglected and only a single structure need be counted, in which case one uses the ordinary vibrational partition function for the torsional vibration, and $\sigma = 2$ for the overall rotations. On the other hand, if interchange of the two equivalent structures by tunneling was more probable, it might be argued that the symmetry number is 4 because the structures

$$
\begin{array}{ccc}
\begin{array}{cc}
B_1 & \quad D_2 \\
\diagdown & \diagup \\
A\!-\!C \\
\diagup & \diagdown \\
B_2 & \quad D_1
\end{array}
& \text{and} &
\begin{array}{cc}
B_2 & \quad D_1 \\
\diagdown & \diagup \\
A\!-\!C \\
\diagup & \diagdown \\
B_1 & \quad D_2
\end{array}
\end{array}
$$

are now allowed. However, each vibrational level is now doubly degenerate[4] because there are two potential minima about which the nuclei can oscillate. Thus, the torsional partition function would contain a factor of 2 due to this degeneracy, and the product of this with $1/\sigma = \frac{1}{4}$ would give the same result as before, namely, $\frac{1}{2}$. Thus, it makes no difference whether tunneling is allowed, conceptually or not. The proper expansion for the overall partition function is obtained by either procedure.

For a molecule in the other extreme case, free rotation ($kT \gg$ barrier height) one uses the usual value of σ for overall rotation ($\sigma = 2$ for B_2ACD_2). The partition function for torsional vibration is replaced by that for internal rotation, Eq. (5-72). Since this expression contains a

---

[4]Almost, at least, if the barrier is high.

factor $1/n$, it can be seen that the overall product $(\sigma n)^{-1}$ for the B_2ACD_2 molecule capable of free internal rotation is $\frac{1}{4}$.   This is due to the fact that four structures can be obtained by free rotations, both overall and internal.

For molecules in the intermediate regime of hindered rotation, the usual value $\sigma = 2$ is used for overall rotation, and Q_{ir} is obtained from tables.[5]

Problems

1. Determine the symmetry numbers of the following molecules:
   (a) CH_4
   (b) CH_3Cl
   (c) NH_3
   (d) C_2H_2
   (e) C_6H_6
   (f) C_2H_6
2. Estimate the temperature at which the equilibrium ratio $[D_2]_0/[D_2]_p = 2.5$. Use data in Appendix IV for H_2, and set $\nu_{D_2} = \nu_{H_2}/\sqrt{2}$, and $I_{D_2} = 2I_{H_2}$.
3. Prepare diagrams similar to Figs. 6-6, 6-7 and 6-8 for a diatomic molecule with ground state symmetry $(g, -)$ composed of nuclei with $i = 2$.   Which rotational states are excluded?

---

[5]K. S. Pitzer, *Quantum Chemistry* (Englewood Cliffs, N.J.: Prentice-Hall, 1953).

chapter 7

CHEMICAL EQUILIBRIUM

So far this book has only considered ensembles of identical particles. If a mixture of molecules in chemical equilibrium is considered, the previous discussions need to be amended. Suppose one has a collection of a large number of atoms of several types, which may exist as the free atoms or as various molecules. If the system is in thermodynamic equilibrium, any arbitrary chemical reaction

$$aA + bB + \cdots \rightleftarrows pP + qQ + \cdots \tag{7-1}$$

has equal forward and backward rates. In this expression, $a, b, \ldots$, $p, q, \ldots$ represent the stoichiometric coefficients for molecules $A, B, \ldots$, $P, Q, \ldots$ in the reaction. Under equilibrium conditions, Eq. (7-1) may be written as a mathematical equation

$$pP + qQ + \cdots - aA - bB - \cdots = 0 \tag{7-2}$$

A more convenient notation is

$$\nu_1 A_1 + \nu_2 A_2 + \cdots = 0 \tag{7-3}$$

where A_j is the jth molecular species, and ν_j is the jth reaction coefficient (negative for reactants and positive for products). Thus, a chemical reaction at equilibrium is written as

$$\sum_{\substack{i=\text{all} \\ \text{species}}} \nu_i A_i = 0 \tag{7-4}$$

It is assumed that each molecular species i has energy levels ϵ_{ji} with degeneracies g_{ji}, relative to a common zero of energy for all species as shown in Figure 7-1. The jth level of molecular species A_i is ϵ_{ji}. The number of particles existing as species A_i in level j is denoted as n_{ji}.

It was shown[1] in Chapter 3 that for a single substance A_i, the thermo-

---

[1] See Table 3-1.

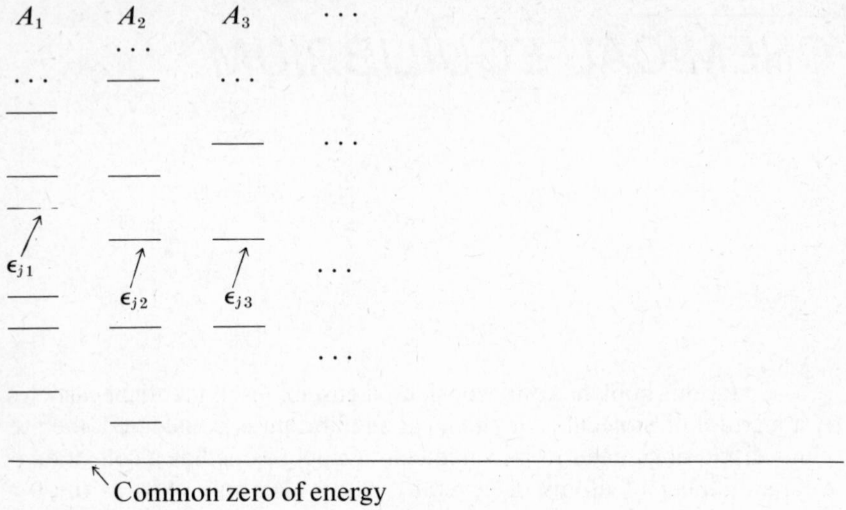

$$A_1 \qquad A_2 \qquad A_3 \qquad \cdots$$

ϵ_{j1} 　　ϵ_{j2} 　　ϵ_{j3} 　　$\cdots$

Common zero of energy

FIGURE 7-1   Energy levels of molecular species

dynamic degeneracy of a corrected Boltzmann gas is

$$W_i = \prod_{\substack{j=\text{all} \\ \text{levels} \\ \text{of } A_i}} g_{ji}{}^{n_{ji}}/n_{ji}! \tag{7-5}$$

In a mixture, any microstate of substance A_i can be associated with any microstate of substance A_k. Therefore, the total thermodynamic degeneracy of a gas mixture is

$$W = \prod_{\substack{j=\text{all} \\ \text{levels} \\ \text{of } A_1}} W_{j1} \prod_{\substack{j=\text{all} \\ \text{levels} \\ \text{of } A_2}} W_{j2} \cdots \tag{7-6}$$

where W_{ji} is given in Eq. (7-5).   Thus,

$$W = \prod_{\substack{i=\text{all} \\ \text{species}}} \prod_{\substack{j=\text{all} \\ \text{levels}}} g_{ji}{}^{n_{ji}}/n_{ji}! \tag{7-7}$$

It should be noted that for any matrix a_{ij},

$$\ln \left\{ \prod_i \prod_j a_{ij} \right\} = \ln (a_{11} a_{12} a_{21} a_{22} a_{13} \ldots) \tag{7-8}$$
$$= \ln a_{11} + \ln a_{12} + \cdots = \sum_i \sum_j \ln a_{ij}$$

Therefore, using Stirling's approximation for $\ln (n_{ji}!)$, it is found that

$$\ln W = \sum_{\substack{i=\text{all} \\ \text{species}}} \sum_{\substack{j=\text{all} \\ \text{levels}}} n_{ji}\{\ln (g_{ji}/n_{ji}) + 1\} \tag{7-9}$$

In the usual way, the n_{ji} are treated as continuous variables and allowed to vary.   Then,

$$d \ln W = \sum_i \sum_j \ln (g_{ji}/n_{ji})\, dn_{ji} \qquad (7\text{-}10)$$

The set of n_{ji} which maximizes W is obtained by setting $d \ln W = 0$.   The restrictive conditions on the n_{ji} are analogous to the case of a single substance.   Conservation of energy requires that

$$dE = 0 = \sum_j \sum_i \epsilon_{ji}\, dn_{ji} \qquad (7\text{-}11)$$

The conservation of particles restriction is a bit more complicated.   Let the total number of atoms of type s in the gas (existing in any form) be $N^{(s)}$.   Let the number of atoms of type s in molecule A_i be q_{is}.   Then conservation of s-atoms requires that

$$\sum_{\substack{i=\text{all} \\ \text{species}}} n_i q_{is} = N^{(s)} \qquad (7\text{-}12)$$

where n_i is the total number of A_i molecules regardless of energy level

$$n_i = \sum_{\substack{j=\text{all} \\ \text{levels}}} n_{ji} \qquad (7\text{-}13)$$

Differentiation of Eq. (7-12), leads to

$$\sum_{\substack{i=\text{all} \\ \text{species}}} q_{is} dn_i = 0 \qquad (7\text{-}14)$$

This may be rewritten

$$\sum_i \sum_j q_{is} dn_{ji} = 0 \qquad s = 1, 2, 3, \dots \qquad (7\text{-}15)$$

for each type of atom s.   The set of Eqs. (7-15) together with Eq. (7-11) form the restrictive conditions on the n_{ji}.   The method of undetermined multipliers is applied by multiplying Eq. (7-11) by $-\beta$ and Eq. (7-15) by α_s, and adding to Eq. (7-10).   The result is

$$\sum_{\substack{i=\text{all} \\ \text{species}}} \sum_{\substack{j=\text{all} \\ \text{levels}}} \left\{ \ln (g_{ji}/n_{ji}) + \sum_{\substack{s=\text{all} \\ \text{atoms}}} \alpha_s q_{is} - \beta \epsilon_{ji} \right\} dn_{ji} = 0 \qquad (7\text{-}16)$$

Since the dn_{ji} may now be treated as independent, the term in braces can be set equal to zero.   Thus, the most probable value for n_{ji} is

$$n_{ji} = g_{ji}\, e^{-\beta \epsilon_{ji}} \exp \left[\sum_s \alpha_s q_{is} \right] \qquad (7\text{-}17)$$

It can be shown by the usual procedure of setting $\partial S/\partial E = 1/T$ that

$\beta = 1/kT$. Thus,

$$n_i = \sum_j n_{ji} = \tilde{Q}_i \exp\left[\sum_s \alpha_s q_{is}\right] \tag{7-18}$$

where $\tilde{Q}_i$ is the partition function of substance A_i,

$$\tilde{Q}_i = \sum_{\substack{j=\text{all}\\ \text{levels}}} g_{ji} e^{-\epsilon_{ji}/kT} \tag{7-19}$$

relative to a common zero of energy for all molecules.   Equation (7-18) gives the most probable value of the total number of molecules of type i. In chemical equilibrium, it is usually desired to determine the relative number of molecules of different species.   Consider the product

$$\prod_{\substack{i=\text{all}\\ \text{species}}} n_i{}^{\nu_i} = \prod_i \tilde{Q}_i{}^{\nu_i} \prod_i \exp\left[\sum_s \alpha_s q_{is}\nu_i\right] \tag{7-20}$$

The second product on the right side of Eq. (7-20) may be expanded as

$$\prod_i \exp\left[\nu_i \sum_s q_{is}\alpha_s\right] = \exp\left\{\nu_1 \sum_s \alpha_s q_{1s} + \nu_2 \sum_s \alpha_s q_{2s} + \cdots\right\} \tag{7-21}$$

$$= \exp\left\{\sum_i \nu_i \sum_s \alpha_s q_{is}\right\} = \exp\left\{\sum_s \alpha_s \sum_i \nu_i q_{is}\right\}$$

In a balanced chemical reaction, the number of atoms of each type s is conserved, so that

$$\sum_i \nu_i q_{is} = 0 \tag{7-22}$$

For example, in the reaction

$$\underbrace{\text{NO}}_{A_1} + \underbrace{2\text{O}}_{A_2} = \underbrace{\text{NO}_3}_{A_3} \tag{7-23}$$

$\nu_1 = -1$, $\nu_2 = -2$, and $\nu_3 = 1$.   If $s = 1$ and 2 denote atoms N and O, respectively, then $q_{12} = 1$, $q_{22} = 1$, $q_{32} = 3$, $q_{11} = 1$, $q_{21} = 0$, and $q_{31} = 1$. It may be verified that Eq. (7-22) holds for $s = 1$ as well as for $s = 2$.   For all balanced equations, Eq. (7-21) reduces to unity.   Therefore, Eq. (7-20) becomes

$$\prod_i n_i{}^{\nu_i} = \prod_i \tilde{Q}_i{}^{\nu_i} \tag{7-24}$$

The product on the left side of Eq. (7-24) is simply the equilibrium constant of a reaction in terms of absolute numbers of molecules.   In reaction (7-23) for example,

$$\prod_i n_i{}^{\nu_i} = \frac{n_{NO_3}}{n_{NO}n_0{}^2} = K_{eq} \tag{7-25}$$

The partition functions $\tilde{Q}_i$ defined in Eq. (7-19) are *not* the usual functions Q_i which are relative to the lowest level of molecule i as the zero of energy. The relation between the two can be obtained by defining ξ_{ji} to be the energy of level j of substance i relative to the lowest level chosen as the zero of energy. It is evident that

$$\epsilon_{ji} = \epsilon_{0i} + \xi_{ji} \tag{7-26}$$

and therefore,

$$\tilde{Q}_i = \sum_j g_{ji} e^{-\xi_{ji}/kT} = Q_i e^{\epsilon_{0i}/kT} \tag{7-27}$$

$$K_{eq} = \prod_i n_i{}^{\nu_i} = \prod_i \tilde{Q}_i{}^{\nu_i} = \left\{ \prod_i Q_i{}^{\nu_i} \right\} \exp \left[-\sum_i \nu_i \epsilon_{0i}/kT \right] \tag{7-28}$$

The quantity $\sum_i \nu_i \epsilon_{0i}$ is the energy of reaction for all reactant and product molecules in the ground energy levels, and is denoted as $\Delta \epsilon^0$. For example, in reaction (7-23),

$$\Delta \epsilon^0 = \epsilon_{NO_3} - \epsilon_{NO} - 2\epsilon_O \tag{7-29}$$

where ϵ in Eq. (7-29) means the energy of the ground level. When Eq. (7-28) is written out for reaction (7-1),

$$K_{eq} = \frac{n_P{}^{\nu_P} n_Q{}^{\nu_Q} \dots}{n_A{}^{\nu_A} n_B{}^{\nu_B} \dots} = \frac{\tilde{Q}_P{}^{\nu_P} \tilde{Q}_Q{}^{\nu_Q} \dots}{\tilde{Q}_A{}^{\nu_A} \tilde{Q}_B{}^{\nu_B} \dots}$$

$$= \frac{Q_P{}^{\nu_P} Q_Q{}^{\nu_Q} \dots}{Q_A{}^{\nu_A} Q_B{}^{\nu_B} \dots} e^{-\Delta \epsilon^0/kT} \tag{7-30}$$

An alternative definition of Q, which is used in Chapter 10, is to use a zero of energy given by the energy of the molecule at rest at the equilibrium position, and then $\Delta \epsilon^0$ is the energy of reaction for all molecules at rest in their equilibrium configurations. Each partition function has a factor of V in it from the translational contribution. If we define a modified partition function per unit volume as

$$Q_i' = Q_i/V, \tag{7-31}$$

$$K_{eq} = V^{\sum \nu_i} \prod_i (Q_i')^{\nu_i} e^{-\Delta \epsilon^0/kT} \tag{7-32}$$

where Q_i' is a function of only the temperature. Thus, the equilibrium constant in terms of concentrations, K_c, is

$$K_c = \prod_{\substack{i=\text{all} \\ \text{species}}} \left(\frac{n_i}{V} \right)^{\nu_i} = \prod_i c_i{}^{\nu_i} = \prod_i (Q_i')^{\nu_i} e^{-\Delta \epsilon^0/kT} \tag{7-33}$$

Problems

1. (Illustrative problem) Calculate the equilibrium constant for the reaction

$$O_2 \rightarrow 2O$$

at 3000°K. Therefore, calculate the percentage dissociation if one started with a container of O_2 at 0.02 atm and 300°K, and heated it to 3000°K.

Answer: According to Eq. (7-33),

$$K_c = \frac{c_O^2}{c_{O_2}} = \frac{(Q_O')^2}{(Q_{O_2}')} e^{-\Delta\epsilon^0/kT}$$

(a) $\Delta\epsilon^0 =$ energy of reaction per molecule based on the "bottom of the well" as the zero of energy

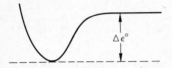

From Appendix IV, $\Delta\epsilon^0 = 5.08 \text{ eV} = 8.14 \times 10^{-12} \text{ erg}$

$$\frac{\Delta\epsilon^0}{k} = 5.89 \times 10^4 \text{°K} \qquad \exp\left(-\frac{\Delta\epsilon^0}{kT}\right) = 2.95 \times 10^{-9}$$

(b) Q_O' is the partition function of O, based on the lowest level as the zero of energy, with V omitted in the translational part of the partition function.

$$Q_O' = Q_{\text{tro}}' Q_{\text{elO}}$$

$$Q_{\text{tro}}' = \left(\frac{2\pi mkT}{h^2}\right)^{3/2} = 1.20 \times 10^{22} T^{3/2} = 1.97 \times 10^{27} \text{ cm}^{-3}$$

$$\begin{aligned}
Q_{\text{elO}} &= (5)(1) + (3) \exp(-232/T) + (1) \exp(-325/T) \\
&\quad + (5) \exp(-22,800/T) + \cdots \\
&= 5 + (3)(0.926) + (1)(0.878) + (5)(0.005) + \cdots \\
&= 8.69
\end{aligned}$$

$$Q_O' = 8.69 \times 1.97 \times 10^{27} = 1.71 \times 10^{28} \text{ cm}^{-3}$$

(c) For calculating Q_{O_2}', we shall assume that only the ground electronic state of O_2 is populated. Thus

$$Q_{O_2}' \cong Q_{\text{tr}O_2}' Q_{\text{rot}O_2} Q_{\text{vib}O_2} Q_{\text{el}O_2}$$

$$\begin{aligned}
Q_{\text{tr}O_2}' &= 2^{3/2} Q_{\text{tro}}' \\
&= 5.57 \times 10^{27} \text{ cm}^{-3}
\end{aligned}$$

$$Q_{\text{rot}O_2} = \frac{T}{\sigma\theta_{\text{rot}}} = \frac{3000}{2 \times 2.01} = 747$$

$$Q_{\text{vib}O_2} = \frac{\exp(-\theta_{\text{vib}}/2T)}{1 - \exp(-\theta_{\text{vib}}/T)} = \frac{0.686}{1 - 0.470} = 1.29$$

$$Q_{el_{O_2}} = 3$$

$$Q'_{O_2} = 1.61 \times 10^{31} \text{ cm}^{-3}$$

(d) According to Eq. (7-33),

$$K_c = \frac{(1.71 \times 10^{28})^2}{(1.61 \times 10^{31})} \times 2.95 \times 10^{-9} = 5.35 \times 10^{16} \text{ cm}^{-3}$$

with concentrations expressed in molecules/cm³. If the gases are assumed to be ideal, we can write the ith partial pressure as

$$p_i = c_i kT \qquad (i = O \text{ or } O_2)$$

Thus,

$$K_c = \frac{c_O{}^2}{c_{O_2}} = \frac{K_p}{kT}$$

where $K_p = p_O{}^2/p_{O_2}$. Therefore, with p in atm, $k = \dfrac{82.07}{6.02 \times 10^{23}} \dfrac{\text{cm}^3\text{-atm}}{\text{molecule-}°K}$

and

$$K_p = (5.35 \times 10^{16})\,(1.365 \times 10^{-22})\,3000$$

$$K_p = 2.19 \times 10^{-2} \text{ atm}$$

(e) If we started with 0.02 atm of O_2 at 300°K and heated it to 3000°K, the total pressure would be 0.2 atm if no dissociation occurred. If a fraction, x, of the O_2 dissociates, the total pressure will be

$$p_{tot} = 0.2(1+x)$$

The partial pressures will be

$$p_O = 0.4x$$

$$p_{O_2} = 0.2(1-x)$$

Thus,

$$K_p = 0.0219 = \frac{(0.4x)^2}{0.2(1-x)}$$

$$\frac{0.16x^2}{0.2(1-x)} = 0.0219$$

Thus, $x = 0.152$

The percent dissociation is, therefore, 15.2%.
It should be noted that there are two major factors which determine the amount of dissociation. If one considers the number of states available to two oxygen atoms when they exist as separate atoms, and when they exist as an oxygen molecule, it will be seen that there are so many translational states that $Q_O{}^2 \gg Q_{O_2}$. This is due to the fact that each state of one O atom can be associated with all states of the other O atom. Therefore, there is a tendency to cause O_2 to dissociate because there are more states available to $(O+O)$ than to O_2. However, the states of $(O+O)$ lie higher in energy (by about 5 eV) than the states of O_2. Thus, the Boltzmann factor $\exp(-\Delta\epsilon^0/kT) =$

3×10^{-9} tends to reduce the fraction of O_2 which dissociates. The net result of these opposing factors is that O_2 is about 15% dissociated at 3000°K, at a total pressure of 0.23 atm. It is also interesting to calculate the populations of the excited vibrational states at this temperature:

$$\frac{n_j}{n_0} = e^{-j\theta_{vib}/T}$$

$$\frac{n_1}{n_0} = 0.470, \qquad \frac{n_2}{n_0} = 0.221, \qquad \frac{n_{10}}{n_0} = 0.005$$

Thus, the upper vibrational states are less populated than the continuum, despite the fact that the continuum lies higher in energy.

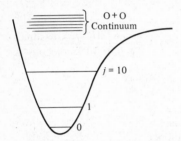

This is due to the fact that there are so many more states available to the continuum.

2. Calculate the fraction of total CO_2 which exists as CO_2, $CO + O$, and $C + O + O$, if one starts with CO_2 at 0·01 atm and 300°K, and heats it to 4000°K.

3. Show that the equilibrium constant for the reaction

$$H_2 + D_2 \rightarrow 2HD$$

is four at high T.   Explain.

chapter 8

THE PERFECT QUANTUM GAS

8.1 THE CONSTANTS α AND β

In Chapter 3, it was shown that the equilibrium distribution of particles of a perfect quantum gas among the energy levels is

$$n_i = \frac{g_i}{e^\alpha e^{\beta\epsilon_i} \pm 1} \tag{8-1}$$

where α and β are dependent on the total number of particles, N, and the temperature, T. The $+$ and $-$ signs refer to Fermi-Dirac and Bose-Einstein substances, respectively. Throughout this chapter, the upper sign of expressions like $\pm$ and $\mp$ will refer to Fermions and the lower sign to Bosons.

For substances that obey corrected Boltzmann statistics,

$$e^\alpha e^{\beta\epsilon_i} \gg 1$$

and the ± 1 in the denominator of Eq. (8-1) can be neglected. In this case, α is readily evaluated as

$$e^\alpha = Q/N \tag{8-1a}$$

where Q is the partition function

$$Q = \sum_{\substack{i=\text{all} \\ \text{levels}}} g_i\, e^{-\beta\epsilon_i} \tag{8-2}$$

The thermodynamic properties were then evaluated by appropriate operations on Q. For example, the energy was obtained as follows:

$$E = \sum_{\substack{i=\text{all} \\ \text{levels}}} n_i\epsilon_i = \frac{N}{Q} \sum_i g_i\epsilon_i\, e^{-\beta\epsilon_i} \tag{8-2a}$$

$$E = \frac{-N}{Q} \left\{ \frac{\partial}{\partial\beta} \sum_i g_i\, e^{-\beta\epsilon_i} \right\} = \frac{-N}{Q} \left(\frac{\partial Q}{\partial\beta} \right)_{V,N} \tag{8-2b}$$

When a similar approach is tried for Fermions or Bosons, the following is found:

$$E = \sum_{\substack{i=\text{all} \\ \text{levels}}} n_i \epsilon_i = \sum_i \frac{g_i \epsilon_i}{(e^\alpha e^{\beta \epsilon_i} \pm 1)} \tag{8-3a}$$

$$E = \sum_i \frac{g_i \epsilon_i e^{-\alpha} e^{-\beta \epsilon_i}}{(1 \pm e^{-\alpha} e^{-\beta \epsilon_i})} \tag{8-3b}$$

This cannot be simply expressed in terms of the partition function. However, a new function may be defined as

$$\Psi = \pm \sum_{\substack{i=\text{all} \\ \text{levels}}} g_i \ln \{1 \pm e^{-\alpha} e^{-\beta \epsilon_i}\} \tag{8-4}$$

On comparison of Eqs. (8-3b) and (8-4), it follows directly that

$$E = -\left(\frac{\partial \Psi}{\partial \beta}\right)_{V,\alpha} \tag{8-5}$$

Therefore, it follows that the function Ψ can be used to evaluate thermodynamic properties of quantum substances. Note that in Eq. (8-4) differentiation is at constant α, whereas in Eq. (8-2b) α varies since N is constant.

A similar result can be shown for the entropy. For Fermions, it is shown in Eq. (3-28) that

$$W_{\text{FD}} = \prod_{\substack{i=\text{all} \\ \text{levels}}} \frac{g_i!}{(g_i - n_i)! n_i!} \tag{8-6}$$

$$\ln W_{\text{FD}} = \sum_{\substack{i=\text{all} \\ \text{levels}}} \left\{ g_i \ln\left(\frac{g_i}{g_i - n_i}\right) + n_i \ln\left(\frac{g_i - n_i}{n_i}\right) \right\} \tag{8-7}$$

But from Eq. (8-1), it follows that

$$\frac{g_i - n_i}{n_i} = e^\alpha e^{\beta \epsilon_i} \tag{8-8}$$

$$\frac{g_i}{g_i - n_i} = 1 + e^{-\alpha} e^{-\beta \epsilon_i} \tag{8-9}$$

Therefore,

$$S_{\text{FD}} = k \ln W_{\text{FD}} = k \sum_i \{ g_i \ln(1 + e^{-\alpha} e^{-\beta \epsilon_i}) + n_i(\alpha + \beta \epsilon_i) \}$$

$$S_{\text{FD}} = k\Psi + Nk\alpha + k\beta E \tag{8-11}$$

In corrected Boltzmann statistics,

$$S_{\text{B}} = Nk \ln Q + k\beta E - k \ln N! \tag{8-12}$$

The relation between this and S_{FD} is obtained by taking the limit of S_{FD} as

$g_i/n_i \to \infty$. For large g_i/n_i, $e^{\alpha} e^{\beta \epsilon_i} \gg 1$, and $n_i/g_i \cong e^{-\alpha} e^{-\beta \epsilon_i} = (N/Q) e^{-\beta \epsilon_i}$. Thus, in this limit, $\ln(1 + e^{-\alpha} e^{-\beta \epsilon_i}) \cong e^{-\alpha} e^{-\beta \epsilon_i} \cong n_i/g_i$. Hence, from Eq. (8-4), $\Psi \cong \Sigma g_i(n_i/g_i) = N$. Since $Nk\alpha \cong kN \ln Q - kN \ln N$, Eq. (8-11) may be written in the form (for large g_i/n_i).

$$S_{FD} \cong kN + Nk \ln Q - kN \ln N + k\beta E$$

$$S_{FD} \cong Nk \ln Q + k\beta E - k(N \ln N - N)$$

If Stirling's approximation for $\ln N!$ is used, Eq. (8-12) is obtained.

Similarly, by using Eq. (3-2) for Bosons, one can show

$$\ln W_{BE} = \sum_i \left\{ (g_i - 1) \ln \left(\frac{g_i + n_i - 1}{g_i - 1} \right) + n_i \ln \left(\frac{g_i + n_i - 1}{n_i} \right) \right\} \quad (8\text{-}13)$$

If the term -1 is neglected compared to g_i in four places in this equation, it follows that Eq. (8-11) also holds for S_{BE} for Bosons.

To evaluate α and β in quantum statistics, the same procedures are used as in Boltzmann statistics. The quantity $(\partial S/\partial E)_{V,N}$ is calculated and set equal to $1/T$ to obtain β. From Eq. (8-11), for quantum substances (FD or BE):

$$\left(\frac{\partial S}{\partial E} \right)_{V,\alpha} = k \left(\frac{\partial \Psi}{\partial E} \right)_{V,\alpha} + k\beta + k \left(\frac{\partial \beta}{\partial E} \right)_{V,\alpha} E + k\alpha \left(\frac{\partial N}{\partial E} \right)_{V,\alpha} \quad (8\text{-}14)$$

Since Ψ depends on β,

$$\left(\frac{\partial \Psi}{\partial E} \right)_{V,\alpha} = \left(\frac{\partial \Psi}{\partial \beta} \right)_{V,\alpha} \left(\frac{\partial \beta}{\partial E} \right)_{V,\alpha} \quad (8\text{-}15)$$

$$\left(\frac{\partial S}{\partial E} \right)_{V,\alpha} = k \left(\frac{\partial \beta}{\partial E} \right)_{V,\alpha} \underbrace{\left\{ E + \left(\frac{\partial \Psi}{\partial \beta} \right)_{V,\alpha} \right\}}_{-E} + k\beta + k\alpha \left(\frac{\partial N}{\partial E} \right)_{V,\alpha} \quad (8\text{-}16)$$

$$\left(\frac{\partial S}{\partial E} \right)_{V,\alpha} = k\beta + k\alpha \left(\frac{\partial N}{\partial E} \right)_{V,\alpha} \quad (8\text{-}17)$$

Now determine $(\partial S/\partial E)_{V,N}$ from $(\partial S/\partial E)_{V,\alpha}$. To do this, note that

$$\left(\frac{\partial S}{\partial E} \right)_{V,\alpha} = \left(\frac{\partial S}{\partial E} \right)_{V,N} + \left(\frac{\partial S}{\partial N} \right)_{E,V} \left(\frac{\partial N}{\partial E} \right)_{\alpha,V} \quad (8\text{-}17a)$$

Thus,

$$\left(\frac{\partial S}{\partial E} \right)_{V,N} = \frac{1}{T} = \left(\frac{\partial S}{\partial E} \right)_{V,\alpha} - \left(\frac{\partial S}{\partial N} \right)_{E,V} \left(\frac{\partial N}{\partial E} \right)_{\alpha,V} \quad (8\text{-}17b)$$

Combining this with Eq. (8-17), the following is found:

$$\frac{1}{T} + \left(\frac{\partial S}{\partial N} \right)_{E,V} \left(\frac{\partial N}{\partial E} \right)_{\alpha,V} = k\beta + k\alpha \left(\frac{\partial N}{\partial E} \right)_{\alpha,V} \quad (8\text{-}17c)$$

It is now necessary to divert to prove

$$\left(\frac{\partial S}{\partial N}\right)_{E,V} = k\alpha$$

It will then follow that $\beta = 1/kT$.   From Eq. (8-11),

$$\left(\frac{\partial S}{\partial N}\right)_{E,V} = k\left(\frac{\partial \Psi}{\partial N}\right)_{E,V} + k\alpha + Nk\left(\frac{\partial \alpha}{\partial N}\right)_{E,V} + kE\left(\frac{\partial \beta}{\partial N}\right)_{E,V} \qquad (8\text{-}18a)$$

But, from the definition of Ψ,

$$\left(\frac{\partial \Psi}{\partial N}\right)_{E,V} = \sum_i \frac{g_i\, e^{-\alpha}\, e^{-\beta\epsilon_i}}{(1 \pm e^{-\alpha}\, e^{-\beta\epsilon_i})} \left\{-\left[\left(\frac{\partial \alpha}{\partial N}\right)_{E,V} + \epsilon_i\left(\frac{\partial \beta}{\partial N}\right)_{E,V}\right]\right\} \qquad (8\text{-}18b)$$

$$\left(\frac{\partial \Psi}{\partial N}\right)_{E,V} = -\sum_i n_i \left\{\left(\frac{\partial \alpha}{\partial N}\right)_{E,V} + \epsilon_i\left(\frac{\partial \beta}{\partial N}\right)_{E,V}\right\} \qquad (8\text{-}18c)$$

$$\left(\frac{\partial \Psi}{\partial N}\right)_{E,V} = -N\left(\frac{\partial \alpha}{\partial N}\right)_{E,V} - E\left(\frac{\partial \beta}{\partial N}\right)_{E,V} \qquad (8\text{-}18d)$$

When Eq. (8-18d) is substituted into Eq. (8-18a), the desired result is obtained:

$$\left(\frac{\partial S}{\partial N}\right)_{E,V} = k\alpha$$

Therefore, from Eq. (8-17c), it follows that as in the case of Boltzmann statistics

$$\beta = \frac{1}{kT} \qquad (8\text{-}18)$$

To calculate α, one must use conservation of particles:

$$\sum_i n_i = N$$

$$\sum_{\substack{i=\text{all} \\ \text{levels}}} \left(\frac{g_i}{e^\alpha\, e^{\beta\epsilon_i} \pm 1}\right) = N \qquad (8\text{-}19)$$

Equation (8-19) must be solved for α.   This is not a trivial problem, and will be discussed at greater length later in this chapter.

8.2   DENSITY OF STATES OF TRANSLATION AND THE EQUATION OF STATE

Consider particles of mass m in a cubical container of side L.   This results in no actual loss of generality and makes the discussion simpler. According to quantum mechanics, the energy levels of a particle in a cubical box are

$$\epsilon_n = \frac{n^2 \pi^2 \hbar^2}{2mL^2} \tag{8-20}$$

where

$$n^2 = n_x^2 + n_y^2 + n_z^2 \tag{8-21}$$

and n_x, n_y, and n_z are the quantum numbers associated with translation in the x, y and z directions. The values of n_x, n_y, and n_z can range from 1, 2, 3, ... A diagram can be constructed to represent the states as shown in Fig. 8-1. Each interior lattice point represents a state. Each point is shared by eight unit cubes, and there are 8 points in a cube. Thus, there is one state per unit cube in the n_x, n_y, n_z space, provided that a volume element is chosen large compared to a unit cube. This discussion has neglected spin. If each particle has spin s, the spin degeneracy associated with each translational state is $2s + 1$, so there are actually $(2s + 1)$ states per unit volume. For very large values of n, the distribution of lattice points may be approximated as a continuum in n_x, n_y, n_z space. The number of states in a thin shell of radius n can be calculated, as shown in Fig. 8-2. The volume of the shell is $\frac{1}{8}(4\pi n^2 \, dn)$, where the factor $\frac{1}{8}$ is due to the fact that only positive values of n_x, n_y, and n_z are allowed, so only one octant of n-space is considered. The number of states in the shell is

$$d\Gamma = (\tfrac{1}{2}\pi n^2 \, dn)(2s + 1) \tag{8-22}$$

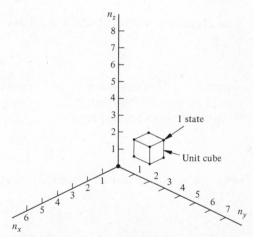

FIGURE 8-1 Plot of translational quantum numbers, n_x, n_y, and n_z. Each interior point ($n_x \geq 1$, $n_y \geq 1$, $n_z \geq 1$) corresponds to a translational state. A unit cube contains 8 points, but each point is shared by 8 cubes, so the density of points is 1 per unit cube.

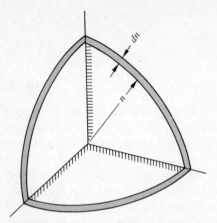

FIGURE 8-2 Expanded view of Fig. 8-1 for large quantum numbers. A spherical shell of radius n and thickness dn is shown.

But if ϵ_n is treated as continuous for large n, and called ϵ, then

$$n^2 = \frac{8mL^2}{h^2}\,\epsilon \tag{8-23}$$

and

$$dn = \left(\frac{2mL^2}{h^2}\right)^{1/2} \frac{d\epsilon}{\epsilon^{1/2}} \tag{8-24}$$

Therefore, the number of states in a thin shell is

$$d\Gamma = 2\pi \left(\frac{2m}{h^2}\right)^{3/2} V(2s+1)\epsilon^{1/2}\,d\epsilon \tag{8-25}$$

where $V = L^3$ is the volume of the container.

The general expression for the equation of state of a gas is obtained from the thermodynamic relation[1]

$$p = -\left(\frac{\partial A}{\partial V}\right)_T \tag{8-26}$$

where $A = E - TS$ is the Helmholtz free energy.   Thus,

$$p = -\left(\frac{\partial E}{\partial V}\right)_T + T\left(\frac{\partial S}{\partial V}\right)_T \tag{8-27}$$

For a perfect gas of noninteracting particles,

$$\left(\frac{\partial E}{\partial V}\right)_T = 0 \tag{8-28}$$

---

[1]All partials in this section are implied to be taken at $N = $ const.

or

$$p = T \left(\frac{\partial S}{\partial V} \right)_T \tag{8-29}$$

In general, for a quantum gas, S is given by Eq. (8-11). Therefore,

$$\frac{\partial S}{\partial V} = k \left(\frac{\partial \Psi}{\partial V} \right)_T + Nk \left(\frac{\partial \alpha}{\partial V} \right)_T + \underbrace{\frac{1}{T} \left(\frac{\partial E}{\partial V} \right)_T}_{= 0} \tag{8-30}$$

and

$$p = kT \left[\left(\frac{\partial \Psi}{\partial V} \right)_T + N \left(\frac{\partial \alpha}{\partial V} \right)_T \right] \tag{8-31}$$

But from Eq. (8-3),

$$\left(\frac{\partial \Psi}{\partial V} \right)_T = -\sum_i \left[\left(\frac{g_i}{e^\alpha e^{\beta \epsilon_i} \pm 1} \right) \left\{ \left(\frac{\partial \alpha}{\partial V} \right)_T + \frac{1}{kT} \left(\frac{\partial \epsilon_i}{\partial V} \right)_T \right\} \right]$$

$$\left(\frac{\partial \Psi}{\partial V} \right)_T = -N \left(\frac{\partial \alpha}{\partial V} \right)_T - \frac{1}{kT} \sum_i n_i \left(\frac{\partial \epsilon_i}{\partial V} \right)_T \tag{8-32}$$

and hence

$$p = -\sum_i n_i \frac{d\epsilon_i}{dV} \tag{8-33}$$

But $\epsilon_i \sim L^{-2} = V^{-2/3}$. Therefore, $\ln \epsilon_i = \ln (\text{const}) - \frac{2}{3} \ln V$, and

$$\frac{d\epsilon_i}{dV} = -\frac{2}{3} \frac{\epsilon_i}{V} \tag{8-34}$$

and hence,

$$p = \frac{2}{3V} \sum_i n_i \epsilon_i = \frac{2E}{3V} \tag{8-35}$$

for a perfect quantum gas.

In the dilute limit where $g_i \gg n_i$, the quantum gas becomes equivalent to a Boltzmann gas, and

$$E = \tfrac{3}{2} NkT$$

so that Eq. (8-35) becomes

$$pV = NkT$$

for the equation of state. If g_i is not very large compared to n_i, Eq. (8-35) must be used.

8.3 EQUATION OF STATE AND THERMODYNAMIC FUNCTIONS OF A QUANTUM GAS—CASE OF WEAK DEGENERACY

For mathematical purposes, it is convenient to re-express the function Ψ in Eq. (8-3) as

$$\Psi = \pm \sum_{\substack{n=\text{all} \\ \text{states}}} \ln \{1 \pm e^{-\alpha} e^{-\beta \epsilon_n}\} \tag{8-36}$$

by removing the factor g_n and letting n refer to states rather than levels. For large values of n a sum may be taken over spherical shells of thickness dn, treating $\ln\{1\pm e^{-\alpha}e^{-\beta\epsilon_n}\}$ as if it varied continuously with n. Thus, Eq. (8-36) may be approximated as

$$\Psi = \pm \int_0^\infty \ln\{1\pm e^{-\alpha}e^{-\beta\epsilon(n)}\}\, d\Gamma(n) \tag{8-37}$$

With Eq. (8-25) used for $d\Gamma(n)$, this becomes

$$\Psi = \pm 2\pi \left(\frac{2m}{h^2}\right)^{3/2} V(2s+1) \int_0^\infty \ln\left[1\pm e^{-\alpha}e^{-\beta\epsilon}\right] \epsilon^{1/2}\, d\epsilon \tag{8-38}$$

If a quantity x is defined by the relation $x^2 = \epsilon/kT$, then $\epsilon^{1/2}\, d\epsilon = 2(kT)^{3/2}x^2\, dx$, and

$$\Psi = \pm 4\pi \left(\frac{2mkT}{h^2}\right)^{3/2} V(2s+1) \int_0^\infty \ln\left[1\pm e^{-\alpha}e^{-x^2}\right] x^2\, dx \tag{8-39}$$

This procedure will only be a good approximation at temperatures high enough that many translational states are excited.

To evaluate α, a similar procedure is used.[2]

$$N = \sum_{\substack{n=\text{all}\\ \text{states}}} n_n = \sum_{\substack{n=\text{all}\\ \text{states}}} \frac{1}{(e^\alpha e^{\beta\epsilon_n}\pm 1)} \cong \int_0^\infty \frac{d\Gamma(n)}{(e^\alpha e^{\beta\epsilon_n}\pm 1)} \tag{8-40}$$

$$N = 2\pi \left(\frac{2m}{h^2}\right)^{3/2} V(2s+1) \int_0^\infty \frac{\epsilon^{1/2}\, d\epsilon}{(e^\alpha e^{\beta\epsilon}\pm 1)} \tag{8-41}$$

$$N = 4\pi \left(\frac{2mkT}{h^2}\right)^{3/2} V(2s+1) \int_0^\infty \frac{x^2\, dx}{(e^\alpha e^{x^2}\pm 1)} \tag{8-42}$$

First, consider the extreme limit of very weak degeneracy corresponding to a dilute gas with $g_i \gg n_i$. In this case it can be shown that $e^\alpha \ggg 1$. Therefore, the factor ± 1 in Eq. (8-42) can be neglected in the denominator of the integrand. Thus, to this approximation, the integral in Eq. (8-42) becomes

$$e^{-\alpha} \int_0^\infty e^{-x^2} x^2\, dx = \frac{\pi^{1/2}}{4} e^{-\alpha}$$

Therefore, in the limit of very weak degeneracy,

$$e^\alpha = \left(\frac{2\pi mkT}{h^2}\right)^{3/2} \frac{V}{N} (2s+1) = \frac{Q}{N} (2s+1) \tag{8-43}$$

where Q is the partition function. This is the first approximation to α for very weak degeneracy.

---

[2]The symbol n_n means the number of particles in quantum state n.

A better approximation for α can be obtained by assuming that e^{α}, while large, is not extremely large compared to unity. The integral in Eq. (8-42) is written

$$I = \int_0^{\infty} \frac{e^{-\alpha} e^{-x^2} x^2 \, dx}{1 \pm e^{-\alpha} e^{-x^2}} \tag{8-44}$$

Since $e^{-\alpha} e^{-x^2}$ is less than unity, the denominator may be expanded using the formula

$$\frac{1}{1 \pm y} = 1 \mp y + y^2 \mp \cdots \tag{8-45}$$

Thus,

$$I = \int_0^{\infty} e^{-\alpha} e^{-x^2} x^2 \, dx \, \{ 1 \mp e^{-\alpha} e^{-x^2} + e^{-2\alpha} e^{-2x^2} \mp \cdots \} \tag{8-46}$$

Since

$$\int_0^{\infty} x^{2n} e^{-ax^2} \, dx = \frac{1 \cdot 3 \cdot 5 \cdots (2n-1)}{2^{n+1} a^n} \sqrt{\frac{\pi}{a}} \tag{8-47}$$

it follows that

$$I = \frac{\sqrt{\pi}}{4} \left\{ e^{-\alpha} \mp \frac{e^{-2\alpha}}{2^{3/2}} + \frac{e^{-3\alpha}}{3^{3/2}} \mp \cdots \right\} \tag{8-48}$$

Thus, an equation is derived which may, in principle, be solved for α.

$$N = \left(\frac{2\pi m k T}{h^2} \right)^{3/2} V(2s+1) \left\{ e^{-\alpha} \mp \frac{e^{-2\alpha}}{2^{3/2}} + \frac{e^{-3\alpha}}{3^{3/2}} \mp \cdots \right\} \tag{8-49}$$

This may be rewritten in terms of the variable

$$\phi = \frac{N}{V} \cdot \frac{1}{2s+1} \cdot \left(\frac{h^2}{2\pi m k T} \right)^{3/2} = \frac{N}{(2s+1)Q} \tag{8-50}$$

[which is the inverse of the first approximation to e^{α}, Eq. (8-43)] as

$$1 = \phi^{-1} \left\{ e^{-\alpha} \mp \frac{e^{-2\alpha}}{2^{3/2}} + \frac{e^{-3\alpha}}{3^{3/2}} \mp \cdots \right\} \tag{8-51}$$

Equation (8-51) is a general expression which determines α. In the case of moderately weak degeneracy, $e^{-\alpha}$ is small, and so $e^{-2\alpha}$ is small compared to $e^{-\alpha}$. The second approximation for α is obtained by using Eq. (8-43) for α in the term $e^{-2\alpha}$, and neglecting higher terms. Thus, the second approximation is

$$1 \cong \phi^{-1} \left\{ e^{-\alpha} \mp \frac{\phi^2}{2^{3/2}} + \cdots \right\}$$

$$e^{-\alpha} \cong \phi \pm \frac{\phi^2}{2^{3/2}} \tag{8-52}$$

Higher order approximations could be obtained by using this expression in successive approximation procedures. In the general case, Eq. (8-51) should be used, provided $e^{-\alpha} < 1$.

In Eq. (8-51), we have calculated a series expression which can be used to evaluate α. A similar procedure for evaluating Ψ can be carried out. From Eq. (8-39),

$$\Psi = \pm \left(\frac{4N}{\pi^{1/2}\phi}\right) \int_0^\infty \ln\left[1 \pm e^{-\alpha}\, e^{-x^2}\right] x^2 \, dx$$

Since $e^{-\alpha}\, e^{-x^2} < 1$, we may expand

$$\pm \ln(1 \pm y) = \mp \sum_{n=1}^\infty (\mp 1)^n \frac{y^n}{n} \tag{8-53}$$

where y represents $e^{-\alpha}\, e^{-x^2}$. Therefore,

$$\Psi = \mp \frac{4N}{\pi^{1/2}\phi} \sum_{n=1}^\infty (\mp 1)^n \frac{e^{-n\alpha}}{n} \int_0^\infty x^2\, e^{-nx^2}\, dx$$

$$\Psi = \mp \frac{N}{\phi} \sum_{n=1}^\infty \frac{e^{-n\alpha}(\mp 1)^n}{n^{5/2}} \tag{8-54}$$

Equation (8-54) is an exact expression for Ψ in terms of α. To the accuracy of the second approximation for α, Eq. (8-52) may be used for $e^{-\alpha}$. Then, the second approximation for Ψ becomes

$$\Psi \cong \mp \frac{N}{\phi}\left\{\mp(\phi \pm \phi^2/2^{3/2}) + \frac{(\phi \pm \phi^2/2^{3/2})^2}{2^{5/2}} + \cdots\right\}$$

$$\Psi = \mp \frac{N}{\phi}\left\{\mp\phi - \frac{\phi^2}{2^{3/2}} + \frac{\phi^2}{2^{5/2}} + \cdots\right\}$$

$$\Psi = N\left\{1 \pm \frac{\phi}{2^{5/2}} + \cdots\right\} \tag{8-55}$$

Equation (8-55) is the second approximation for Ψ, corresponding to use of the second approximation for α in Eq. (8-54).

Next, calculate the second approximation to the energy of a quantum gas. The following relation will be used:

$$E = -\left(\frac{\partial \Psi}{\partial \beta}\right)_{V,\alpha} \tag{8-56}$$

derived in Eq. (8-4). There is an algebraic relationship between α, N, and T which may be written $\alpha(N, T) = 0$ [see Eq. (8-51)]. In order to take $\partial/\partial\beta$ with α held constant, one must allow N to vary in such a way that α remains unchanged as T varies.   To the second approximation,

$$\Psi \cong N \left(1 \pm \frac{\phi}{2^{5/2}} + \cdots \right)$$

$$\phi = (\text{const}) N\beta^{3/2}$$

$$\left(\frac{\partial \Psi}{\partial \beta} \right)_{V,\alpha} = \left(\frac{\partial N}{\partial \beta} \right)_{V,\alpha} \pm \left[\frac{\beta^{3/2}}{2^{3/2}} N \left(\frac{\partial N}{\partial \beta} \right)_{V,\alpha} (\text{const}) + \frac{N^2 \cdot 3/2}{2^{5/2}} \beta^{1/2} (\text{const}) + \cdots \right]$$

(8-57)

If Eq. (8-51) is multiplied by N, we obtain an equation which determines α:

$$N = \frac{N}{\phi} \left\{ e^{-\alpha} \mp \frac{e^{-2\alpha}}{2^{3/2}} + \cdots \right\}$$

(8-58)

This may be written

$$N = (\text{const}')\beta^{-3/2} f(\alpha)$$

where $f(\alpha)$ is the term in braces in Eq. (8-58), and $(\text{const}') = (\text{const})^{-1}$. Thus,

$$\left(\frac{\partial N}{\partial \beta} \right)_{V,\alpha} = -\frac{3}{2} \beta^{-5/2} (\text{const}') f(\alpha) = -\frac{3}{2} NkT$$

(8-59)

Equation (8-57) may now be rewritten

$$E = -\left(\frac{\partial \Psi}{\partial \beta} \right)_{V,\alpha} = +\frac{3}{2} NkT \left\{ 1 \pm \frac{\phi}{2^{5/2}} + \cdots \right\}$$

(8-60)

This is the second approximation for E.

A similar result can be obtained for S from Eq. (8-11).   The second approximation for α is obtained from Eq. (8-52) as

$$e^{-\alpha} = \phi \pm \frac{\phi^2}{2^{3/2}} + \cdots$$

$$-\alpha = \ln \left(\phi \pm \phi^2/2^{3/2} + \cdots \right)$$

$$-\alpha \cong \ln \phi + \ln \left(1 \pm \phi/2^{3/2} \right)$$

$$\alpha \cong -\ln \phi \mp \frac{\phi}{2^{3/2}} + \cdots$$

(8-61)

Thus,

$$S = Nk \left\{ 1 \pm \frac{\phi}{2^{5/2}} + \cdots \right\} + Nk \left\{ -\ln \phi \mp \frac{\phi}{2^{3/2}} + \cdots \right\} + \frac{3}{2} Nk \left\{ 1 \pm \frac{\phi}{2^{5/2}} + \cdots \right\}$$

$$S = Nk \left\{ \frac{5}{2} - \ln \phi \pm \frac{\phi}{2^{7/2}} + \cdots \right\}$$

(8-62)

From Eq. (8-35), the equation of state is

$$pV = \frac{2}{3}E = NkT\left\{1 \pm \frac{\phi}{2^{5/2}} + \cdots\right\}$$

(8-63)

The effect of weak degeneracy of a quantum gas on E, S, and the equation of state has been shown to be

$$E = \frac{3}{2}NkT\left\{1 \pm \frac{\phi}{2^{5/2}} + \cdots\right\}$$

(8-64)

$$S = Nk\left\{\frac{5}{2} - \ln\phi \pm \frac{\phi}{2^{7/2}} + \cdots\right\}$$

(8-65)

$$pV = NkT\left\{1 \pm \frac{\phi}{2^{5/2}} + \cdots\right\}$$

(8-66)

where $\phi = N/[(2s+1)Q]$. In each case where a double sign appears, the upper sign corresponds to Fermi-Dirac statistics and the lower sign to Bose-Einstein statistics. An examination of Eq. (8-1) shows that when compared to Boltzmann statistics, Bose-Einstein statistics tends to put extra particles in the lower states, and Fermi-Dirac statistics tends to put extra particles in the upper states.

The distribution functions for weakly degenerate Bose-Einstein and Fermi-Dirac substances are illustrated in Fig. 8-3. Thus, it is not surprising that the correction to E is $+$ for Fermi-Dirac statistics and $-$ for Bose-Einstein statistics. Similarly, Fermi-Dirac substances have higher S and p than Bose-Einstein substances, because more states are populated, and the particles tend to strike the walls with higher momenta.

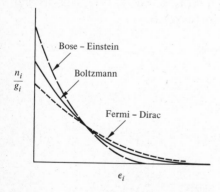

FIGURE 8-3 Comparison of populations of levels in Bose-Einstein, Boltzmann, and Fermi-Dirac statistics.

8.4 THE PERFECT BOSE-EINSTEIN GAS — CASE OF STRONG DEGENERACY

This section deals with a cubical container containing a perfect Bose-Einstein gas composed of molecules which do not interact with one another. In the usual procedure, the zero of energy is defined as the potential inside the container. The translational energy levels of a particle in a three-dimensional box with respect to this zero are $\epsilon_1, \epsilon_2, \epsilon_3, \ldots$, with $\epsilon_1 > 0$ due to the zero-point energy from the uncertainty principle. Alternatively, the zero of energy can be defined as ϵ_1, the energy of the lowest quantum state. With this definition, the energy levels are

$$\xi_j = \epsilon_j - \epsilon_1 \tag{8-67}$$

where $\xi_1 = 0$, $\xi_2 = 3\pi^2\hbar^2/2mL^2$ and L is the dimension of the cubical box. If we define

$$\gamma(T, N) = \alpha(T, N) + \epsilon_1/kT \tag{8-68}$$

then Eq. (8-1) takes the form

$$n_j = \frac{g_j}{e^\gamma\, e^{\xi_j/kT} - 1} \tag{8-69}$$

In particular,

$$n_1 = \frac{g_1}{e^\gamma - 1} \quad \text{and} \quad n_2 = \frac{g_2}{e^\gamma\, e^{\xi_2/kT} - 1} \tag{8-70}$$

Thus,

$$\frac{n_2}{n_1} = \frac{g_2}{g_1} \cdot \frac{(e^\gamma - 1)}{(e^\gamma\, e^{\xi_2/kT} - 1)} \tag{8-71}$$

Furthermore, $\gamma > 0$, so $e^\gamma > 1$. Three regions of temperature are distinguished:

(I) Very low T such that $\xi_2 \gg kT$. Since $e^{\xi_2/kT} \gg 1$, the factor -1 can be neglected in the denominator of Eq. (8-71). It will be shown later that the qualitative variation of γ with T at fixed N is as illustrated in Fig. 8-4. For low T, e^γ is only slightly greater than one, and the numerator of Eq. (8-71) may be expanded as

$$e^\gamma - 1 \cong (1 + \gamma + \cdots) - 1 = \gamma + \cdots$$

Therefore, in this temperature range,

$$\frac{n_2}{n_1} \cong \frac{g_2}{g_1} e^{-\xi_2/kT}\gamma \tag{8-72}$$

with $\gamma \ll 1$. Similarly, if we calculate n_3/n_2, it is found that

$$\frac{n_3}{n_2} \cong \frac{g_3(e^\gamma\, e^{\xi_2/kT} - 1)}{g_2(e^\gamma\, e^{\xi_3/kT} - 1)} \cong \frac{g_3}{g_2} e^{(\xi_2 - \xi_3)/kT} \tag{8-73}$$

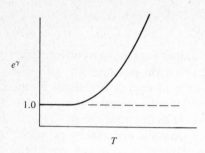

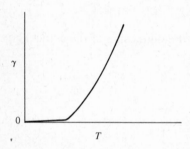

FIGURE 8-4 The functions γ and e^γ vs T.

The expression for n_3/n_2 is the same as one would obtain from Boltzmann statistics. The same holds true for all ratios n_i/n_j not involving the ground state. However, the ratio n_2/n_1 is much smaller than would be predicted from Boltzmann statistics due to the factor γ. Therefore, at very low T, there is a greater relative population of the ground energy level in Bose-Einstein statistics than in Boltzmann statistics. This may be put in the following way. In Boltzmann statistics at low T,

$$n_1 \gg n_2 \gg n_3 \gg n_4 \gg \cdots$$

whereas in Bose-Einstein statistics at low T,

$$n_1 \ggg n_2 \gg n_3 \gg n_4 \gg \cdots$$

(II) Very high T. At very high T, $e^\gamma \to e^\alpha \to (Q/N)(2s+1)$, which is large compared to 1, and the distribution over states becomes Boltzmann.

(III) Intermediate T such that $T \gg \xi_2/k$ but γ is small enough that $\xi_2/k\gamma \gg T$. Somewhere between the extremes of regions I and II, the temperature will be high enough that $kT \gg \xi_2$, yet low enough that $\xi_2 \gg kT\gamma$. This statement will be justified later. In this region, the exponentials in the numerator and the denominator of Eq. (8-71) may both be expanded in a series and only the leading terms retained. Thus,

$$\frac{n_2}{n_1} \cong \frac{g_2}{g_1} \frac{[(1+\gamma+\cdots)-1]}{[(1+\gamma+\xi_2/kT+\cdots)-1]} = \frac{g_2}{g_1}\left(\frac{\gamma}{\gamma+\xi_2/kT}\right)$$

$$\frac{n_2}{n_1} \cong \frac{g_2}{g_1}\left(\frac{1}{1+\xi_2/kT\gamma}\right) \cong \frac{g_2}{g_1}\left(\frac{kT\gamma}{\xi_2}\right) \tag{8-74}$$

This ratio would be simply g_2/g_1 in Boltzmann statistics since $\xi_2 \ll kT$. Thus, an enhancement of the population of the ground state can remain in Bose-Einstein statistics even when $kT \gg \xi_2$. In this temperature range, it is possible that $n_2, n_3, n_4, \ldots$ may all be of comparable magnitude, but n_1 is very much larger than any of the other n_j. Thus,

$$n_1 \ggg n_2 \sim n_3 \sim n_4 \ldots$$

provided that γ is small as assumed. If kT is large compared to a large number of ξ_j, then many states will be populated even though the ground state has the highest population.

Consider a gas of Bosons in temperature region III such that $n_1 \ggg n_2$, but a very large number of states beginning with n_2 are nearly equally populated. Then an attempt may be made to evaluate γ by replacing the sum

$$N = \sum_{\substack{j=\text{all}\\\text{states}}} \left(\frac{1}{e^\gamma e^{\xi_j/kT}-1}\right) \tag{8-75}$$

by the integral

$$N \cong \int_0^\infty \frac{d\Gamma(n)}{e^\gamma e^{\xi(n)/kT}-1} \tag{8-76}$$

This procedure works well when there is a broad distribution over many quantum states and the upper quantum states dominate the overall population. At moderately high T, the inaccurate representation of the lower quantum states in this procedure produces only a very small error. However, in the range of T considered here, n_1 is so large that it must be specifically treated or serious error will result. Since $d\Gamma(n)$ is very small for small n [see, for example, Eq. (8-22)] the use of Eq. (8-76) essentially neglects n_1 altogether. To specifically include n_1 in the distribution, write

$$N \cong n_1 + \int_0^\infty \frac{d\Gamma(n)}{e^\gamma e^{\xi(n)/kT}-1} \tag{8-77}$$

At high T, n_1 will be negligible, but in the intermediate temperature range it can be larger than the integral. The integral was evaluated previously in Eq. (8-49). Thus,

$$N \cong \frac{g_1}{e^\gamma - 1} + (2s+1)V \left(\frac{2\pi mkT}{h^2}\right)^{3/2} \left\{e^{-\gamma} + \frac{e^{-2\gamma}}{2^{3/2}} + \frac{e^{-3\gamma}}{3^{3/2}} + \cdots\right\} \qquad (8\text{-}78)$$

In general, $e^{-\gamma}$ is less than one, so the largest that $\sum\limits_{j=2}^{\infty} n_j$ can be is

$$\sum_{j=2}^{\infty} n_j = (2s+1)V \left(\frac{2\pi mkT}{h^2}\right)^{3/2} \underbrace{\left\{1 + \frac{1}{2^{3/2}} + \frac{1}{3^{3/2}} + \cdots\right\}}_{= 2.612}$$

Thus, if the assumption that $\gamma \ll 1$ is correct, the maximum number of particles that can be in states $2, 3, 4, \ldots, \infty$ is

$$N_m = (2s+1)V \left(\frac{2\pi mkT}{h^2}\right)^{3/2} (2.612) \qquad (8\text{-}79)$$

As long as $\gamma \ll 1$, $\sum\limits_{j=2}^{\infty} n_j$ will be only slightly less than N_m.

Now examine the consequences of our assumption that $\gamma \ll 1$:

$$N \cong n_1 + N_m$$

or

$$n_1 = N - N_m = N \left\{1 - \left(\frac{T}{T_0}\right)^{3/2}\right\} \qquad (8\text{-}80)$$

where T_0 is defined by

$$N = (2s+1)V \left(\frac{2\pi mkT_0}{h^2}\right)^{3/2} (2.612) \qquad (8\text{-}81)$$

Thus, at constant V and N, the variation of n_1 with T is as shown in Fig. 8-5. This behavior is known as "Einstein condensation." Of course, n_1 does not really go to 0 at $T = T_0$, but it goes to a very small value compared to N. For a gas of Bosons, there is a critical temperature below which n_1/N begins to rise with decreasing T to far greater values than predicted by Boltzmann statistics. Einstein condensation can also be

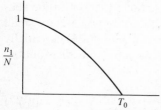

FIGURE 8-5 The ratio of populations of the ground state to all states vs T in Bose-Einstein statistics at fixed V below V_0.

produced by increasing the density N/V at constant T. By defining V_0 so that

$$N = (2s+1)\left(\frac{2\pi mkT}{h^2}\right)^{3/2}(2.612)V_0, \qquad (8\text{-}82)$$

there results from Eq. (8-80),

$$n_1 = N - N_m = N\left\{1 - \frac{V}{V_0}\right\} \qquad (8\text{-}83)$$

This is illustrated in Fig. 8-6.

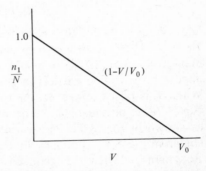

FIGURE 8-6 The ratio of populations of the ground state to all states vs V for fixed T below T_0 for a perfect gas of Bosons.

Now consider the validity of the approximations. All the previous discussions were based on the assumptions that $\gamma(T,N) \ll 1$, and that $T \gg \theta_{tr} \cong 3 \times 10^{-15\circ}\text{K}$ so that many excited translational states are populated. The assumption of many translational states populated is certainly valid for $T \gg 10^{-15\circ}\text{K}$. Thus, for temperatures greater than, say, $10^{-5\circ}\text{K}$, the replacement of the sums over states by integrals such as in Eqs. (8-75) and (8-76) is certainly justified. The question still remains about the magnitude of γ. It has been shown that if $\gamma \ll 1$, $e^{-\gamma} \cong 1$ and

$$N \cong \frac{1}{e^{\gamma}-1} + N_m \qquad (8\text{-}84)$$

since $g_1 = 1$ for the ground state of a particle in a box. Therefore, if $\gamma \ll 1$,

$$e^{\gamma} = 1 + \frac{1}{N-N_m}$$

$$1 + \gamma + \cdots = 1 + \frac{1}{N-N_m}$$

$$\gamma \cong \frac{1}{N-N_m}$$

Thus, γ will be $\ll 1$ if $N - N_m \gg 1$. It has been shown that

$$N_m = (2s+1)Q(2.612)$$

where $Q = (2\pi mkT/h^2)^{3/2}V$. If He atoms are considered at $1°K$ in 1 cm^3 of volume, $Q = 2 \times 10^{21}$, and $N_m = 0.5 \times 10^{22}$ particles. The value of N depends on the density. The empirical value of the density at this temperature is $\sim 0.15 \text{ g/cm}^3$. Therefore,

$$N = 0.15 \text{ g/cm}^3 \times \frac{6.023 \times 10^{23}}{4(\text{g/molecule})} \cong 2 \times 10^{22} \text{ particles}$$

and $N - N_m \cong 1.5 \times 10^{22}$ particles. Thus, γ is approximately $(1.5 \times 10^{22})^{-1} \cong 10^{-22}$, justifying our assumption that $\gamma \ll 1$. If T were increased to about $3.2°K$, N would be equal to N_m. At this temperature, almost no particles are left in the ground state, and thus $T_0 \cong 3.2°K$. When T is raised above T_0, the only way for the left and right sides of Eq. (8-78) to be equal is if $e^{-\gamma}$ decreases. Therefore, γ must vary with T as shown in Fig. 8-7. The variation of the distribution over states is shown in Fig. 8.8. For $T > T_0$, the distribution over states is essentially Boltzmann. When T is reduced below T_0, Einstein condensation produces a very large number of systems in the ground state. It is important to realize that this is entirely different than what happens at $\sim 10^{-15}°K$ when kT becomes comparable to the separation between energy levels. At a temperature of $0.01°K$, more than 99% of the He atoms are in the ground state due to Einstein condensation, whereas the Boltzmann distribution would put less than $10^{-18}\%$ of the particles in the ground state! Of course, at these low temperatures, the interatomic forces

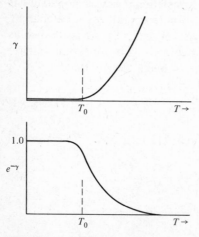

FIGURE 8-7 The functions γ and $e^{-\gamma}$ vs T.

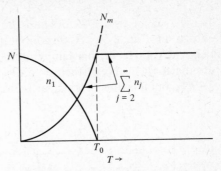

FIGURE 8-8 The population of the ground state and the sum of populations of all excited states vs T for $V < V_0$ for a perfect gas of Bosons.

cannot be neglected, and He is actually a liquid, not a perfect gas. Nevertheless, the qualitative conclusions drawn from this simplified discussion are basically correct.

Now calculate the energy of a degenerate Bose-Einstein gas. As shown in Eq. (8-54) (replacing α by γ),[3]

$$\Psi = (2s+1)\left(\frac{2\pi mkT}{h^2}\right)^{3/2} V \sum_{n=1}^{\infty} \frac{e^{-n\gamma}}{n^{5/2}}$$ (8-85)

For $T < T_0$, $e^{-n\gamma}$ may be replaced by unity and since

$$1 + \frac{1}{2^{5/2}} + \frac{1}{3^{5/2}} + \cdots = 1.341$$ (8-86)

then

$$\Psi = (2s+1)\left(\frac{2\pi mkT}{h^2}\right)^{3/2} V(1.341)$$

From Eq. (8-56), therefore,

$$E = \frac{3}{2}kT\left\{(2s+1)(1.341)\left(\frac{2\pi mkT}{h^2}\right)^{3/2} V\right\}$$

$$E = \frac{3}{2}kT\left\{\frac{1.341}{2.612}N_m(T)\right\}$$ (8-86a)

Since $\xi_1 = 0$, there is no contribution to the energy from particles in the ground state, so E is proportional to N_m. The specific heat is

$$c_v = \left(\frac{\partial E}{\partial T}\right)_V = \frac{15}{4}(1.341)(2s+1)\left(\frac{2\pi m}{h^2}\right)^{3/2} Vk^{5/2}T^{3/2}$$ (8-86b)

At $T = T_0$, $c_v = \frac{15}{4}\left(\frac{1.341}{2.612}\right)Nk = 1.926\,Nk.$ (8-86c)

---

[3] It is assumed here that $kT \gg \epsilon_1$.

For $T \gg T_0$, $c_v = 1.5Nk$. Thus, the temperature dependence of c_v is as shown in Fig. 8-9. A discontinuity is predicted at T_0. The actual behavior of He in this temperature range is complicated by the inter-atomic forces. Nevertheless, a discontinuity in c_v does in fact occur at 2.19°K, but it is due, at least partly, to a phase transition.

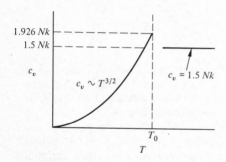

FIGURE 8-9 The specific heat vs T of a perfect gas of Bosons. In all actual gases, liquification obscures the discontinuity at $T = T_0$.

The entropy of a degenerate Boson gas may also be calculated. From Eqs. (8-55) and (8-60), for $T < T_0$,

$$k\Psi = \frac{2}{3}\frac{E}{T}$$

Therefore, from Eq. (8-11), noting that $\alpha \cong 0$, we obtain

$$S = \frac{5}{3}\frac{E}{T} = (\text{const})T^{3/2} \qquad (8\text{-}86d)$$

8.5 THE PERFECT FERMI-DIRAC GAS – CASE OF STRONG DEGENERACY

The distribution function for a gas of Fermions is

$$n_i = \frac{g_i}{e^{\alpha}\,e^{\epsilon_i/kT} + 1}$$

As $T \to 0$, all particles tend to go into the lowest quantum states available. However, since no two particles can be in the same quantum state, particles "pack in" to the lowest N quantum states if there are N particles. Thus, as $T \to 0$, the distribution of particles among the quantum states is as shown in Fig. 8-10. The maximum energy level that is filled at $T = 0$ is called the Fermi energy and is denoted ϵ_{m0}. A Boltzmann or

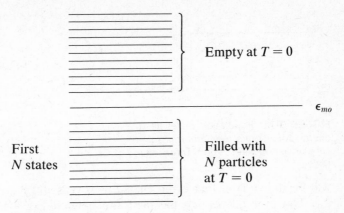

FIGURE 8-10 Filled and unfilled energy levels of a substance obeying Fermi-Dirac statistics at $T = 0$.

Bose-Einstein substance would have all particles in the ground state as $T \to 0$. Therefore, a gas of Fermions has relatively high energy at low T. To study the behavior of α as a function of T for low T, it is required that $n_i = g_i$ for all $\epsilon_j < \epsilon_{m0}$ at $T = 0$. Thus, as $T \to 0$,

$$e^\alpha e^{\epsilon_i/kT} \to 0 \qquad \text{for } \epsilon_i \leqq \epsilon_{m0}$$

$$\to \infty \qquad \text{for } \epsilon_i > \epsilon_{m0} \tag{8-87}$$

This can only happen if $\alpha \to -\epsilon_{m0}/kT$ as $T \to 0$. Thus α is a large negative quantity at low T, as shown in Fig. 8-11. The distribution function can then be written

$$n_i = \frac{g_i}{e^{(\epsilon_i - \epsilon_{m0})/kT} + 1} \tag{8-88}$$

This function satisfies the requirements of Eq. (8-87). Another way of depicting Eq. (8-88) is shown as the solid line in Fig. 8-12 for $T = 0$. It

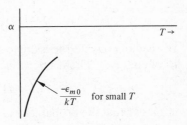

FIGURE 8-11 The temperature dependence of α for a substance obeying Fermi-Dirac statistics.

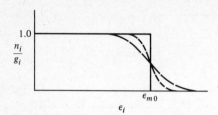

FIGURE 8-12 Population of levels vs energy for various temperatures for a substance obeying Fermi-Dirac statistics.

will be shown later that even when $T \neq 0$, α is still $\cong -\epsilon_{mo}/kT$ for small T. Thus, for $T \neq 0$, n_i/g_i has the form shown in Fig. 8-12 as the broken lines. Only in the limit of very high T does n_i/g_i approach the limit predicted by corrected Boltzmann statistics.

Now calculate the magnitude of ϵ_{mo}. Let the quantum number associated with level ϵ_{mo} be n_{mo}. The definition of ϵ_{mo} is such that

$$\int_0^{\epsilon_{mo}} d\Gamma(\epsilon) = N$$

$$\int_0^{\epsilon_{mo}} 2\pi \left(\frac{2m}{h^2}\right)^{3/2} V(2s+1)\epsilon^{1/2} \, d\epsilon = N$$

$$\epsilon_{mo} = \left(\frac{h^2}{2m}\right)\left(\frac{3N}{4\pi V}\right)^{2/3}\left(\frac{1}{2s+1}\right)^{2/3} \tag{8-89}$$

For electrons in a metal, the use of N/V corresponding to the atom density (assume one electron per atom) and m and the electron mass, leads to the result $\epsilon_{mo} \cong 6\,\text{eV}$. This is indeed a very high energy compared to kT (~ 0.03 eV at room temperature).

In order to calculate the properties of a degenerate gas of Fermions, the following analysis is resorted to. The probability that a state corresponding to energy level ϵ is populated is

$$f(\epsilon) = \frac{\text{population of level } \epsilon}{\text{degeneracy of level } \epsilon} \tag{8-90}$$

Instead of considering a single state, consider all the states in a band width $d\epsilon$ of energy. Then, since the total population of these states is

$$d\Gamma/(e^\alpha e^{\epsilon/kT} + 1)$$

it follows that the probability of any one state in the group being populated is

$$f(\epsilon) = \left(\frac{d\Gamma(\epsilon)}{e^\alpha e^{\epsilon/kT} + 1}\right)\left(\frac{1}{d\Gamma(\epsilon)}\right) = \frac{1}{e^\alpha e^{\epsilon/kT} + 1} \tag{8-91}$$

A digression is necessary at this point to evaluate integrals of the form

$$I = \int_0^\infty f(\epsilon) \frac{dF(\epsilon)}{d\epsilon} d\epsilon \tag{8-92}$$

These integrals come about in calculating the average properties of a substance, because if the quantity of interest is written in the form $dF(\epsilon)/d\epsilon$, then I is the average value of this quantity over the entire population of states. To evaluate I, Eq. (8-92) is integrated by parts, obtaining

$$I = f(\epsilon)F(\epsilon) \Big]_0^\infty - \int_0^\infty F(\epsilon) \frac{df}{d\epsilon} d\epsilon \tag{8-93}$$

The calculation will be restricted to functions $F(\epsilon)$ for which $F(0) = 0$. Furthermore, since $f(\infty) = 0$, it follows that

$$I = -\int_0^\infty F(\epsilon) \frac{df}{d\epsilon} d\epsilon \tag{8-94}$$

The function $f(\epsilon)$ has the qualitative form shown in Fig. 8-13. Therefore, $df/d\epsilon$ has the form shown in Fig. 8-14. Hence, if $F(\epsilon)$ is a smoothly varying function of ϵ, the integrand in Eq. (8-94) is appreciable only in the region near ϵ_m. Therefore, $F(\epsilon)$ can be expanded about $\epsilon = \epsilon_m$ as follows:

$$F(\epsilon) = \sum_{n=0}^\infty \left(\frac{d^n F}{d\epsilon^n}\right)_{\epsilon=\epsilon_m} (\epsilon - \epsilon_m)^n / n! \tag{8-95}$$

Since

$$\frac{df}{d\epsilon} = \frac{-(1/kT) e^\alpha e^{\epsilon/kT}}{(1 + e^\alpha e^{\epsilon/kT})^2} \tag{8-96}$$

it follows that

$$I = \frac{1}{kT} \sum_{n=0}^\infty \left(\frac{d^n F}{d\epsilon^n}\right)_{\epsilon=\epsilon_m} \frac{1}{n!} \int_0^\infty \frac{(\epsilon - \epsilon_m)^n e^y d\epsilon}{(1 + e^y)^2} \tag{8-97}$$

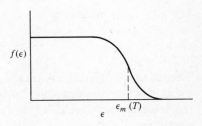

FIGURE 8-13 The distribution function $f(\epsilon)$ at temperature T. A point of inflection occurs at $\epsilon_m(T)$.

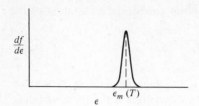

FIGURE 8-14 The slope of the curve in Fig. 8-13. The peak occurs at $\epsilon_m(T)$.

where $y = \alpha + \epsilon/kT$. Since the integration over $d\epsilon$ is carried out for constant T and α,

$$d\epsilon = kT \, dy \tag{8-98}$$

In general, α is some function of T, which may be defined as

$$\alpha = -\frac{\epsilon_m(T)}{kT} \tag{8-99}$$

without loss of generality. Therefore,

$$\epsilon - \epsilon_m = kTy \tag{8-100}$$

Equation (8-97) may then be rewritten

$$I = \sum_{n=0}^{\infty} \left(\frac{d^n F}{d\epsilon^n}\right)_{\epsilon_m} \frac{(kT)^n}{n!} \int_{\alpha}^{\infty} \frac{y^n e^y \, dy}{(1 + e^y)^2} \tag{8-101}$$

For many systems at low and moderate temperatures, it turns out that $\epsilon_m \gg kT$, so that $-\alpha \gg 1$. Therefore, the lower limit of the integral in Eq. (8-101) can be replaced by $-\infty$ without significant error. The integral in Eq. (8-101) for $n = 0$ is

$$\frac{1}{4} \int_{-\infty}^{\infty} \operatorname{sech}^2\left(\frac{y}{2}\right) dy = \frac{1}{4} \times 4 = 1$$

For $n \geq 1$, the integral is

$$\int_{-\infty}^{\infty} \frac{y^n \, dy}{(e^{-y/2} + e^{y/2})^2}$$

The numerator of this integrand is an even (odd) function of y if n is even (odd), whereas the denominator is always an even function of y. Thus, the integral vanishes if $n = $ odd. Equation (8-101) can, therefore, be rewritten

$$I = F(\epsilon_m) + \sum_{j=1}^{\infty} \left(\frac{d^{2j} F}{d\epsilon^{2j}}\right)_{\epsilon_m} \frac{(kT)^{2j}}{(2j)!} K_j \tag{8-102}$$

where

$$K_j = \int_{-\infty}^{\infty} \frac{y^{2j} \, dy}{(e^{-y/2} + e^{y/2})^2} = 2 \int_0^{\infty} \frac{y^{2j} e^{-y} \, dy}{(1 + e^{-y})^2} \qquad (8\text{-}103)$$

The denominator of this integrand can be expanded, using the formula

$$(1+x)^{-2} = 1 - 2x + 3x^2 - 4x^3 + \cdots = \sum_{l=1}^{\infty} l(-1)^{l-1} x^{l-1}$$

which is valid for $x < 1$. Since $e^{-y} < 1$ for $y > 0$, this is justified in our case for $x = e^{-y}$. Therefore,

$$K_j = 2 \sum_{l=1}^{\infty} l(-1)^{l-1} \underbrace{\int_0^{\infty} y^{2j} e^{-ly} \, dy}_{= \dfrac{(2j)!}{l^{2j+1}}} \qquad (8\text{-}104)$$

Thus, Eq. (8-102) becomes

$$I = F(\epsilon_m) + 2 \sum_{j=1}^{\infty} \left(\frac{d^{2j} F}{d\epsilon^{2j}} \right)_{\epsilon_m} (kT)^{2j} \sum_{l=1}^{\infty} \frac{(-1)^{l-1}}{l^{2j}} \qquad (8\text{-}105)$$

$$I = F(\epsilon_m) + 2(kT)^2 \left(\frac{d^2 F}{d\epsilon^2} \right)_{\epsilon_m} \sum_{l=1}^{\infty} \frac{(-1)^{l-1}}{l^2} + \text{terms in } j = 2, 3, \ldots \qquad (8\text{-}106)$$

If the higher derivatives of F are small, one may use Eq. (8-106) and neglect the higher order terms. It can be shown that

$$\sum_{l=1}^{\infty} \frac{(-1)^{l-1}}{l^2} = \frac{\pi^2}{12} \qquad (8\text{-}107)$$

Thus,

$$I = \int_0^{\infty} f(\epsilon) \left(\frac{dF}{d\epsilon} \right) d\epsilon = F(\epsilon_m) + \frac{\pi^2}{6} \left(\frac{d^2 F}{d\epsilon^2} \right)_{\epsilon_m} (kT)^2 + \cdots \qquad (8\text{-}108)$$

Now proceed to evaluate thermodynamic functions of a degenerate gas of Fermions using Eq. (8-108). First consider the conservation of particles:

$$N = \int_0^{\infty} f(\epsilon) \, d\Gamma(\epsilon) \qquad (8\text{-}109)$$

The total number of particles is the integral of probability of population of a state times the density of states in an interval of energy. Upon comparison with Eq. (8-108), it follows that F must be defined by

$$d\Gamma(\epsilon) = \frac{dF}{d\epsilon} \, d\epsilon$$

Therefore, in this case,

$$\frac{dF}{d\epsilon} = \frac{d\Gamma(\epsilon)}{d\epsilon} = 2\pi \left(\frac{2m}{h^2}\right)^{3/2} (2s+1) V\epsilon^{1/2} \tag{8-110}$$

and by integration,

$$F(\epsilon) = 2\pi \left(\frac{2m}{h^2}\right)^{3/2} V(2s+1)\epsilon^{3/2}\left(\frac{2}{3}\right) \tag{8-111}$$

By differentiation of Eq. (8-110),

$$\frac{d^2F}{d\epsilon^2} = 2\pi \left(\frac{2m}{h^2}\right)^{3/2} V(2s+1) \left(\frac{1}{2}\right) \epsilon^{-1/2} \tag{8-112}$$

Thus,

$$N = 2\pi \left(\frac{2m}{h^2}\right)^{3/2} V(2s+1) \left(\frac{2}{3}\right) \epsilon_m^{3/2}\left\{1 + \frac{\pi^2}{8}\left(\frac{kT}{\epsilon_m}\right)^2 + \cdots\right\} \tag{8-113}$$

As $T \to 0$, $\epsilon_m \to \epsilon_{m0}$, and

$$N = 2\pi \left(\frac{2m}{h^2}\right)^{3/2} V(2s+1) \frac{2}{3} \epsilon_{m0}^{3/2} \tag{8-114}$$

If Eq. (8-114) is solved for ϵ_{m0}, the result is the same as obtained previously in Eq. (8-89). By eliminating N between Eqs. (8-113) and (8-114), the following result is obtained:

$$\epsilon_m = \frac{\epsilon_{m0}}{\left\{1 + \frac{\pi^2}{8}\left(\frac{kT}{\epsilon_m}\right)^2 + \cdots\right\}^{2/3}} \tag{8-115}$$

Since

$$\frac{1}{(1+x)^{2/3}} = 1 - \frac{2}{3}x + \cdots, \qquad \text{for } x < 1,$$

$$\epsilon_m = \epsilon_{m0}\left\{1 - \frac{\pi^2}{12}\left(\frac{kT}{\epsilon_m}\right)^2 + \cdots\right\} \tag{8-116}$$

If $\epsilon_m \gg kT$, ϵ_m can be replaced by ϵ_{m0} on the right side of Eq. (8-116) without much error. One then finds that at $T = 300°K$, for an electron gas in a metal, $(\pi kT/\epsilon_{m0})^2/12 \cong 2 \times 10^{-5}$. Therefore, at room temperature, $\epsilon_m \cong 0.99998\,\epsilon_{m0}$. Note that ϵ_m decreases as T increases. Therefore, Fig. 8-12 should really be redrawn with ϵ_m shifted slightly to the left as T increases.

Now perform a similar calculation for the energy.

$$E = \int_0^\infty f(\epsilon)\epsilon\, d\Gamma(\epsilon) \tag{8-117}$$

$$\left(\frac{dF}{d\epsilon}\right) d\epsilon = \epsilon\, d\Gamma(\epsilon) = 2\pi \left(\frac{2m}{h^2}\right)^{3/2} V(2s+1)\epsilon^{3/2}\, d\epsilon$$

$$F(\epsilon) = 2\pi \left(\frac{2m}{h^2}\right)^{3/2} V(2s+1) \cdot \frac{2}{5} \epsilon^{5/2}$$

$$\frac{d^2F}{d\epsilon^2} = 2\pi \left(\frac{2m}{h^2}\right)^{3/2} V(2s+1) \frac{3}{2} \epsilon^{1/2}$$

From Eq. (8-108), the following is obtained:

$$E = 2\pi \left(\frac{2m}{h^2}\right) V(2s+1) \cdot \frac{2}{5} \epsilon_m^{5/2} \left\{1 + \frac{5\pi^2}{8}\left(\frac{kT}{\epsilon_m}\right)^2 + \cdots\right\} \qquad (8\text{-}118)$$

The approximate value, ϵ_{m0} from Eq. (8-89), can be used for ϵ_m inside the braces because the second term is small compared to unity. In the factor $\epsilon_m^{5/2}$, however, Eq. (8-116) must be used. Therefore,

$$E = \left[2\pi \left(\frac{2m}{h^2}\right) V(2s+1) \frac{2}{3} \epsilon_{m0}^{3/2}\right]\left(\frac{3}{2}\right)\left(\frac{2}{5}\right) \epsilon_{m0} \left\{1 - \frac{5\pi^2}{24}\left(\frac{kT}{\epsilon_{m0}}\right)^2 + \cdots\right\}$$

$$\cdot \left\{1 + \frac{5\pi^2}{8}\left(\frac{kT}{\epsilon_{m0}}\right)^2 + \cdots\right\} \qquad (8\text{-}119)$$

and the following has been used:

$$(1-y)^{5/2} = 1 - \tfrac{5}{2}y + \cdots$$

For small values of a and b,

$$(1-a)(1+b) = 1 - a + b - ab \cong 1 - a + b$$

Therefore, Eq. (8-119) becomes

$$E = N \cdot \frac{3}{5} \epsilon_{m0} \left\{1 + \frac{5\pi^2}{12}\left(\frac{kT}{\epsilon_{m0}}\right)^2 + \cdots\right\} \qquad (8\text{-}120)$$

As $T \to 0, E \to \frac{3}{5}N\epsilon_{m0}$, which is very large compared to $\frac{3}{2}NkT$.

The specific heat can be obtained simply by taking

$$c_v = \left(\frac{\partial E}{\partial T}\right)_v = \frac{\pi^2}{2} Nk \left(\frac{kT}{\epsilon_{m0}}\right) + \cdots \qquad (8\text{-}121)$$

At room temperature, $c_v \cong 0.02Nk$ for an electron gas in a metal. Since $c_v \cong 3Nk$ for metals due to the vibrations of the lattice, the relative contribution to the specific heat of a metal by the electrons is small despite the high energy of the electrons.

The entropy of an electron gas is

$$S = \int_0^T c_v \frac{dT'}{T'} = \frac{\pi^2}{2} Nk \left(\frac{kT}{\epsilon_{m0}}\right) = c_v \qquad (8\text{-}122)$$

The results of this section will now be reviewed. If kT is large compared to the spacing between translational energy levels, we may treat the levels as if they constituted a continuum. The number of particles in a range of energy between ϵ and $\epsilon + d\epsilon$ is $dn(\epsilon)$. The sum over all energy is

$$N = \int_0^\infty dn(\epsilon) \qquad (8\text{-}123)$$

The number of particles in a band width of energy is the product of the number of translational states in the interval, $d\Gamma(\epsilon)$, and the probability of population of a state, $f(\epsilon)$. It has been shown that

$$d\Gamma(\epsilon) = (\text{const})\epsilon^{1/2}\, d\epsilon \qquad (8\text{-}124)$$

and

$$f(\epsilon) = \frac{1}{e^{(\epsilon - \epsilon_m)/kT} + 1} \qquad (8\text{-}125)$$

and ϵ_m is a very slowly varying function of T. Therefore,

$$dn(\epsilon) = \frac{(\text{const})\, \epsilon^{1/2}\, d\epsilon}{e^{(\epsilon - \epsilon_m)/kT} + 1} \qquad (8\text{-}126)$$

The distribution function $dn(\epsilon)/\epsilon^{1/2}\, d\epsilon$ is shown in Fig. 8-15. The difference between ϵ_m and ϵ_{m0} has been greatly exaggerated. A plot of $dn(\epsilon)/d\epsilon$ is given in Fig. 8-16. The average energy per particle is

$$\langle \epsilon \rangle = \frac{E}{N} = \frac{\int_0^\infty \epsilon\, dn(\epsilon)}{\int_0^\infty dn(\epsilon)}$$

At $T = 0°K$,

$$\langle \epsilon \rangle = \frac{\int_0^{\epsilon_{m0}} (1)\epsilon(\text{const})\epsilon^{1/2}\, d\epsilon}{\int_0^{\epsilon_{m0}} (1)(\text{const})\epsilon^{1/2}\, d\epsilon} = \frac{3}{5}\epsilon_{m0}$$

Were it not for the factor of $\epsilon^{1/2}$ in $dn(\epsilon)$, $\langle \epsilon \rangle$ would be equal to $\epsilon_{m0}/2$. Although $\langle \epsilon \rangle$ is large compared to kT for an electron gas in a metal, it

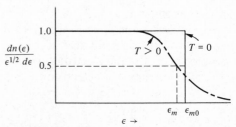

FIGURE 8-15 Distribution function vs energy level for a gas of Fermions. ϵ_{m0} is the value of ϵ_m at $T = 0$.

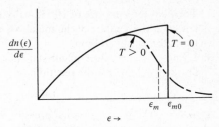

FIGURE 8-16 Distribution function vs. energy level for a gas of Fermions. ϵ_{m0} is the value of ϵ_m at $T = 0$.

does not vary much as T is changed. This is why the contribution to the specific heat of a metal by the electrons is small despite the fact that the electrons have most of the energy. The velocity distribution of a gas of Fermions is obtained by writing $\epsilon = mv^2/2$ and $d\epsilon = mvdv$.

$$dn(v) = \frac{(\text{const}')v^2\,dv}{e^{-(mv^2/2-\epsilon_m)/kT} + 1}$$

The result is plotted in Fig. 8-17, with the differences between ϵ_m and ϵ_{m0} greatly exaggerated.

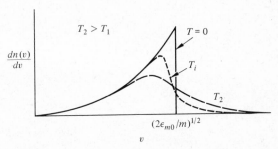

FIGURE 8-17 Distribution function vs velocity for a gas of Fermions. ϵ_{m0} is the value of ϵ_m at $T = 0$.

Problems

1. Derive the third approximation for α from Eq. (8-49) by using Eq. (8-52) for α in $e^{-2\alpha}$, and Eq. (8-43) in $e^{-3\alpha}$. Similarly, derive the third approximations for $\Psi, E, S,$ and pV/NkT.

2. Using Eqs. (8-50) and (8-63), calculate the percent deviation of He gas from the $pV = NkT$ due to quantum effects at 300°K, 200°K, and 30°K.

3. Derive the equation of state of a degenerate ideal gas of Bosons from Eqs. (8-86a) and (8-35). Note that $N_m(T)$ is given by Eq. (8-79) for $T > T_0$ and for $T < T_0$.

4. Compare ϵ_{m0} for an electron gas confined to a metal block by the coulomb attractive forces, with an electron gas in a hot plasma at a pressure of 10^{-4}

Torr and temperature of $10^{4\circ}$K.   Assume there is a metal atom in every 10^{-23} cm³, and the volume of the metal is 1 cm³.   Which statistics does the electron gas in the plasma follow?

5. Using corrected Boltzmann statistics for the electron gas, calculate the equilibrium constant for the reaction

$$Li(gas) \rightarrow Li^+(gas) + e^-$$

at $T = 5000°$K.   If the total pressure is 10^{-4} Torr, what is the concentration of electrons per cubic centimeter?

chapter 9

DEPENDENT NONLOCALIZED SYSTEMS – THE IMPERFECT GAS

9.1 INTRODUCTION

This chapter deals with a collection of N identical point particles that are indistinguishable and that exert forces on each other. It is assumed that kT is large compared to the spacing between translational levels, and $g_i \gg n_i$ for each pseudo-level. Therefore, corrected Boltzmann statistics is used. Since a great many closely packed translational levels are populated, classical statistical mechanics is employed.

In the discussion given in Chapter 4, the partition function and distribution function for a single particle was described. Each particle could be represented by a point in phase space, and the density of points in phase space gives the probability of occurrence of a configuration. This is a "microscopic" formulation, and is usually called the micro-canonical ensemble. An equivalent but different procedure is to use a multidimensional phase space in which the coordinates and momenta of all N particles form a $6N$ dimensional space ($3N$ coordinates and $3N$ momenta). The configuration of the entire system of N particles may be represented by a single point in this phase space. Now the existence of a very large number of complete ensembles of N particles can be conjectured, each ensemble identical with the one under consideration. If the configuration of each of these ensembles is represented by a point in the multidimensional phase space, then the density of points in any region is interpreted as the probability of any arbitrary ensemble being in that configuration. The multidimensional partition function for a collection of N interacting particles is

$$Q = \frac{1}{N!} h^{-3N} \int \cdots \int_{6N} e^{-E/kT} \, dq_1 \cdots dq_{3N} \, dp_1 \cdots dp_{3N} \qquad (9\text{-}1)$$

where

$$E = \frac{1}{2m} \sum_{i=1}^{3N} p_i^2 + U(q_1, q_2, \ldots q_{3N}) \qquad (9\text{-}2)$$

and U is the total potential energy.   Thus,

$$Q = \frac{1}{N!} \left(\frac{2\pi mkT}{h^2} \right)^{3N/2} \left\{ \underset{3N}{\int \cdots \int} e^{-U/kT} \, dq_1 \, dq_2 \ldots dq_{3N} \right\} \qquad (9\text{-}3)$$

The quantity in braces in Eq. (9-3) is denoted as Z and is called the configuration integral.

As a trivial example of the use of this formulation of statistical mechanics, consider the case where no forces act between the molecules, so that

$$U = 0 \text{ inside a container}$$
$$U = \infty \text{ outside the container}$$

Then the configurational integral is

$$Z = \iiint dq_1 \, dq_2 \, dq_3 \iiint dq_4 \, dq_5 \, dq_6 \int \cdots = V^N \qquad (9\text{-}4)$$

where V is the volume of the container in which the molecules exist. Thus

$$Q = \frac{1}{N!} \left[\left(\frac{2\pi mkT}{h^2} \right)^{3/2} V \right]^N = \frac{1}{N!} [\text{partition function for one molecule}]^N$$

$$(9\text{-}5)$$

The partition function is $1/N!$ times the Nth power of the partition function for a single molecule. In this formulation, there is only a single hypothetical system with $6N$-fold degrees of freedom. Therefore, the thermodynamic properties are obtained from Q by means of the usual formulas, but with the number of particles taken as one. Thus, for a single system with $6N$ degrees of freedom,

$$S = (1)k \ln Q + E/T \qquad (9\text{-}6)$$

$$E = (1)kT^2 \frac{d \ln Q}{dT} \qquad (9\text{-}7)$$

where Q is given by Eq. (9-5). The results are the same as those obtained in previous chapters.

In order to include intermolecular forces in this description, it will be assumed that the total potential energy of the ensemble of N interacting molecules depends on a sum of pairwise interactions between molecules:

$$U = \underset{\substack{i \quad j \\ \text{all pairs} \\ \text{of molecules}}}{\sum \sum} u(r_{ij}) \qquad (9\text{-}8)$$

where r_{ij} is the distance between molecules i and j, and $u(r_{ij})$ is the interaction energy between molecules i and j. It is assumed that the interaction between molecules i and j, $u(r_{ij})$, is not affected if a third molecule (k) approaches. This is illustrated in Fig. 9-1, where $u(r_{21})$ is independent of coordinates r_{32} and r_{31}.

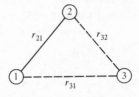

FIGURE 9-1 Illustration of the assumption that the potential between molecules 1 and 2 is independent of the presence of molecule 3.

Equation (9-8) may be rewritten

$$U = \sum_{i>j}^{i=N} \sum_{j=1}^{N-1} u(r_{ij}) \tag{9-9}$$

in which i varies from 2 to N and is always greater than j, and j varies from 1 to $N-1$, since r_{ij} and r_{ji} are not to be counted as different. There are a total of $\frac{1}{2}N(N-1)$ different pairs of molecules. The term $\exp(-U/kT)$ is given by

$$e^{-U/kT} = \exp\left\{-\sum_{i>j}^{N} \sum_{j=1}^{N-1} u(r_{ij})/kT\right\} = \prod_{N\geq i>j\geq 1} \exp\left[-u(r_{ij})/kT\right] \tag{9-10}$$

In general, the intermolecular potential between a pair of molecules has the qualitative form shown in Fig. 9-2. At large distances there is a

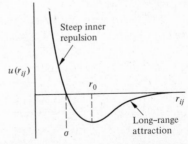

FIGURE 9-2 Typical potential energy of interaction between molecules i and j, consisting of a long-range attraction and a short-range repulsion.

weak Van der Waals attraction, and at small distances there is a steep inner repulsion. The function $\exp[-u(r_{ij})/kT]$, therefore, has the form shown in Fig. 9-3. As T increases, this curve spreads, whereas it narrows when T is reduced. This is illustrated in Fig. 9-4. As $T \to 0$, all the molecules tend to go to the most stable configuration, a distance r_0 apart from one another. As $T \to \infty$, the interactions become unimportant and the interacting gas closely approximates perfect gas behavior.

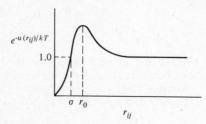

FIGURE 9-3   The function $\exp[-u(r_{ij})/kT]$ vs r_{ij} for a $u(r_{ij})$ function as shown in Fig. 9-2.

Define a function

$$f_{ij} \equiv f(r_{ij}) = e^{-u(r_{ij})/kT} - 1 \qquad (9\text{-}11)$$

which is almost the same as $e^{-u(r_{ij})/kT}$ except that f_{ij} goes to 0 as $r_{ij} \to \infty$. This function is plotted in Fig. 9-5. Then

$$e^{-U/kT} = \prod_{N \geq i > j \geq 1} (1+f_{ij}) = (1+f_{21})(1+f_{31})(1+f_{32}) \ldots (1+f_{N,N-1}) \quad (9\text{-}12)$$

and thus

$$e^{-U/kT} = 1 + \sum_{\substack{\text{all pairs} \\ i>j}} f_{ij} + \sum_{\substack{\text{all pairs} \\ k>l \\ (k,l \neq i,j)}} \sum_{\substack{\text{all pairs} \\ i>j}} f_{ij}f_{kl} + \cdots \qquad (9\text{-}13)$$

If there were no forces between the molecules, all the f_{ij} would be zero. Then $\exp(-U/kT)$ would be unity, and Z would be V^N.

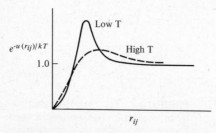

FIGURE 9-4   Variation of the function $\exp[-u(r_{ij})/kT]$ with T.

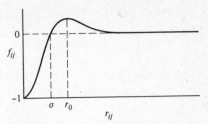

FIGURE 9-5 The function $f_{ij} = \exp[-u(r_{ij})/kT] - 1$ vs r_{ij}.

9.2 CONSIDERATION OF ELEMENTARY CALCULATIONS OF EQUATION OF STATE

Some authors attempt to calculate the effect of intermolecular forces on Z to first order by assuming the f_{ij} are small and neglecting terms like $f_{ij}f_{kl}$ compared to f_{ij}. The double sum in Eq. (9-13) is therefore neglected and the approximation is made that

$$e^{-U/kT} \cong 1 + \sum_{\substack{\text{all pairs} \\ i>j}} f_{ij} \tag{9-14}$$

Although each typical term $f_{ij}f_{kl}$ is indeed small compared to a typical term f_{ij}, there are many more terms in the double sum over $f_{ij}f_{kl}$ than in the single sum over f_{ij}. It will be shown in a later part of this chapter that the third, fourth, and additional terms in Eq. (9-13) are generally not negligible, and, in fact, they exceed the first two terms. Therefore, Eq. (9-14) is grossly incorrect. Nevertheless, use of Eq. (9-14) in the expression for Z,

$$Z = \int \cdots_{6N} \int e^{-U/kT} \, dq_1 \ldots dq_{3N} \, dp_1 \ldots dp_{3N} \tag{9-15}$$

does (by compounding the felony) give the correct first-order correction due to intermolecular forces. This incorrect procedure is illustrated here. When Eq. (9-14) is put into Eq. (9-15), the result is

$$Z \cong V^N + \int \cdots \int \{f_{21} + f_{31} + f_{32} + \cdots\} \, d\tau_1 \ldots d\tau_N \tag{9-16}$$

where $d\tau_j$ is the volume element for particle j. Thus,

$$Z \cong V^N + \int \cdots \int f_{21} \, d\tau_1 \ldots d\tau_N + \int \cdots \int f_{32} \, d\tau_1 \ldots d\tau_N$$

$$+ \cdots + \int \cdots \int f_{N,N-1} \, d\tau_1 \ldots d\tau_N \tag{9-17}$$

Each of the integrals in Eq. (9-17) is equal because all the molecules are identical and it does not matter which pair is taken for evaluation of the integral. There are $\frac{1}{2}N(N-1)$ such integrals. Thus,

$$Z \cong V^N + \frac{1}{2}N(N-1) \int \cdots \int f_{ij}\, d\tau_1 \ldots d\tau_N \qquad (9\text{-}18)$$

The integral in Eq. (9-18) may be integrated over $N-2$ volume elements, not including i and j. A factor of V is obtained for each such integration. Thus,

$$Z \cong V^N + \frac{1}{2}N^2 V^{N-2} \int\int f_{ij}\, d\tau_i\, d\tau_j \qquad (9\text{-}19)$$

in which $N(N-1)$ has been replaced by N^2 since $N \gg 1$. The integral in Eq. (9-19) may be reduced further by conceptually placing molecule i in some place in the container, and referring the coordinates of molecule j relative to molecule i. The interaction potential between i and j is independent of the position of molecule i in the container since it only depends on the relative position of j to i. It is assumed that surface effects are not important, so that configurations where i is near a wall occur rarely. One then integrates over $d\tau_i$ and the result is

$$Z \cong V^N + \frac{N^2}{2} V^{N-2} V \int f_{ij}\, d\tau_j \qquad (9\text{-}20)$$

If $u(r_{ij})$ is spherically symmetric, we may put $d\tau_j = 4\pi r^2\, dr$ and $r_{ij} = r$. Therefore,

$$Z \cong V^N(1 + N^2 b/2V) \qquad (9\text{-}21)$$

where b is given by

$$b = \int_0^\infty f(r)4\pi r^2\, dr \qquad (9\text{-}22)$$

The quantity b has a magnitude of the order of a molecular volume, namely, about 10^{-23}cm^3. For a gas in a 1 cm^3 container under ordinary conditions, N is about 10^{19} particles. Thus,

$$\frac{N^2 b}{2V} \cong \frac{10^{38}10^{-23}}{2} \cong 10^{15} \qquad (9\text{-}23)$$

Since this term greatly outweighs the first term (unity) in Eq. (9-21), it is evident that the first-order assumption Eq. (9-14) is very poor. Succeeding terms in Eq. (9-13) also produce non-negligible contributions to Z, and therefore, Eq. (9-21) is grossly in error. Nevertheless, if Eq. (9-21) is used to calculate Q from Eq. (9-3), the use of this Q in standard formulas results in the correct first-order correction to the equation of state of an ideal gas due to intermolecular forces. This result occurs because the assumption that $N^2 b/2V \ll 1$ is made several times. This will be discussed further, after examining a more rigorous calculation. The approximate approach will then be compared with the correct calculation.

Before proceeding with a more rigorous calculation, briefly review two aspects of the simplified procedure involving the use of Eq. (9-21) for Z. First, estimate the magnitude of b. (In the previous discussion, a magnitude of $\sim 10^{-23}\ cm^3$ was merely stated.) Then, examine the procedure by which the equation of state is obtained from Z.

The magnitude of b is determined from Eq. (9-22). Since $f(r) = \exp(-u(r)/kT) - 1$, it follows that the intermolecular potential $u(r)$ determines the magnitude of b. The intermolecular potential generally has the form shown in Fig. 9-2. As a very simple qualitative illustration, assume the molecules can be represented by rigid spheres. Then $u(r)$ and $f(r)$ are as shown in Figs. 9-6 and 9-7. The calculation of b is then very simple.

$$b = \int_0^{r_0} (-1)4\pi r^2\, dr = -\tfrac{4}{3}\pi r_0^3$$

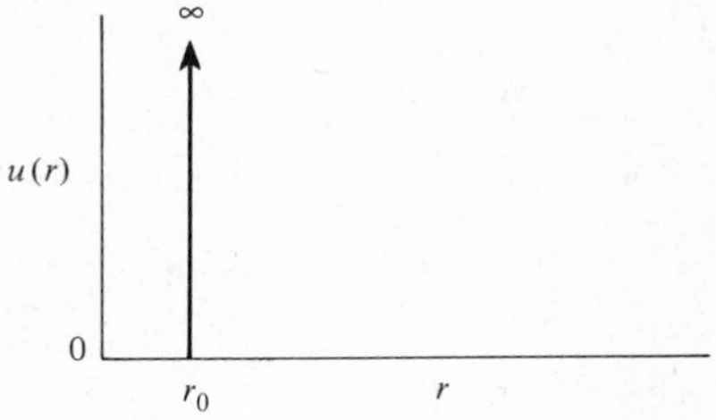

FIGURE 9-6 The potential function $u(r)$ representing a hard sphere interaction.

If one considers the molecules to be hard spheres of diameter r_0 (see Fig. 9-8), then

$$b = -8[\tfrac{4}{3}\pi (r_0/2)^3] = -8V_m$$

where V_m is the volume of a molecule. If $V_m \cong 10^{-24}\ cm^3$, $|b|$ is $\cong 10^{-23}$ cm^3.

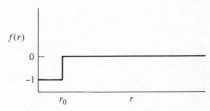

FIGURE 9-7 The function $f(r)$ corresponding to the potential $u(r)$ given in Fig. 9.6.

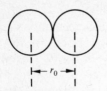

FIGURE 9-8 The molecular diameter r_0 for a hard-sphere interaction.

The determination of the equation of state of a gas of interacting particles follows from the thermodynamic equations

$$p = -\,(\partial A/\partial V)_{T,N} \tag{9-24}$$

$$A = E - TS = -kT \ln Q \tag{9-25}$$

where Q is the partition function for N molecules.   Then

$$p = kT\,(\partial \ln Q/\partial V)_{T,N} \tag{9-26}$$

Since

$$Q = \frac{1}{N!}\left(\frac{2\pi mkT}{h^2}\right)^{3N/2} Z \tag{9-27}$$

then

$$\frac{\partial \ln Q}{\partial V} = \frac{\partial \ln Z}{\partial V} \tag{9-28}$$

For an ideal gas, where no forces exist between molecules, $Z = V^N$, and $\partial \ln Z/\partial V = N/V$.   Therefore, for an ideal gas,

$$p = kT\,(N/V) \tag{9-29}$$

When intermolecular forces act, Eq. (9-15) should be used for Z.

Now proceed to a more rigorous calculation of Z.   Reexamine Eq. (9-13) in order to estimate the effect of higher terms in contributing to Z.   It has already been shown that the first term (1) in Eq. (9-13) contributes V^N to Z.   The second term contributes $(N^2 b/2V)V^N$.   To estimate the effect of the third term in the sum, let us count up the number of equivalent factors of the type $f_{ij}f_{kl}$ with $i > j$, $k > l$ and $k,l \ne i,j$.   There are $N(N-1)/2$ terms f_{ij} with $i > j$.   For each such term, there are $N(N-1)/2$ factors f_{kl}.   After eliminating the one term, for which $i,j = k,l$, there are a total of $(N(N-1)/2 - 1)$ factors f_{kl} for which $k,l \ne i,j$.   However, this calculation is based on treating terms like $f_{75}f_{32}$ to be different than $f_{32}f_{75}$.   Therefore, the result must be divided through by 2.   Finally, there are

$$\frac{1}{2}\left[\frac{N(N-1)}{2}\right]\left[\frac{N(N-1)}{2} - 1\right] \tag{9-29a}$$

different terms $f_{ij}f_{kl}$ with $i > j, k > l$ and $i,j \neq k,l$. If $N \gg 1$, this is nearly equal to $N^4/8$. In a similar way, it can be shown that there are $\sim N^6/48$ terms $f_{ij}f_{kl}f_{mn}$ in the fourth term of the sum in Eq. (9-13).

In taking the integral

$$\underset{N}{\int \cdots \int} f_{ij}f_{kl}\, d\tau_1 \ldots d\tau_N$$

great simplification results if none of the indices i, j, k, l are the same, because

$$\underset{N}{\int \cdots \int} f_{ij}f_{kl}\, d\tau_1 \ldots d\tau_N = \int f_{ij}\, d\tau_j \int f_{kl}\, d\tau_l \underset{N-2}{\int \cdots \int} d\tau_a\, d\tau_b \ldots$$

$$= V^{N-2}b^2$$

It can be shown[1] that the overwhelming majority of terms $f_{ij}f_{kl}$ involve no common indices. The number of such terms without common indices is

$$\frac{1}{2}\left[\frac{N(N-1)}{2}\right]\left[\frac{(N-2)(N-3)}{2}\right] \tag{9-29b}$$

If $N \gg 1$, this is essentially equal to $N^4/8$, which is the total number of terms including common indices.

Equations (9-13) and (9-15), which are exact, may be combined to yield the exact equation

$$Z = V^N\left\{1 + \left(\frac{N^2b}{2V}\right) + \frac{1}{2}\left(\frac{N^2b}{2V}\right)^2 + \frac{1}{6}\left(\frac{N^2b}{2V}\right)^3 + \cdots\right\} \tag{9-30}$$

Since $N^2b/2V \gg 1$, it can be seen that each succeeding term in the series outweighs the previous term.

Now examine the result obtained in the crude approximation of Eq. (9-21) for Z.

When Eq. (9-21) is used for Z,

$$\ln Z \cong \ln\left\{V^N\left(1 + \frac{N^2b}{2V}\right)\right\} = N \ln V + \ln\left(1 + \frac{N^2b}{2V}\right) \tag{9-31}$$

$$\frac{\partial \ln Z}{\partial V} \cong \frac{N}{V} + \frac{-(N^2b/2V^2)}{(1+N^2b/2V)} = \frac{N}{V}\left\{1 - \frac{Nb/2V}{(1+N^2b/2V)}\right\}$$

$$p \cong \frac{NkT}{V}\left\{1 - \frac{Nb/2V}{(1+N^2b/2V)}\right\} \tag{9-32}$$

---

[1] The ratio of terms without common indices to all terms is Eq. (9-29b) divided by Eq. (9-29a):

$$\frac{(N-2)(N-3)}{N(N-1)-2} = \frac{N^2-5N+6}{N^2-N-2} = \frac{1-5/N+6/N^2}{1-1/N-2/N^2} \cong 1 - \frac{4}{N} + \cdots$$

It is at this point that the ultrasimplified calculation goes astray. One must assume that $N^2b/2V \ll 1$, and set $1 + N^2b/2V \cong 1$. This is, of course, absurd because $N^2b/2V \cong 10^{15}$. Nevertheless, when this is done, the result is

$$p \cong \frac{NkT}{V}\left\{1 - \frac{Nb}{2V}\right\} \tag{9-33}$$

Note that $Nb/2V \ll 1$. When a rigorous calculation is made (see Sec. 9.3), it is found that

$$p = \frac{NkT}{V}\left\{1 - \frac{Nb}{2V} + \text{higher order terms in } \frac{N}{V}\right\} \tag{9-34}$$

Only the fortunate cancellation of large errors caused by using Eq. (9-21) for Z and setting $1 + 10^{15} = 1$ produces the proper first-order term in N/V. If Eq. (9-30) were used for Z, it would be found that

$$\ln Z = N \ln V + \ln\left\{1 + \frac{N^2b}{2V} + \frac{1}{2}\left(\frac{N^2b}{2V}\right)^2 + \frac{1}{6}\left(\frac{N^2b}{2V}\right)^3 + \cdots\right\} \tag{9-35}$$

$$\frac{\partial \ln Z}{\partial V} = \frac{N}{V}\left\{1 - \left(\frac{Nb}{2V}\right)\left[\frac{1 + \theta + \theta^2/2 + \theta^3/12 + \cdots}{1 + \theta + \theta^2/2 + \theta^3/6 + \cdots}\right]\right\} \tag{9-36}$$

where $\theta = N^2b/2V$ is of the order of 10^{15}. It is not evident from Eq. (9-36) what the form of the equation of state is, and recourse has to be made to a more rigorous procedure.

9.3 THE DESCRIPTION OF NONIDEAL GASES IN TERMS OF CLUSTERS

It should be readily evident from the previous paragraphs that the quantity that makes a gas of noninteracting molecules nonideal is the configuration integral Z. This in turn depends primarily on the factor

$$e^{-U/kT} = 1 + \sum f_{ij} + \sum \sum f_{ij}f_{kl} + \sum \sum \sum f_{ij}f_{kl}f_{mn} + \cdots \tag{9-37}$$

Each term in each of the sums in Eq. (9-37) may be represented by a picture. The term f_{ij} involves an interaction between molecules i and j, but with no interactions involving the other molecules. Thus, a description of the interactions implied by this term is as shown in Fig. 9-9a. The term $f_{ij}f_{kl}$, where none of the coefficients are common, implies the interactions shown in Fig. 9-9b. The term $f_{ij}f_{ik}$ is as shown in Fig. 9-9c. It is more convenient to adopt a scheme whereby only the molecules listed in the subscripts of the f terms are included in the pictorial representations. When this is done, a systematic representation of all the possible terms involving three molecules is shown in Fig. 9-10. The term 1 is represented by a noninteracting system. Each conglomerate of interacting

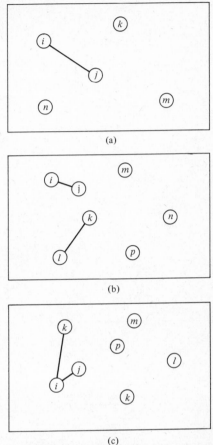

FIGURE 9-9 Diagrams illustraing terms in the sum in Eq. (9-37): (a) f_{ij}; (b) $f_{ij}f_{kl}$; (c) $f_{ij}f_{ik}$.

molecules is called a *cluster*. For example, in f_{31}, there is one cluster of 2 molecules and one cluster of one molecule. In $f_{21}f_{31}$, there is a single cluster of three molecules.

Now a quantity, $S_{ijk}\dots$, is defined, known as the *cluster sum*, as the sum of all possible terms in Eq. (9-37) which connect molecules $i, j, k, \dots$ in a cluster, with none of the remaining molecules connected in a cluster to $i, j, k, \dots$. For example, S_{21} is the sum of all clusters containing *only* molecules 1 and 2, and therefore,

$$S_{21} = f_{21}$$

Similarly,

$$S_{321} = f_{21}f_{32}f_{31} + f_{21}f_{32} + f_{31}f_{32} + f_{21}f_{31}$$

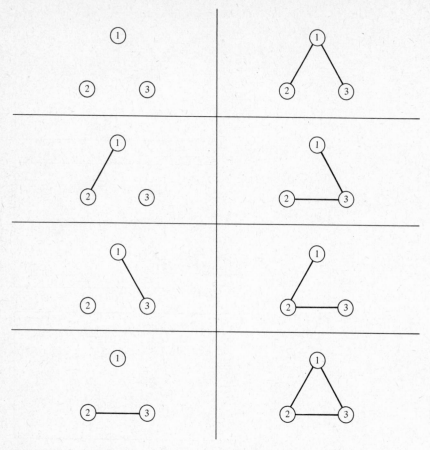

FIGURE 9-10   All the possible interaction terms for three interacting particles.

To complete the definition, we put $S_i = 1$ so that

$$S_1 = S_2 = S_3 = 1$$

It can be shown that $e^{-U/kT}$ is equal to the sum of all products of $S_{ijk} \ldots$ such that any particular index appears only once in each product. For a hypothetical gas of 3 molecules, this is illustrated below:

$$e^{-U/kT} = 1 + (f_{21} + f_{31} + f_{32}) + (f_{21}f_{31}f_{32} + f_{21}f_{31} + f_{32}f_{21} + f_{31}f_{32})$$

$$= S_1 S_2 S_3 + S_1 S_{32} + S_2 S_{31} + S_3 S_{21} + S_{321}$$

$$= 1 + S_{32} + S_{21} + S_{31} + S_{321}$$

For a gas of N molecules, $e^{-U/kT}$ will be a large sum over products of a large number of $S_{ijk} \ldots$ terms.

Primary interest is in the configurational integral

$$Z = \int \cdots \int_{3N} e^{-U/kT} \, dq_1 \ldots dq_{3N} \tag{9-38}$$

Therefore, the *cluster integral* is defined as

$$b_l = \frac{1}{l!V} \int \cdots \int_{3l} S_{123\ldots l} \, dq_1 \ldots dq_{3l} \tag{9-39}$$

where l is the number of molecules in the cluster, and the integrals in Eqs. (9-38) and (9-39) are taken over the entire volume of the container. For example,

$$b_1 = \frac{1}{V} \iiint (1) \, dx_1 \, dy_1 \, dz_1 = 1 \tag{9-40}$$

$$b_2 = \frac{1}{2V} \int \cdots \int_6 f_{21} \, dx_1 \, dy_1 \, dz_1 \, dx_2 \, dy_2 \, dz_2 \tag{9-41}$$

It is evident that b_2 is the same as b as previously defined in Sec. 9.2. By integrating over the coordinates of particle 1, and over the possible orientations of particle 2, one obtains[2]

$$b_2 = 2\pi \int_0^\infty f(r) r^2 \, dr \tag{9-42}$$

The factor $f(r)$ depends only on the potential between two particles and the temperature. It is independent of the volume of the container. This holds in general for all f_{ij}. The b_l do not depend on *which* molecules are chosen, only on how many. In performing the integral

$$b_l = \frac{1}{l!V} \int \cdots \int_{3l} S_{123\ldots l} \, dq_1 \ldots dq_{3l} \tag{9-42a}$$

there is freedom to choose the position of particle 1 anywhere in the container (provided surface and capillary effects can be neglected). Then $S_{123\ldots l}$ may be taken as independent of the coordinates of particle 1 since the coordinates of all the other particles 2, 3, . . . l may be taken relative to particle 1. Therefore,

$$b_l = \frac{1}{l!V} \iiint dx_1 \, dy_1 \, dz_1 \int \cdots \int_{3l-3} S_{123\ldots l} \, dq_4 \ldots dq_{3l} \tag{9-43}$$

$$b_l = \frac{1}{l!} \int \cdots \int_{3l-3} S_{12\ldots l} \, dq_4 \ldots dq_{3l} \tag{9-44}$$

---

[2]Ignoring wall effects.

The integral in Eq. (9-44) depends only on temperature and how the forces depend on relative distance between molecules, but does not depend on volume.   Thus, the b_l are independent of V.

The configurational integral can be expressed in terms of the b_l. Temporarily consider the simple case where there are only 3 particles in the ensemble.   Since

$$e^{-U/kT} = S_1 S_2 S_3 + S_1 S_{32} + S_2 S_{31} + S_3 S_{21} + S_{321}$$

the configurational integral is

$$Z = \int \cdots \int_9 e^{-U/kT} \, d\tau_1 \, d\tau_2 \, d\tau_3 = \int S_1 \, d\tau_1 \int S_2 \, d\tau_2 \int S_3 \, d\tau_3$$

$$+ \int S_1 \, d\tau_1 \int\int S_{32} \, d\tau_2 \, d\tau_3 + \int S_2 \, d\tau_2 \int\int S_{31} \, d\tau_1 \, d\tau_3$$

$$+ \int S_3 \, d\tau_3 \int\int S_{21} \, d\tau_1 \, d\tau_2 + \int\int\int S_{321} \, d\tau_1 \, d\tau_2 \, d\tau_3$$

According to the definition of b_l,

$$\int S_j \, d\tau_j = (1!Vb_1)$$

$$\int S_{ij} \, d\tau_i \, d\tau_j = (2!Vb_2)$$

$$\int S_{ijk} \, d\tau_i \, d\tau_j \, d\tau_k = (3!Vb_3)$$

Thus, for a system of 3 interacting particles

$$Z = (1!Vb_1)^3 + 3(1!Vb_1)(2!Vb_2) + (3!Vb_3) \qquad (9\text{-}44a)$$

In general, when calculating Z for N interacting particles, Z is a sum of terms, each of the following type (with each index appearing once and only once in the entire product):

$$\int \cdots \int_{3N} \underbrace{[S_i S_j \ldots S_q]}_{\substack{n_1 \text{ clusters} \\ \text{of one molecule}}} \underbrace{[S_{kl} S_{mn} \ldots S_{op}]}_{\substack{n_2 \text{ clusters} \\ \text{of two molecules}}} \ldots \underbrace{[S_{abc\ldots l} \ldots]}_{\substack{n_l \text{ clusters} \\ \text{of } l \text{ molecules}}} \ldots \, d\tau_1 \ldots d\tau_N$$

$$= (1!Vb_1)^{n_1}(2!Vb_2)^{n_2} \ldots (l!Vb_l)^{n_l} \ldots \qquad (9\text{-}45)$$

Since the total number of molecules in all the clusters is equal to N, when summing the n_l clusters of l molecules over all l, one finds

$$N = \sum_{l=1}^{N} l n_l \qquad (9\text{-}46)$$

Thus, Z will be given by a sum of terms of type in Eq. (9-45) such that Eq. (9-46) is satisfied for each term.   It remains only to count the number of times a term of the type: n_1 clusters of 1, n_2 clusters of 2, ..., n_l clusters of l molecules, ... occurs in the sum. Therefore, it is necessary to determine the number of different ways N particles can be arranged in clusters: n_1 clusters of 1, ..., n_l clusters of l, ..., subject to the restrictive condition Eq. (9-46). Since corrected Boltzmann statistics are being used with a factor of $(N!)^{-1}$ in the expression for Q, the molecules can be treated as if they were distinguishable. Now imagine that the N particles are labeled $a, b, c, ...$, and arrange them along a line. There are $N!$ different permutations possible. Suppose one of these arrangements is subdivided into clusters as

$$|\,abc\ldots\quad|\quad (g\,h)\,(i\,j)\,\ldots\quad|\quad (n\,o\,p)\,(q\,r\,s)\,\ldots\quad|\quad\ldots$$

$$
\begin{array}{ccc}
n_1 \text{ clusters} & n_2 \text{ clusters} & n_3 \text{ clusters} \\
\text{of 1 molecule} & \text{of 2 molecules} & \text{of 3 molecules}
\end{array}
$$

Vertical bars are inserted to denote the separation between groups of clusters of different size. There are still $N!$ possible permutations of the particles. However, there is no physical difference between the arrangements

$$|a\,b\ldots|\,(gh)\,(ij)\,\ldots|\ldots$$

and

$$|\,a\,b\ldots|\,(ij)\,(gh)\,\ldots|\ldots$$

In order to eliminate permutations corresponding merely to shifting clusters within the vertical bars, divide through by $\prod_l n_l!$. In addition, there is no physical difference between the arrangements

$$|\,a\,b\ldots|\,(gh)\,(ij)\,\ldots|\ldots$$

and

$$|\,a\,b\ldots|\,(hg)\,(ij)\,\ldots|\ldots$$

To eliminate permutations corresponding merely to shifting molecules within clusters, divide through by $\prod_l (l!)^{n_l}$. Therefore, there are a total of

$$W(n_1, n_2, \ldots) = \frac{N!}{\prod\limits_l n_l!(l!)^{n_l}} \tag{9-47}$$

different ways to divide N particles into n_1 clusters of 1 molecule, n_2 clusters of 2 molecules, and so forth. The contribution to Z made by any one of these arrangements is given in Eq. (9-45). Therefore, the total contribution to Z of all arrangements with n_1 clusters of 1, n_2 clusters of 2, ... is

$$W(n_1, n_2, \ldots) \cdot (1!Vb_1)^{n_1}(2!Vb_2)^{n_2}\ldots \tag{9-48}$$

Thus, after summing over all possibilities for sets of $n_1, n_2, \ldots,$

$$Z = \sum_{\substack{\text{all possible} \\ \text{sets of } n_l \\ \text{such that} \\ \Sigma\, l n_l = N}} \left(\frac{N!}{\prod\limits_{l=1}^{N} n_l!(l!)^{n_l}} \right) \left(\prod_{l=1}^{N} (l!Vb_l)^{n_l} \right) \qquad (9\text{-}49)$$

Therefore,

$$Z = N! \sum_{\substack{\text{all sets} \\ \text{of } n_l \text{ with} \\ \Sigma\, l n_l = N}} \left(\prod_{l=1}^{N} \frac{(Vb_l)^{n_l}}{n_l!} \right) \qquad (9\text{-}50)$$

As an example of the use of Eq. (9-50), consider again the case where $N = 3$ particles. There are three possible distributions:

Distribution	n_1	n_2	n_3	$\Sigma\, l n_l$	$W(n_1, n_2, n_3)$
1	3	0	0	3	1
2	1	1	0	3	3
3	0	0	1	3	1

Therefore,

$$Z = 3! \left[\frac{(Vb_1)^3}{3!} + \frac{(Vb_1)^1}{1!} \frac{(Vb_2)^1}{1!} + \frac{(Vb_3)^1}{1!} \right]$$

$$Z = 3! \left[\frac{(Vb_1)^3}{3!} + (Vb_1)(Vb_2) + (Vb_3) \right]$$

which is in agreement with the result obtained previously in Eq. (9-44a).

The partition function of a nonideal gas is obtained by combining Eqs. (9-3) and (9-50). The result is

$$Q = \left(\frac{2\pi mkT}{h^2} \right)^{3N/2} \sum_{\substack{\text{all sets of} \\ n_l \text{ such that} \\ \Sigma\, l n_l = N}} \left(\prod_{l=1}^{N} \frac{(Vb_l)^{n_l}}{n_l!} \right) \qquad (9\text{-}51)$$

The first factor may be brought under the summation sign, and N replaced by $\sum_l n_l l$ for each term representing a set of n_l. Thus,

$$Q = \sum_{\substack{\text{all sets of } n_l \\ \text{such that} \\ \Sigma\, n_l l = N}} \left\{ \left(\frac{2\pi mkT}{h^2} \right)^{(3/2)\sum_l n_l} \prod_{l=1}^{N} \frac{(Vb_l)^{n_l}}{n_l!} \right\} \qquad (9\text{-}52)$$

A factor $(\alpha)^{(3/2)\Sigma\, l n_l}$ can be expressed as

$$\alpha^{(3/2)\Sigma\, l n_l} = \alpha^{(3/2)(1 \cdot n_1)} \alpha^{(3/2)(2 \cdot n_2)} \ldots = \prod_{l=1}^{N} (\alpha^{3l/2})^{n_l}$$

Thus, Eq. (9-52) may be rewritten

$$Q = \sum_{\substack{\text{all } n_l \\ \text{such that} \\ \Sigma \, l n_l = N}} \prod_{l=1}^{N} \left\{ \frac{[(2\pi m k T / h^2)^{3l/2}(V b_l)]^{n_l}}{n_l!} \right\} \qquad (9\text{-}53)$$

Let $(2\pi m k T / h^2)^{3l/2} b_l$ be denoted as λ_l, which has the dimensions of a length. Then

$$Q = \sum_{\substack{\text{all sets of } n_l \\ \text{such that} \\ \Sigma \, l n_l = N}} \prod_{l=1}^{N} \left\{ \frac{(\lambda_l V)^{n_l}}{n_l!} \right\} \qquad (9\text{-}54)$$

Let $\mathscr{F}_{n_l}$ be defined by

$$\mathscr{F}_{n_l} = \frac{\displaystyle\prod_{l=1}^{N} (V\lambda_l)^{n_l}}{\displaystyle\prod_{l=1}^{N} n_l!} \qquad (9\text{-}55)$$

for a particular set of n_l. Then

$$Q = \sum_{\substack{\text{all sets of } n_l \\ \text{such that} \\ \Sigma \, l n_l = N}} \mathscr{F}_{n_l} \qquad (9\text{-}56)$$

For each possible set of $n_1, n_2, n_3, \ldots$, there is a particular value of $\mathscr{F}_{n_l}$. The sum of such terms over all sets of n_l is Q. When $N = 3$ particles were dealt with, there were three possible sets of the n_l, and the set which was most diverse in the n_l was most probable. When N is a very large number, the most probable set of n_l becomes overwhelmingly probable, and is to all intents and purposes the only set that occurs. This is illustrated schematically in Fig. 9-11. It is assumed that this particular

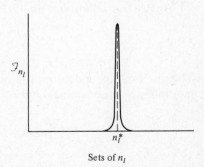

FIGURE 9-11 Schematic plot of $\mathscr{F}_{n_l}$ vs various sets of n_l, showing a sharp maximum at the set $n_1^*, n_2^*, \ldots$.

set of n_l, identified as $n_l{}^*$, leads to a $\mathscr{F}_{n_l}{}^*$ which is so large compared to all the other $\mathscr{F}_{n_l}$ that $\ln Q$ may be calculated as if Q were equal to $\mathscr{F}_{n_l}{}^*$. The argument here is the same as in Sec. 1.7, in which $\ln \Omega$ was replaced by $\ln W^*$. Since $\mathscr{F}_{n_l}$ is an extremely large number, $\ln \Sigma \, \mathscr{F}_{n_l} \cong \ln \mathscr{F}_{n_l}^*$. To find the set of n_l which maximizes $\mathscr{F}_{n_l}$, treat the n_l as if they were continuous variables and set

$$d \ln \mathscr{F}_{n_l} = 0 \qquad (9\text{-}57)$$

subject to the restrictive condition

$$\sum_l l n_l = N \qquad (9\text{-}58)$$

From Eq. (9-55),

$$\ln \mathscr{F}_{n_l} = \sum_{l=1}^{N} \underbrace{\{ n_l \ln \lambda_l + n_l \ln V - \ln n_l! \}}_{n_l \ln (\lambda_l V)} \qquad (9\text{-}59)$$

It is assumed that $n_l \gg 1$, so Stirling's approximation can be used for $\ln n_l!$. Then

$$d \ln \mathscr{F}_{n_l} = 0 = \sum_{l=1}^{N} \left\{ \ln (\lambda_l V) \, dn_l - \ln n_l{}^* \, dn_l \right.$$

$$\left. - n_l{}^* \frac{dn_l}{n_l{}^*} + dn_l \right\} \qquad (9\text{-}60)$$

and hence

$$\sum_{l=1}^{N} \ln (\lambda_l V / n_l{}^*) \, dn_l = 0$$

The restrictive condition, Eq. (9-58), can be written

$$\sum_{l=1}^{N} l \, dn_l = 0 \qquad (9\text{-}61)$$

The method of undetermined multipliers is applied and Eq. (9-61) is multiplied by a constant α, and added to Eq. (9-60). The result is

$$\sum_{l=1}^{N} \{ \ln (\lambda_l V / n_l{}^*) + \alpha l \} \, dn_l = 0 \qquad (9\text{-}62)$$

Since the dn_l in Eq. (9-62) may be treated as independent, the term in braces may be set equal to zero. Thus,

$$n_l{}^* = \lambda_l V \, e^{\alpha l} = \lambda_l V a^l \qquad (9\text{-}63)$$

where $a = e^\alpha$ will turn out later to be the absolute activity.   Thus,

$$\ln \mathscr{F}_{n_l*} = \sum_{l=1}^{N} \{(\lambda_l V a^l) \ln (\lambda_l V) - (n_l^* \ln n_l^* - n_l^*)\}$$

$$\ln \mathscr{F}_{n_l*} = \sum_{l=1}^{N} \{-n_l^* l \ln a + n_l^*\}$$

$$\downarrow$$
(pull out of summation sign)

$$\ln \mathscr{F}_{n_l*} = -N \ln a + \sum_{l=1}^{N} n_l^* \tag{9-64}$$

According to the basic assumption, $\ln Q$ may be calculated by approximating the sum in Eq. (9-56) by $\mathscr{F}_{n_l*}$.   Therefore,

$$\ln Q \cong \ln \mathscr{F}_{n_l*} = -N \ln a + \sum_{l=1}^{N} n_l^* \tag{9-65}$$

According to Eq. (9-26),

$$p = kT (\partial \ln Q / \partial V)_{T,N} \tag{9-66}$$

When Eq. (9-65) is used for $\ln Q$, the result is

$$p = -\frac{NkT}{a} \left(\frac{\partial a}{\partial V}\right)_T + kT \sum_l \left(\frac{\partial n_l^*}{\partial V}\right)_T \tag{9-67}$$

Note that λ_l is independent of V and depends only on T.   For, by definition, $\lambda_l = (2\pi m k T/h^2)^{3l/2} b_l$, and b_l is independent of volume.   Since $n_l^* = \lambda_l V a^l$, and λ_l is independent of V,

$$\left(\frac{\partial n_l^*}{\partial V}\right)_T = \lambda_l a_l + \lambda_l V l a^{l-1} \left(\frac{\partial a}{\partial V}\right)_{T,N} \tag{9-68}$$

Thus,

$$p = -\frac{NkT}{a} \left(\frac{\partial a}{\partial V}\right)_{T,N} + kT \sum_{l=1}^{N} (n_l^*/Va^l)a^l + kT \sum_{l=1}^{N} \lambda_l V l a^{l-1} \left(\frac{\partial a}{\partial V}\right)_{T,N} \tag{9-69}$$

The third term in Eq. (9-69) may be written

$$kT \sum_{l=1}^{N} \frac{(\lambda_l a^l)}{a} lV \left(\frac{\partial a}{\partial V}\right)_{T,N} = \frac{kT}{a} \left(\frac{\partial a}{\partial V}\right)_{T,N} \underbrace{\sum l n_l^*}_{\text{equals } N} \tag{9-70}$$

Therefore, the first and third terms in Eq. (9-69) cancel, and

$$pV = kT \sum_{l=1}^{N} n_l^* \tag{9-71}$$

Thus, a gas of interacting molecules acts thermodynamically like a perfect gas of noninteracting particles where each cluster in the most probable distribution of clusters is to be treated as a particle.

To show that a is the absolute activity, calculate the Gibbs free energy:

$$G = H - TS = E + PV - TS = A + PV$$

$$G = \left\{ NkT \ln a - kT \sum n_l^* \right\} + kT \sum n_l^* \tag{9-72}$$

$$G = NkT \ln a$$

Equation (9-72) is the definition of activity; a is the quantity necessary to make the right side equal to G. The equation of state of a nonideal gas is

$$p = kT \sum_{l=1}^{N} \lambda_l a^l \tag{9-73}$$

Although this gives p/kT in terms of a, this is not very useful because a cannot be measured directly. It would be much more valuable to obtain p/kT as a function of gas density ρ.

Since $\sum ln_l^* = N$, it follows that

$$V \sum_l l\lambda_l a^l = N$$

or

$$\rho = \frac{N}{V} = \sum_l l\lambda_l a^l \tag{9-74}$$

This may be expanded for clarity:

$$\rho = \lambda_1 a + 2\lambda_2 a^2 + 3\lambda_3 a^3 + \cdots \tag{9-75}$$

Each coefficient λ_l is independent of V, and at constant N, λ_l depends only on T, and is independent of ρ. Now take the inverse of Eq. (9-75), and obtain a as a function of ρ. Before doing this, consider the limiting case when $\rho \to 0$. In this case, $a \ll 1$, and therefore all higher terms on the right side of Eq. (9-75), may be neglected. Thus, as $\rho \to 0$,

$$\rho \cong \lambda_1 a$$

$$a \cong \frac{\rho}{\lambda_1} = \frac{\rho}{(2\pi mkT/h^2)^{3/2} b_1} = \frac{N}{Q_0} \tag{9-76}$$

where Q_0 is the partition function for an ideal gas. For ordinary gases, at, say, one atmosphere and 300°K, $N \cong 10^{19}$ particles/cm³ and $Q_0 \cong 10^{25}$ states/cm³, so $a \cong 10^{-6}$. If $2\lambda_2 a^2 \ll \lambda_1 a$, the gas is nearly ideal. λ_2 will be calculated later.

Assume that a can be expanded as a power series in the density

$$a = c_1\rho + c_2\rho^2 + \cdots$$

Then Eq. (9-75) can be written

$$\rho = \lambda_1(c_1\rho + c_2\rho^2 + c_3\rho^3 + \cdots) + 2\lambda_2(c_1\rho + c_2\rho^2 + c_3\rho^3 + \cdots)^2$$
$$+ 3\lambda_3(c_1\rho + c_2\rho^2 + c_3\rho^3 + \cdots)^3 + \cdots$$

The leading terms are

$$\rho = \lambda_1(c_1\rho + c_2\rho^2 + c_3\rho^3 + \cdots) + 2\lambda_2(c_1^2\rho^2 + 2c_1c_2\rho^3 + \cdots)$$
$$+ 3\lambda_3(c_1^3\rho^3 + \cdots)$$

If the leading powers of ρ are grouped,

$$\rho = (\lambda_1 c_1)\rho + (\lambda_1 c_2 + 2\lambda_2 c_1^2)\rho^2 + (\lambda_1 c_3 + 4c_1c_2\lambda_2 + 3\lambda_3 c_1^3)\rho^3 + \cdots \quad (9\text{-}77)$$

In order for the left and right sides of Eq. (9-77) to be equal, the coefficient of ρ must be unity, and the coefficients of higher powers of ρ are 0.   Thus,

$$\lambda_1 c_1 = 1$$
$$(\lambda_1 c_2 + 2\lambda_2 c_1^2) = 0 \qquad\qquad (9\text{-}78)$$
$$(\lambda_1 c_3 + 4c_1c_2\lambda_2 + 3\lambda_3 c_1^3) = 0$$

$$\cdots$$

Hence,

$$c_1 = \frac{1}{\lambda_1} = (2\pi mkT/h^2)^{-3/2} \qquad\qquad (9\text{-}79)$$

$$c_2 = -\frac{2\lambda_2}{\lambda_1}c_1^2 = -2\lambda_2/\lambda_1^3 \qquad\qquad (9\text{-}80)$$

$$c_3 = -4c_1c_2\frac{\lambda_2}{\lambda_1} - 3\frac{\lambda_3}{\lambda_1}c_1^3 = \frac{8\lambda_2^2}{\lambda_1^5} - \frac{3\lambda_3}{\lambda_1^4} \qquad\qquad (9\text{-}81)$$

Therefore, Eq. (9-73) may be expanded as

$$p = kT\left\{\lambda_1[\lambda_1^{-1}\rho - \frac{2\lambda_2}{\lambda_1^3}\rho^2 + \left(\frac{8\lambda_2^2}{\lambda_1^5} - \frac{3\lambda_3}{\lambda_1^4}\right)\rho^3 + \cdots\right]$$
$$+ \lambda_2\left[\lambda_1^{-2}\rho^2 - 4\frac{\lambda_2}{\lambda_1^4}\rho^3 + \cdots\right] \qquad\qquad (9\text{-}82)$$
$$+ \lambda_3[\lambda_1^{-3}\rho^3 + \cdots + \cdots]\right\}$$

or,

$$p = kT\left\{\rho - \frac{\lambda_2}{\lambda_1^2}\rho^2 + \left(\frac{4\lambda_2^2}{\lambda_1^4} - \frac{2\lambda_3}{\lambda_1^3}\right)\rho^3 + \cdots\right\} \qquad\qquad (9\text{-}83)$$

This is usually written as a *virial expansion*

$$p = kT \{B_1\rho + B_2\rho^2 + B_3\rho^3 + \cdots\}$$

where the virial coefficients are

$$B_1 = 1$$
$$B_2 = -\lambda_2/\lambda_1^2 = -b_2 \qquad\qquad (9\text{-}84)$$
$$B_3 = \frac{4\lambda_2^2}{\lambda_1^4} - \frac{2\lambda_3}{\lambda_1^3} = 4b_2 - 2b_3$$

Before proceeding on to some examples, let us review this chapter briefly to put the material in perspective. The potential energy of a gas of interacting molecules is expanded as

$$e^{-U/kT} = 1 + \sum f_{ij} + \sum \sum f_{ij}f_{kl} + \cdots$$

Each $f_{ij} = e^{-u(r_{ij})/kT} - 1$, where $u(r_{ij})$ is the interaction potential between any two molecules. To determine the configurational integral, $e^{-U/kT}$ must be integrated over all space. Therefore, the (volume-independent) cluster integrals are defined as

$$b_l = \frac{1}{l!V} \int \cdots \int_{3l} S_{123\ldots l}\, d\tau_1 \ldots d\tau_l$$

where $S_{123\ldots l}$ is the sum of all terms consisting of products of f_{ij} which contain *all* the indices $1, 2, 3, \ldots l$, but no others. Since $e^{-U/kT}$ is equal to the sum of all products of the S's such that each index appears once and only once in any particular product, it follows that

$$Z = \int e^{-U/kT}\, d\tau_1 \ldots d\tau_N = \sum_{\substack{\text{all distinguishable} \\ \text{sets of } n_l \text{ such} \\ \text{that } \sum_l ln_l = N}} \left\{ \prod_{l=1}^{N} (l!Vb_l)^{n_l} \right\}$$

By counting up the number of ways any particular set of n_l can be formed, $W(n_1, n_2, \ldots)$, this becomes

$$Z = \sum_{\substack{\text{all sets of} \\ n_l \text{ such that} \\ \sum_l n_l l = N}} W(n_1, n_2, \ldots) \prod_{l=1}^{N} (l!Vb_l)^{n_l}$$

After evaluating $W(n_1, n_2, \ldots)$, it can be shown that the partition function is

$$Q = \frac{1}{N!} \left(\frac{2\pi mkT}{h^2} \right)^{3N/2} Z = \sum_{\substack{\text{all sets of} \\ n_l \text{ such that} \\ \sum ln_l = N}} \mathscr{F}_{n_l}$$

where

$$\mathscr{F}_{n_l} = \prod_{l=1}^{N} \frac{(\lambda_l V)^{n_l}}{n_l!}$$

for any set of n_l, and $\lambda_l = (2\pi mkT/h^2)^{3l/2} b_l$. The equation of state of a gas depends on $(\partial \ln Q/\partial V)_{T,N}$. In taking the log of a sum of terms $(\mathscr{F}_{n_l})$ which have a very sharp maximum at $\mathscr{F}_{n_l^*}$, a good approximation is obtained by replacing the sum by the single term $\mathscr{F}_{n_l^*}$. Thus we seek the maximum value $\mathscr{F}_{n_l^*}$ by taking $d \ln \mathscr{F}_{n_l} = 0$ with respect to variations in the n_l. This procedure is entirely analogous to the procedure for evaluating the properties of a noninteracting gas, the difference being that n_l refers to the number of clusters of l molecules in state l. The most probable set of n_l is $n_l^* = \lambda_l V a^l$, where a is the absolute activity. Therefore, when $\ln Q$ is approximated as $\ln \mathscr{F}_{n_l^*}$, the equation of state becomes

$$pV = kT \left(\frac{\partial \ln Q}{\partial V}\right)_{T,N} = kT \sum_{l=1}^{N} n_l^*$$

This shows that each cluster in the most probable set of cluster arrangements acts like a single perfect gas molecule. Since n_l^* is known in terms of absolute activity, it is not in very useful form. It is more convenient to express a as a power series in the density, ρ, thus obtaining the virial expansion

$$p = kT [B_1\rho + B_2\rho^2 + B_3\rho^3 + \cdots]$$

where $B_1 = 1, B_2 = -b_2, B_3 = 4b_2^2 - 2b_3, \ldots$. Thus,

$$p = \frac{NkT}{V}\{1 - b_2\rho + (4b_2^2 - 2b_3)\rho^2 + \cdots\} \qquad (9\text{-}85)$$

Evidently, at low gas densities, the equation of state of an imperfect gas approaches the limiting case of a perfect gas. At higher densities, the higher powers of ρ become increasingly important in Eq. (9-85). The leading first order correction term due to molecular interactions is $-b_2\rho$.

9.4 THE RELATIONSHIP BETWEEN INTERMOLECULAR POTENTIALS AND EQUATIONS OF STATE

This section deals with calculation of the equation of state of an imperfect gas based on various models for the intermolecular potential between two molecules. At moderate gas densities, it can be shown that only the leading nonideal term $-b_2$ in Eq. (9-85) is of significance. Therefore, we need to evaluate b_2.

The simplest case to consider is when the molecules can be approximated as rigid spheres. The $u(r)$ and $f(r)$ are as shown in Figs. 9-6 and 9-7. The calculation of b_2 is very simple. Since f_{ij} depends only on the *relative* positions of particles i and j, we can integrate over $d\tau_i$ to obtain a factor of V.

$$b_2 = \frac{1}{2!V} \iint f_{ij}\, d\tau_i\, d\tau_j = \frac{1}{2!} \int f_{ij}\, d\tau_j$$

Since $d\tau = 4\pi r^2\, dr$, this becomes

$$b_2 = 2\pi \int_0^\infty f(r) r^2\, dr \tag{9-86}$$

$$b_2 = 2\pi \left\{ \int_0^{r_0} (-1) r^2\, dr + \int_{r_0}^\infty (0) r^2\, dr \right\} = -2\pi r_0^3/3$$

Because r_0 is the *diameter* of a molecule, the volume of the molecule is

$$\gamma = \frac{4}{3}\pi \left(\frac{r_0}{2}\right)^3 = \frac{1}{6}\pi r_0^3$$

and

$$b_2 = -4\gamma$$

$$B_2 = 4\gamma$$

Thus, the equation of state for a gas of rigid spheres is

$$p = \frac{NkT}{V}\left[1 + 4\gamma\rho + \cdots\right] \tag{9-87}$$

Next, estimate the magnitude of $4\gamma\rho$. The volume of a molecule is $\sim 10^{-23}\,cm^3$, and at standard temperature and pressure, $\rho \cong 3 \times 10^{19}$ molecules/cm³. Thus, $4\gamma\rho \cong 10^{-3}$. Higher terms in $\rho^2, \rho^3, \ldots$ can be shown to be smaller. Thus,

$$p = \frac{NkT}{V}\left[1 - b_2\frac{N}{V} + \cdots\right] \cong \frac{NkT}{V}\left\{\frac{1}{1 + b_2 N/V - \cdots}\right\} \tag{9-88}$$

since $b_2 N/V \ll 1$. This may be rewritten as

$$p[V + b_2 N] \cong NkT$$

or

$$p[V - 4N\gamma] \cong NkT \tag{9-89}$$

Evidently, the equation of state is the same as for an ideal gas, if the container volume is replaced by a reduced volume obtained by subtracting four times the volume of all the molecules.

As a second example, consider a gas where the molecules interact with a long-range attractive potential and there is a hard-sphere inner core repulsion as shown in Fig. 9-12. The functions $e^{-u/kT}$ and $f(r)$ are

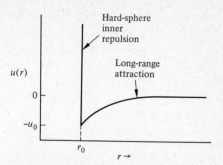

FIGURE 9-12 Potential function consisting of a long range attraction plus a hard-sphere inner-core repulsion.

illustrated in Figs. 9-13 and 9-14. To calculate b_2, the following is used:

$$b_2 = 2\pi \left\{ \int_0^{r_0} (-1)r^2 \, dr + \int_{r_0}^{\infty} f(r)r^2 \, dr \right\}$$

$$b_2 = -2\pi r_0^3/3 + 2\pi I$$

where $I = \int_{r_0}^{\infty} f(r)r^2 \, dr$. Consider the special case where

$$u(r) = -u_0(r_0/r)^m \qquad \text{for } r > r_0$$

with $m > 3$. Then

$$I = \int_{r_0}^{\infty} \{\exp(u_0 r_0^m r^{-m}/kT) - 1\}r^2 \, dr$$

For moderately high T, when the well-depth u_0 is not too large, it may be assumed that[3]

$$u_0/kT \ll 1$$

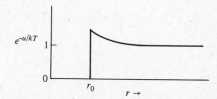

FIGURE 9-13 The function $e^{-u/kT}$, where u is given in Fig. 9-12.

---

[3]For actual gases, u_0/k varies from about 100°K to ~ 500°K. Thus, the use of this approximation implies $T > 1000°K$ if $u_0 \cong 300°K$.

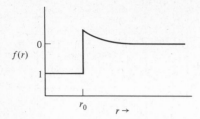

FIGURE 9-14 The function $f(r) = e^{-u(r)/kT} - 1$, where $u(r)$ is given in Fig. 9-12.

Then $e^x \cong 1 + x + \cdots$ may be expanded and the higher order terms may be neglected. Hence,

$$I \cong \int_{r_0}^{\infty} \left(\frac{u_0 r_0^m}{kT}\right) r^{2-m}\, dr = \left(\frac{u_0 r_0^m}{kT}\right) \frac{r_0^{3-m}}{m-3} = \frac{u_0 r_0^3}{kT(m-3)}$$

and therefore,

$$b_2 = -4\gamma + \frac{12\gamma u_0}{(m-3)kT} \tag{9-90}$$

where $\gamma = \frac{1}{6}\pi r_0^3$ is, as before, the volume of one molecule. It is clear from Eq. (9-90) that the inner repulsive forces (due to the nonzero volume of the molecules) tend to produce a negative b_2, whereas the attractive forces contribute a positive quantity to b_2. Since the repulsive forces have been assumed to be given by a simple hard sphere model, they lead to a contribution to b_2 which is temperature-independent. The effect of the attractive forces is diminished, however, as T is increased, because the average kinetic energy of the molecules is greater and they can overcome the attractive forces more easily. A plot of b_2 vs T is shown in Fig. 9-15.

A gas of molecules with a hard-sphere inner core repulsion and an inverse power attraction is called a Van der Waals gas. The equation of

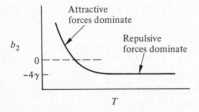

FIGURE 9-15 The second virial coefficient vs T for molecules interacting as shown in Fig. 9-12.

state for such a substance is

$$p = \frac{NkT}{V}\left\{1 + \left[4\gamma - \frac{12\gamma u_0}{(3-m)kT}\right]\frac{N}{V} + \cdots\right\}$$

If we denote $4\gamma N/V = \delta_1$ and $(12\gamma u_0 N)/((3-m)kTV) = \delta_2$, and if $\delta_1 \ll 1$, and $\delta_2 \ll 1$, then

$$1 + \delta_1 - \delta_2 \cong \frac{1}{(1-\delta_1)}\cdot\frac{1}{(1+\delta_2)}$$

Therefore, in this case (corresponding to moderately high T and moderately low u_0),

$$p\left[1 + \frac{12\gamma u_0 N}{(3-m)VkT}\right]V\left[1 - \frac{4\gamma N}{V}\right] \cong NkT$$

This may be expanded to

$$\left[p + \frac{12\gamma u_0}{(3-m)}\left(\frac{p}{kT}\right)\frac{N}{V}\right][V - 4N\gamma] \cong NkT$$

Since the entire term $12\gamma u_0/(3-m)(p/kT)$ is small compared to p, an approximate expression $(p/kT) \cong N/V$ can be used in this term without serious error. Thus, we obtain

$$(p + \alpha N^2/V^2)(V - \beta) = NkT$$

the Van der Waals equation of state, with

$$\alpha = \frac{12u_0\gamma}{(3-m)} \text{ and } \beta = 4N\gamma$$

From an experimental determination of α and β, values for u_0 and r_0 may be inferred. There are theoretical reasons for believing that for many nonpolar molecules the best value for m is 6. The degree of accuracy of this model can be ascertained for any substance by measuring α and β vs T, and determining how independent of T the experimental values are.

For more accurate calculations, the more realistic Lennard-Jones potential,

$$u(r) = 4\epsilon\left\{\left(\frac{\sigma}{r}\right)^{12} - \left(\frac{\sigma}{r}\right)^6\right\} = \epsilon\left\{\left(\frac{r_0}{r}\right)^{12} - 2\left(\frac{r_0}{r}\right)^6\right\} \tag{9-91}$$

is often employed.[4] This potential is plotted in Fig. 9-16. With this potential, it can be shown that $b_2(T)$ has the general form shown in Fig. 9-17. At high T, the repulsive forces dominate in determining the

---

[4]The relation between σ and r_0 is $r_0 = 2^{1/6}\sigma$. At $r = r_0$, $u(r)$ is a minimum, and at $r = \sigma, u(r) = 0$.

magnitude of b_2; whereas at low T, b_2 becomes positive due to the attractive forces. The second virial coefficient B_2 vs T is shown in Fig. 9-18 for fixed ϵ and various r_0. In Fig. 9-19, B_2 is shown for fixed r_0 and various ϵ. By fitting the observed $B_2(T)$ to these curves, best values of ϵ and r_0 can be found for any gas. Some results are shown in Table 9-1.

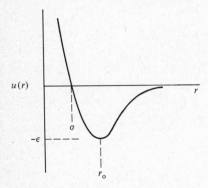

FIGURE 9-16 The Lennard-Jones potential.

In practice, a different procedure is used. The second virial coefficient is written

$$B_2 = -2\pi \int_0^\infty \left[\exp\left(-\frac{u}{kT}\right) - 1 \right] r^2 \, dr \tag{9-92}$$

and this is integrated by parts using

$$\int_a^b U \, dV = UV \Big]_a^b - \int_a^b V \, dU \tag{9-93}$$

with $U = \exp\left(-u/kT\right) - 1$ and $dV = r^2 \, dr$. Thus, $dU = -(kT)^{-1}$ $(du/dr) \exp\left(-u/kT\right) dr$, and $V = r^3/3$. If $u(r) \to 0$ faster than r^{-3} as $r \to \infty$, $UV \to 0$ as $r \to \infty$. Furthermore, UV is clearly zero at $r = 0$.

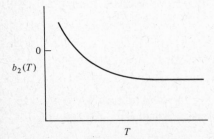

FIGURE 9-17 Second virial coefficient for a Lennard-Jones potential.

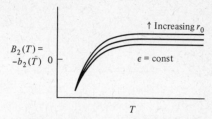

FIGURE 9-18 Second virial coefficient vs T for various r_0 at fixed ϵ, for Lennard-Jones potential.

Thus $UV]_0^\infty = 0$ in this case, and

$$B_2 = \frac{-2\pi}{3kT} \int_0^\infty \left(\frac{du}{dr}\right) e^{-u/kT} r^3 \, dr \qquad (9\text{-}94)$$

For a generalized Lennard-Jones potential

$$u(r) = \frac{\alpha}{r^a} - \frac{\beta}{r^b} \qquad (9\text{-}95)$$

with $a > 3, b > 3, B_2$ takes the form

$$B_2 = \frac{-2\pi}{3kT} \int_0^\infty \left[-\frac{a\alpha}{r^{a+1}} + \frac{b\beta}{r^{b+1}} \right] \exp\left(\frac{-\alpha}{r^a kT}\right) \exp\left(\frac{+\beta}{r^b kT}\right) r^3 \, dr \qquad (9\text{-}96)$$

This integral can be evaluated[5] as follows. The second exponential is

TABLE 9-1[a] Constants for the Lennard-Jones (6-12) Potential Determined from Second Virial Coefficients

Gas	$\epsilon/k(°K)$	$\sigma(\text{Å})$
Ne	34.9	2.78
A	119.8	3.405
Kr	171	3.60
Xe	221	4.10
N_2	95.05	3.698
O_2	118	3.46
CH_4	148.2	3.817
CO_2	189	4.486

[a]J. O. Hirschfelder, C. F. Curtiss, and R. L. Bird, *Molecular Theory of Gases and Liquids*, copyright © 1954, by permission of John Wiley & Sons, Inc.

[5]J. E. Jones, *Proc. Roy. Soc.* **A106**, 463 (1924). The relation between notations is: Jones (Rapp), $n-1$ (a), $m-1$ (b), $\lambda_n(\alpha a)$, $\lambda_m(\beta b)$, $2j(1/kT)$, $\lambda_n/[n-1](\alpha)$, $\lambda_m/[m-1](\beta)$.

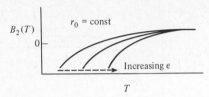

FIGURE 9-19 Second virial coefficient vs T for various ϵ at fixed r_0, for Lennard-Jones potential.

expanded in a power series

$$\exp\left(\frac{\beta}{r^b kT}\right) = \sum_{j=0}^{\infty} \left(\frac{\beta}{kT}\right)^j r^{-jb} \tag{9-97}$$

When this is inserted into Eq. (9-96), the result has the form:

$$B_2 = \frac{-2\pi}{3kT}\left\{ \sum_{j=0}^{\infty} \frac{(-a\alpha)}{j!}\left(\frac{\beta}{kT}\right)^j \int_0^{\infty} r^{-(jb+a-2)} \exp\left[-r^{-a}\left(\frac{\alpha}{kT}\right)\right] dr \right.$$

$$\left. + \sum_{j=0}^{\infty} \frac{(b\beta)}{j!}\left(\frac{\beta}{kT}\right)^j \int_0^{\infty} r^{-(jb+b-2)} \exp\left[-r^{-a}\left(\frac{\alpha}{kT}\right)\right] dr \right\} \tag{9-98}$$

where the integrals can be expressed in terms of gamma functions from the relation

$$q^{-1} c^{(1-t)/q}\Gamma\left(\frac{t-1}{q}\right) = \int_0^{\infty} r^{-t} \exp\left(-cr^{-q}\right) dr \tag{9-99}$$

Therefore,

$$B_2 = \frac{-2\pi}{3kT}\left\{ \sum_{j=0}^{\infty} \frac{(-a\alpha)}{j!}\left(\frac{\beta}{kT}\right)^j a^{-1}\left(\frac{\alpha}{kT}\right)^{(1-jb-a+2)/a} \Gamma\left(\frac{jb+a-3}{a}\right) \right.$$

$$\left. + \sum_{j=0}^{\infty} \frac{(b\beta)}{j!}\left(\frac{\beta}{kT}\right)^j a^{-1}\left(\frac{\alpha}{kT}\right)^{(1-jb-a+2)/a} \Gamma\left(\frac{jb+b-3}{a}\right) \right\} \tag{9-100}$$

By combining terms, this can be re-expressed as

$$B_2 = \frac{2\pi}{3}\left(\frac{\alpha}{kT}\right)^{3/a} \sum_{j=0}^{\infty} \frac{1}{j!} y^j \left\{ \Gamma\left(\frac{jb+a-3}{a}\right) - \frac{yb}{a}\Gamma\left(\frac{jb+b-3}{a}\right) \right\} \tag{9-101}$$

where y is the group of terms

$$y = \left(\frac{\beta}{kT}\right)\left(\frac{\alpha}{kT}\right)^{-b/a} \tag{9-102}$$

By taking out the term in y^0 from the first sum, letting $j = l$ for the rest of the terms in this sum, and letting $j + 1 = l$ for the second sum, Eq. (9-101) can be put in the form

$$B_2 = \frac{2\pi}{3}\left(\frac{\alpha}{kT}\right)^{3/a}\left\{\Gamma\left(\frac{a-3}{a}\right) - \sum_{l=1}^{\infty}\left[\frac{b}{a}\frac{1}{(l-1)!}\Gamma\left(\frac{b(l-1)+b-3}{a}\right)\right.\right.$$
$$\left.\left. -\frac{1}{l!}\Gamma\left(\frac{lb+a-3}{a}\right)\right]y^l\right\} \qquad (9\text{-}103)$$

Since $\Gamma(1 + z) = z\Gamma(z)$, it follows that

$$\Gamma\left(\frac{bl-3+a}{a}\right) = \Gamma\left(1 + \frac{bl-3}{a}\right) = \frac{bl-3}{a}\Gamma\left(\frac{bl-3}{a}\right) \qquad (9\text{-}104)$$

$$B_2 = \frac{2\pi}{3}\left(\frac{\alpha}{kT}\right)^{3/a}\left\{\Gamma\left(\frac{a-3}{a}\right) - \sum_{l=1}^{\infty}\frac{3}{al!}\Gamma\left(\frac{bl-3}{a}\right)y^l\right\} \qquad (9\text{-}105)$$

For the special case of the Lennard-Jones 6–12 potential given in Eq. (9-91), $\alpha = 4\epsilon\sigma^{12}$, $\beta = 4\epsilon\sigma^6$, $a = 12$, $b = 6$, and $y = 2(kT/\epsilon)^{-1/2}$. It is convenient to define reduced temperature and virial coefficients:

$$T^* = kT/\epsilon$$

$$B_2^* = \frac{B_2}{\frac{2}{3}\pi\sigma^3} \qquad (9\text{-}106)$$

Thus, for the 6–12 potential, Eq. (9-105) becomes

$$B_2^* = \sqrt{2}(T^*)^{-1/4}\Gamma(3/4) - \sum_{l=1}^{\infty}\frac{2^{l+1/2}}{4l!}\Gamma\left(\frac{2l-1}{4}\right)(T^*)^{-(2l+1)/4} \qquad (9\text{-}107)$$

Furthermore, since the first term on the right side of Eq. (9-107) fits the definition of the general term under the summation for $l = 0$ because $\Gamma(3/4) = -(1/4)\Gamma(-1/4)$, this equation can be written

$$B_2^* = \sum_{l=0}^{\infty} -\left\{\frac{2^{l+1/2}}{4l!}\Gamma\left(\frac{2l-1}{4}\right)\right\}T^{*-(2l+1)/4} \qquad (9\text{-}108)$$

The factors in braces in Eq. (9-108) are tabulated by Hirschfelder, Curtiss, and Bird[6]. It is clear from Eq. (9-108), that all substances which have intermolecular potentials which can be approximated by Lennard-Jones 6–12 potentials, have the same value of B_2^* at the same T^*.[7] If B_2 is measured at two temperatures, the ratio $B_2(T_1)/B_2(T_2)$ is equal to $B_2^*(T_1)/B_2^*(T_2)$ because the factors $\frac{2}{3}\pi\sigma^3$ cancel. Thus, by trial and error, the

---

[6]J. O. Hirschfelder, C. F. Curtiss and R. B. Bird. *Molecular Theory of Gases and Liquids* (New York: J. Wiley & Sons, 1954).

[7]$B_2^*(T^*)$ is tabulated in Table I-B of Hirschfelder, Curtiss, and Bird, p. 114.

value of ϵ can be found which is necessary to make $B_2^*(T_1)/B_2^*(T_2)$ equal to the experimental ratio $B_2(T_1)/B_2(T_2)$. In practice, ϵ is varied, and a best fit is sought to the shape of the B_2^* vs T^* curve shown in Fig. 9-20. The value of σ is obtained from the ratio

$$\frac{B_2}{B_2^*} = \frac{2}{3}\pi\sigma^3 = 1.2615\sigma^3 \tag{9-109}$$

at any temperature. In Xe, the experimental values of B_2 vs T are given in Table 9-2.

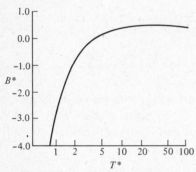

FIGURE 9-20 Reduced virial coefficient vs reduced temperature for a Lennard-Jones interaction potential.

TABLE 9-2[a]

$T\,(^\circ K)$	B_2 (cm^3/molecule) (Experimental)	B_2 (Calculated)
298.2	21.6×10^{-23}	21.3×10^{-23}
348.2	15.7×10^{-23}	15.7×10^{-23}
373.2	13.5×10^{-23}	13.5×10^{-23}
498.2	6.49×10^{-23}	6.47×10^{-23}
548.3	4.65×10^{-23}	4.65×10^{-23}
573.3	3.90×10^{-23}	3.88×10^{-23}

[a]J. O. Hirschfelder, C. F. Curtiss, and R. L. Bird, *Molecular Theory of Gases and Liquids*, copyright© 1954, by permission of John Wiley & Sons, Inc.

By taking various pairs of temperatures, a trial-and-error best fit to ϵ can be made.[8] For example, if $T_1 = 298.2$ and $T_2 = 498.2$ are chosen, then the experimental value of $B_2(T_2)/B_2(T_1)$ is 0.3003. For various trial values

---

[8]This example is taken directly from p. 167 of Hirschfelder, Curtiss, and Bird.

of $\epsilon/k°K$, the calculated ratios $B_2{}^*(T_2{}^*)/B_2{}^*(T_1{}^*)$ (from tables in Hirschfelder, Curtiss, and Bird) are as shown in Table 9-3.

TABLE 9-3[a]

Trial ϵ/k	Calculated $B^*_2(T^*_2)/B^*_2(T^*_1)$
223	0.3060
220	0.3010
219	0.2993

[a]J. O. Hirschfelder, C. F. Curtiss, and R. L. Bird, *Molecular Theory of Gases and Liquids*, copyright © 1954, John Wiley & Sons, Inc. By permission of John Wiley & Sons, Inc.; *J. Chem. Phys.* and J. A. Beattie, R. J. Barrieault, and J. S. Brierly.

It is clear that the best value of ϵ/k is just slightly under 220°K. For other pairs of temperatures, the best fits to ϵ/k vary by only 1 or 2°. This indicates that the Lennard-Jones 6–12 potential is an adequate description of the intermolecular potential for calculating the second virial

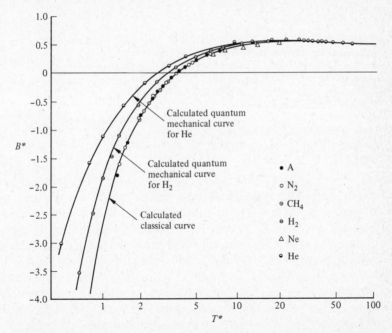

FIGURE 9-21 Reduced second virial coefficient vs reduced temperature for a variety of gases. The results for a wide variety of substances fall on the single curve predicted from classical statistical mechanics. Quantum corrections (not covered in this book) are important only for H_2 and He. From J. O. Hirschfelder, C. F. Curtiss, and R. L. Bird, *Molecular Theory of Gases and Liquids*, copyright © 1954. By permission of John Wiley & Sons, Inc.

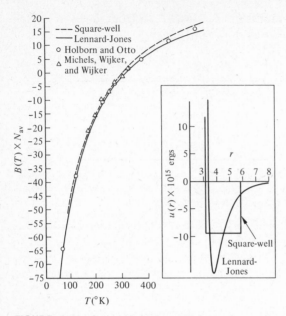

FIGURE 9-22 Second virial coefficients for *argon* calculated for several molecular models. The potential functions obtained from the experimental *B(T)* data are also shown. The experimental data are those of L. Holborn and J. Otto, *Z. Physik* **33**, 1 (1925), and A. Michels, Hub. Wijker, and Hk. Wijker, *Physica* **15**, 627 (1949). From J. O. Hirschfelder, C. F. Curtiss, and R. L. Bird, *Molecular Theory of Gases and Liquids*, copyright © 1954. By permission of John Wiley & Sons, Inc.

coefficient. The average of best values of ϵ/k for various pairs of temperatures is $\epsilon/k = 221°K$. With this value, Eq. (9-109) can be used to evaluate $\sigma = 4.10$ Å. Then, for the set of potential parameters $\epsilon/k = 221°K$, $\sigma = 4.10$ Å, the second virial coefficient of Xe can be calculated from tables of $B^*(T^*)$ for all T. The results are given in Table 9-2.

For a series of gases, one can plot $B^*_2(T^*)$ on the same axes. If the theory is correct, the results for various substances should fall on the same curve. Such a plot[9] is shown in Fig. 9-21. It is clear that except for H_2 and He, the relationship does indeed hold. The deviations in H_2 and He are due to quantum mechanical effects that are beyond the scope of this book.

---

[9]Taken from Fig. 3.6-1 of Hirschfelder, Curtiss and Bird, *Molecular Theory of Gases and Liquids* (New York, J. Wiley & Sons, 1954).

It can be shown that the temperature dependence of the second virial coefficient does not uniquely determine the form of the intermolecular potential. This is illustrated[10] in Figs. 9-22 and 9-23 for Ar and N_2 where a good fit of B_2 vs T is obtained for both square-well and Lennard-Jones potentials. Higher virial coefficients are more sensitive to the exact form of the intermolecular potential.

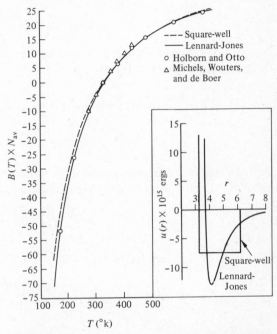

FIGURE 9-23 Second virial coefficient for *nitrogen* calculated for several molecular models. The potential functions obtained from experimental $B(T)$ data are also shown. The experimental data are those of L. Holborn and J. Otto, *Z. Physik* **33**, 1 (1925), and A. Michels, H. Wouters, and J. de Boer, *Physica* **1**, 587 (1934). From J. O. Hirschfelder, C. F. Curtiss, and R. L. Bird, *Molecular Theory of Gases and Liquids*, copyright © 1954. By permission of John Wiley & Sons, Inc.

9.5 PRINCIPLE OF CORRESPONDING STATES

It has been shown that the equation of state of a nonideal gas follows from the use of Eqs. (9-66) and Eq. (9-3). Since the pre-integral factors

---

[10]Taken from Figs. 3.9-3 and 3.9-5 of Hirschfelder, Curtiss and Bird, *Molecular Theory of Gases and Liquids* (New York, J. Wiley & Sons, 1954).

in Eq. (9-3) do not depend on volume, it follows that

$$p = -kT \ (\partial \ln Z/\partial V)_{T,N} \tag{9-110}$$

where

$$Z = \int \cdots \int e^{-U/kT} \, dq_1 \, dq_2 \ldots dq_{3N} \tag{9-111}$$

and

$$U = \sum_{i>j}^{N} \sum_{j=1}^{N} u(r_{ij}) \tag{9-112}$$

Suppose the intermolecular potential has the form (note that the Lennard-Jones potential is of this form):

$$u(r_{ij}) = \text{(energy parameter)} \times \text{function}\left(\frac{r_{ij}}{\substack{\text{diameter} \\ \text{parameter}}}\right)$$

$$= \epsilon\phi(r_{ij}/\sigma) \tag{9-113}$$

where ϵ and σ are the energy and diameter parameters, and ϕ is a function of r_{ij}/σ. Then U may be written in the form

$$U = \epsilon \sum_{i>j}^{N} \sum_{j=1}^{N-1} \phi(r_{ij}/\sigma) \tag{9-114}$$

The ratio kT/ϵ is denoted as T^*. Thus, Eq. (9-111) may be expressed as

$$Z = \sigma^{3N} \int \cdots \int \exp\left\{-T^{*-1} \sum \sum \phi(r_{ij}/\sigma)\right\} \frac{dq_1}{\sigma} \frac{dq_2}{\sigma} \ldots \frac{dq_{3N}}{\sigma} \tag{9-115}$$

Since all distances in the integral in Eq. (9-115) are measured in units of σ, the integral depends only on the parameter T^*, and the volume over which the integral is taken in units of σ^3. Since each particle can range over the full volume of the container, and distances are measured in units of σ, the integral depends on the ratio V/σ^3. Therefore, the integral depends only on T^* and V/σ^3. Let the integral in Eq. (9-115) be denoted $I(T^*, V/\sigma^3)$. Now, let Eq. (9-110) be multiplied through by $(\sigma^3/\sigma^3)(\epsilon/\epsilon)$. The following is obtained:

$$\frac{p\sigma^3}{\epsilon} = \frac{-kT}{\epsilon}\left[\frac{\partial \ln Z}{\partial (V/\sigma^3)}\right]_{T,N} \tag{9-116}$$

Since

$$Z = \sigma^{3N} I(T^*, V/\sigma^3) \tag{9-117}$$

it follows that

$$\left[\frac{\partial \ln Z}{\partial (V/\sigma^3)}\right]_{T,N} = \left[\frac{\partial \ln I}{\partial (V/\sigma^3)}\right]_{T,N} \equiv J(T^*, V/\sigma^3) \tag{9-118}$$

where $J(T^*, V/\sigma^3)$ is defined in this equation. Thus, it has been shown
that

$$\frac{p\sigma^3}{\epsilon} = -\frac{kT}{\epsilon} J(T^*, V/\sigma^3) \qquad (9\text{-}119)$$

If dimensionless variables are defined, such that

$$p^* = p\sigma^3/\epsilon \qquad (9\text{-}120)$$

and

$$V^* = V/\sigma^3 \qquad (9\text{-}121)$$

it follows that

$$p^* = -T^*J(T^*, V^*) \qquad (9\text{-}122)$$

Since J is a function only of T^* and V^*, it follows that regardless of the
values of ϵ and σ, a universal relation exists between p^*, V^*, and T^* for
all gases with intermolecular potential terms of the form in Eq. (9-113).
The variables p^*, V^*, and T^* are called *reduced* variables of state. The
fact that a universal relation exists between p^*, V^*, and T^* is called the
law of corresponding states.

It may be concluded that, to the extent that Eq. (9-113) describes
a series of substances, all the substances obey essentially the same equation of state.

Problems

1. Calculate the second virial coefficient of a gas with an interaction potential
   in the form of a square well:

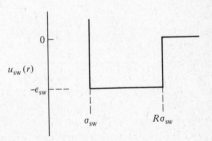

Answer: $B_2(T) = (\frac{2}{3}\pi\sigma_{s\text{-}w\,\text{pot}}^3)\{1 - (e^{\epsilon_{s\text{-}w\,\text{pot}}/kT} - 1)(R^3 - 1)\}$

2. Fit a square-well potential to a Lennard-Jones 6–12 potential by requiring

   (a) The total area of $u(r)$ below zero is the same for both potentials:

$$\underbrace{\int_{\sigma_{s\text{-}w\,\text{pot}}}^{\infty} u_{s\text{-}w\,\text{pot}}(r)\, dr}_{\phi_{s\text{-}w\,\text{pot}}} = \underbrace{\int_{\sigma}^{\infty} u_{L\text{-}J\,\text{pot}}(r)\, dr}_{\phi_{L\text{-}J\,\text{pot}}}$$

(b) The centroid of the two potentials below zero is the same:

$$\phi_{s\text{-}w\,\text{pot}}^{-1} \int_{\sigma_{s\text{-}w\,\text{pot}}}^{\infty} u_{s\text{-}w\,\text{pot}}(r)\, r\, dr = \phi_{L\text{-}J\,\text{pot}}^{-1} \int_{\sigma}^{\infty} u_{L\text{-}J\,\text{pot}}(r)\, r\, dr$$

These relations yield the equations

$$(R-1)\sigma_{s\text{-}w\,\text{pot}}\epsilon_{s\text{-}w\,\text{pot}} = \tfrac{24}{55}\epsilon\sigma$$

$$(R+1)\sigma_{s\text{-}w\,\text{pot}} = \tfrac{55}{20}\sigma$$

Make the arbitrary (but reasonable) assumption that $\sigma_{s\text{-}w\,\text{pot}} = 0.94\sigma$. Show that

$$R = 1.92$$
$$\epsilon_{s\text{-}w\,\text{pot}} = 0.505\epsilon$$

Plot the Lennard-Jones and square-well potentials on the same axes.

3. For Ar, the best values of the Lennard-Jones 6–12 coefficients are $\epsilon/k = 120°\text{K}$ and $\sigma = 3.41$ Å. The resulting calculated values of $B_2(T)$ are shown below.

T $°K$	T^*	B^* cm³/molecule	B cm³/molecule
0.70	84	−4.710	−3.91 × 10⁻²⁴
0.80	96	−3.734	−3.10 × 10⁻²⁴
1.00	120	−2.538	−2.11 × 10⁻²⁴
1.25	150	−1.704	−1.42 × 10⁻²⁴
1.80	216	−0.812	−0.674 × 10⁻²⁴
3.00	360	−0.115	−0.096 × 10⁻²⁴
4.00	480	0.115	+0.096 × 10⁻²⁴
6.00	720	0.323	0.269 × 10⁻²⁴
8.00	960	0.413	0.343 × 10⁻²⁴
10.00	1200	0.461	0.383 × 10⁻²⁴
20.00	2400	0.525	0.436 × 10⁻²⁴

Calculate $B_2(T)$ for the square-well potential using the results of problems 1 and 2. Construct graphs from the above data and plot the square-well calculations on the same axes. Does the second virial coefficient give detailed information on the exact form of the intermolecular potential?

chapter 10

THE MOLECULAR PARTITION FUNCTION IN TERMS OF LOCAL BOND PROPERTIES

10.1 INTRODUCTION

This chapter considers the partition function of a polyatomic molecule in the approximation that the rotations may be treated as rigid rotations with the atoms at their equilibrium positions and the vibrations are small-amplitude harmonic motions about equilibrium. This is the usual approximation that is made in most calculations. In Chapter 5, it was shown that the partition function for a molecule (to this approximation) is

$$Q = Q_{\text{tr}} \sum_{\substack{n=\text{all} \\ \text{electronic} \\ \text{states}}} g_n^{\text{el}} \, e^{-\epsilon_n^{\text{el}}/kT} \, \{Q_{\text{rot}}Q_{\text{vib}}Q_{\text{int rot}}\}_n \qquad (10\text{-}1)$$

where $\{Q_{\text{rot}}Q_{\text{vib}}Q_{\text{int rot}}\}_n$ is the product of rotational, vibrational and internal rotational partition functions for the molecule in electronic state n. These partition functions will generally change with the electronic state of the molecule. For many molecules, the spacing between electronic levels is sufficiently large compared to kT that the excited electronic states can be neglected. In this case,

$$Q \cong g_0^{\text{el}} Q_{\text{tr}} Q_{\text{rot}} Q_{\text{vib}} Q_{\text{int rot}} \qquad (10\text{-}2)$$

and $Q_{\text{rot}}Q_{\text{vib}}Q_{\text{int rot}}$ is to be evaluated for the ground electronic state. This is the approximation used in this chapter. The nature of the partition functions in Eq. (10-2) is such that they depend on overall properties of the molecule. For example, the rotational partition function depends on moments of inertia for overall rigid rotation, and the vibrational partition function depends on frequencies of vibration of normal modes of the molecule. In this chapter, an alternate formulation for Q is derived, which depends on local bond properties such as bond lengths, force constants, and masses rather than overall properties of the molecule. This will only be done in the approximation of classical mechanics. Quantum correction factors for the vibrational modes must be included.

The partition function for a molecule is written as

$$Q = \frac{g_0^{el} Q_{qu}}{\sigma} = \frac{g_0^{el}}{\sigma} \left(\frac{Q_{qu}}{Q_{cl}}\right) Q_{cl} \tag{10-3}$$

where Q_{qu} is the quantum mechanical partition function excluding the electronic contribution and symmetry factors, Q_{cl} is the classical partition function, σ is the net product of symmetry numbers for the rotations, and (Q_{qu}/Q_{cl}) is a quantum correction factor. Since the translational partition function is very nearly exactly classical, and the rotational partition function is generally closely approximated by the classical result, Q_{qu}/Q_{cl} will usually depend only on the vibrational degrees of freedom.

The classical partition function of a polyatomic molecule containing N atoms is

$$Q_{cl} = h^{-3N} \underset{6N}{\int_{-\infty}^{\infty} \cdots \int_{-\infty}^{\infty}} e^{-E/kT} \, dp_{x_1} \ldots dp_{z_N} \, dx_1 \ldots d_{z_N} \tag{10-4}$$

in terms of cartesian coordinates x_i, y_i, z_i and momenta p_{x_i}, p_{y_i}, p_{z_i} for particle i. The total energy E is given by

$$E = \frac{p_{x_1}^2}{2m_1} + \frac{p_{y_1}^2}{2m_i} + \frac{p_{z_1}^2}{2m_1} + \frac{p_{x_2}^2}{2m_2} + \cdots + \frac{p_{z_N}^2}{2m_N} + U(x_1, y_1, z_1, \ldots, z_N) \tag{10-5}$$

and U is the potential energy relative to the equilibrium configuration chosen as zero. The $3N$ integrations over the momenta can be carried out immediately, leading to

$$Q_{cl} = \prod_{\alpha=1}^{N} \Lambda_\alpha^{-3} Z \tag{10-6}$$

where Λ_α is a quantity with the dimensions of a length given by

$$\Lambda_\alpha^{-1} = \int_{-\infty}^{\infty} e^{-p^2/2m_\alpha kT} \, dp/h = \left(\frac{2\pi m_\alpha kT}{h^2}\right)^{1/2} \tag{10-7}$$

and m_α is the mass of atom α. The quantity Z is the *configurational integral*

$$Z = \underset{3N}{\int \cdots \int} e^{-U(x_1, \ldots, z_n)/kT} \, dx_1 \ldots dz_N \tag{10-8}$$

(Note the similarity between this procedure and that of Chapter 9 for nonideal gases.) If there were no forces acting between the atoms of a molecule, the molecule would fall apart and the atoms would be free to roam around the entire container of volume V. In this case, the potential function U would simply be $U = 0$ inside the container, and $U = \infty$

outside the container.   Therefore, one would obtain

$$Z = \int \cdots \int_{\text{container}} (1)\, dx_1 \ldots dz_N = V^N \tag{10-9}$$

$$Q_{\text{cl}} = \prod_{\alpha=1}^{N} \left\{ \left(\frac{2\pi m_\alpha kT}{h^2} \right)^{3/2} V \right\} \tag{10-10}$$

The quantity in braces in Eq. (10-10) is simply the partition function for atom α. This is exactly what should be expected for the partition function of N noninteracting atoms.

For a real molecule, where forces do exist between the atoms, the effective volume available to an atom is less than V because it is restrained by the other atoms. If surface effects are neglected, one may conceptually place the atoms of the molecule in the container serially and determine how U depends on the coordinates of each atom. The potential energy of the molecule is independent of where the molecule is in the container, but only depends on the relative positions of the atoms with one another. The position of the molecule can be specified by placing any arbitrary atom in the container and referencing all other coordinates to this atom as origin. The potential energy is independent of the position of this atom, which is free to wander around the container. If we denote this as atom 1, then $U(x_2, y_2, \ldots z_n)$ is independent of $x_1, y_1,$ and z_1. Thus,

$$Z = V \int \cdots \int_{3N-3} e^{-U(x_2,\ldots,z_n)/kT}\, dx_2 \ldots dz_n \tag{10-11}$$

In this treatment, atom 1 may be thought of as wandering freely around the container, "dragging the rest of the molecule with it."

To determine how U depends on the coordinates of the other atoms, temporarily assume that the molecule has the form of a *chain* for atoms 1, 2, and 3, although *branching* can occur further away. That is, some general form like that shown below is assumed.

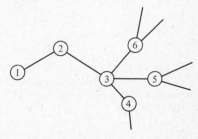

The cartesian coordinates x_2, y_2, and z_2 for particle 2 are relative to particle 1 chosen as the origin, with any arbitrary direction chosen as

z_2 axis. Transform to spherical polar coordinates r_2, θ_2, and ϕ_2 as shown in Fig. 10-1. To locate particle 2 relative to particle 1, either specify x_2, y_2 and z_2, or r_2, θ_2 and ϕ_2. Any change in θ_2 or ϕ_2 with fixed r_2 constitutes a rigid rotation of the molecule because only the orientation of the molecule in space is changed. The potential energy cannot depend on how the molecule is oriented in space; therefore U is a function of r_2 and the coordinates of atoms 3, 4, ... N, but not of θ_2 or ϕ_2. The volume element for atom 2 is $dx_2\, dy_2\, dz_2$. In making a change of variables $(x_2, y_2, z_2) \rightarrow (r_2, \theta_2, \phi_2)$ the new volume element is determined by the expression[1]

$$dx_2\, dy_2\, dz_2 = |J|\, dr_2\, d\theta_2\, d\phi_2 \tag{10-12}$$

where $|J|$ is the *Jacobian determinant* given by

$$|J| = \begin{vmatrix} \dfrac{\partial x_2}{\partial r_2} & \dfrac{\partial x_2}{\partial \theta_2} & \dfrac{\partial x_2}{\partial \phi_2} \\[2mm] \dfrac{\partial y_2}{\partial r_2} & \dfrac{\partial y_2}{\partial \theta_2} & \dfrac{\partial y_2}{\partial \phi_2} \\[2mm] \dfrac{\partial z_2}{\partial r_2} & \dfrac{\partial z_2}{\partial \theta_2} & \dfrac{\partial z_2}{\partial \phi_2} \end{vmatrix} = r_2{}^2 \sin \theta_2 \tag{10-13}$$

In general, in making a transformation from coordinates $x_1, x_2, \ldots$ to q_1, $q_2, \ldots$, it can be shown[2] that

$$dx_1\, dx_2 \ldots = |J|\, dq_1\, dq_2 \ldots \tag{10-14}$$

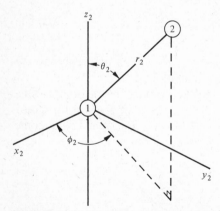

FIGURE 10-1 Spherical polar coordinates used to locate particle 2 with respect to particle 1 as a reference origin of coordinates.

---

[1] P. Franklin, *Treatise on Advanced Calculus* (New York: John Wiley & Sons, 1940), p. 371; also Appendix VI.
[2] See Appendix VI.

where $J_{ij} = \partial x_i / \partial q_j$. Thus, Eq. (10-11) becomes

$$Z = V \int_{3N-3} \cdots \int e^{-U(r_2, x_3, \ldots, z_n)/kT} r_2^2 \, dr_2 \sin \theta_2 \, d\theta_2 \, d\theta_2 \, d\phi_2 \, dx_3 \ldots dz_N \quad (10\text{-}15)$$

One may immediately integrate over $d\theta_2$ and $d\phi_2$ from 0 to π, and 0 to 2π, respectively, to obtain $2(2\pi) = 4\pi$. Thus, in general

$$Z = 4\pi V \int_{3N-5} \cdots \int e^{-U(r_2, x_3, \ldots z_n)/kT} r_2^2 \, dr_2 \, dx_3 \ldots dz_n \quad (10\text{-}16)$$

10.2 TREATMENT OF LINEAR AND NONLINEAR MOLECULES

At this point in the calculation, it is necessary to distinguish between molecules which are linear and nonlinear in the equilibrium configuration. As shown in Chapter 5, linear molecules have only 2 rotational degrees of freedom and $3N - 5$ vibrational modes, whereas nonlinear molecules have 3 rotations and $3N - 6$ vibrational or internal rotational modes. Since we have already included two free rotational degrees of freedom (coordinates θ_2 and ϕ_2), it is at this point that a distinction must be drawn between the two cases. The assumption that U is independent of θ_2 and ϕ_2 means that atom 2 can rotate around a full sphere relative to atom 1 without changing U. For linear molecules, this defines the entire range of rotational motions that are possible. For nonlinear molecules, one more degree of free rotation must be included.

First consider a general linear molecule as shown in Fig. 10-2. The

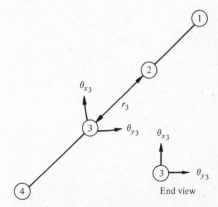

FIGURE 10-2 The linear molecule with many atoms. Atom 3 is located relative to atom 2 as a reference origin by means of coordinates r_3, θ_{x_3} and θ_{y_3}, representing linear stretching, and orthogonal bendings, respectively.

coordinates of atom 3 relative to atom 2 are specified by the distance from atom 2 to atom 3, r_3, and two arbitrary perpendicular angles θ_x and θ_y which specify the angular motion of atom 3 out of line from the equilibrium configuration.   For small displacements, it can be shown that[3]

$$dx_3\, dy_3\, dz_3 = r_3{}^2\, dr_3\, d\theta_{x_3}\, d\theta_{y_3}$$

and since U depends on r_3, θ_{x_3} and θ_{y_3}, the general expression for the configurational integral of a linear molecule is

$$Z = 4\pi V \int \cdots \int_{3N-5} \exp\{-U(r_2, r_3, \theta_{x_3}, \theta_{y_3}, x_4, \ldots z_n)/kT\}$$

$$\times r_2{}^2\, dr_2\, r_3{}^2\, dr_3\, d\theta_{x_3}\, d\theta_{y_3}\, dx_4 \ldots dz_N \tag{10-17}$$

For nonlinear molecules, a set of spherical polar coordinates r_3, θ_3, and ϕ_3 are used to locate particle 3 relative to particle 2 as origin, with the polar axis chosen as the direction from atom 1 to atom 2.   This is shown in Fig. 10-3.   When angle ϕ_3 changes with θ_3 and r_3 fixed, the molecule undergoes a rigid rotation as shown in Fig. 10-4.   This motion is free, and U cannot depend on coordinate ϕ_3.   Thus,

$$Z = 4\pi V \int \cdots \int_{3N-5} \exp\{-U(r_2, r_3, \theta_3, x_4, \ldots z_n)/kT\}$$

$$\times r_2{}^2\, dr_2\, r_3{}^2\, dr_3 \sin\theta_3\, d\theta_3\, d\phi_3\, dx_4 \ldots dz_N \tag{10-18}$$

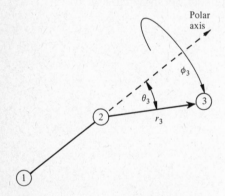

FIGURE 10-3 The bent triatomic molecule.   Atom 3 is located relative to atom 2 as a reference origin by coordinates r_3, θ_3, and ϕ_3, the polar axis being chosen along the $1 \rightarrow 2$ bond axis.

[3]See Appendix VI.

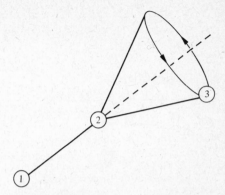

FIGURE 10-4 The cone of motion generated by variations in angle ϕ_3 with r_3 and θ_3 fixed.

Upon integration over $d\phi_3$ from 0 to 2π, one obtains the general expression for the configurational integral of a *nonlinear polyatomic molecule*,

$$Z = 8\pi^2 V \underset{3N-6}{\int \cdots \int} e^{-U/kT} r_2{}^2 \, dr_2 \, r_3{}^2 \, dr_3 \sin\theta_3 \, d\theta_3 \, dx_4 \ldots dz_N \qquad (10\text{-}19)$$

10.3 USE OF THE HARMONIC APPROXIMATION WITH NEGLECT OF THE VIBRATIONAL-ROTATIONAL INTERACTION

In this section, Eqs. (10-17) and (10-19) will be evaluated in greater detail by assuming that the atoms in a molecule are bound to their equilibrium positions by harmonic restoring forces and that the vibrational amplitudes are small compared to the equilibrium bond distances. The procedure will be illustrated by assuming that U depends on the distance between atoms 1 and 2 as $\frac{1}{2}f_{r_2}(r_2 - r_2{}^0)^2$ where f_{r_2} is the force constant for stretching of the 1-2 bond, and $r_2{}^0$ is the equilibrium value of r_2. The integral

$$\int_0^\infty \exp\left\{-f_{r_2}(r_2 - r_2{}^0)^2/2kT\right\} r_2{}^2 \, dr_2 \qquad (10\text{-}20)$$

may then be factored out of the configurational integral. A plot of the integrand functions is shown in Fig. 10-5. An approximate integration can be effected by taking the pre-exponential factor $r_2{}^2$ out of the integral sign as $(r_2{}^0)^2$. The remaining integral is

$$(r_2{}^0)^2 \int_{-r_2{}^0}^\infty e^{-f_{r_2}\xi^2/2kT} \, d\xi$$

where $\xi = r_2 - r_2{}^0$. Since the integrand is very small at the lower limit, the lower limit can be replaced by $-\infty$ to a good approximation. The

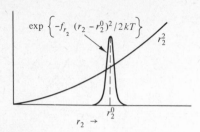

FIGURE 10-5 The functions exp $\{-f_{r_2}(r_2-r_2{}^0)^2/2kT\}$ and $r_2{}^2$ which appear in Eq. (10-20). If the exponential function is sufficiently narrowly peaked around $r_2 = r_2{}^0$, the factor $r_2{}^2$ in the integrand may be taken out in front of the integral sign as a factor $(r_2{}^0)^2$.

integral then becomes

$$(r_2{}^0)^2 \left(\frac{2\pi kT}{f_{r_2}}\right)^{1/2}$$

An exactly analogous procedure can be used for coordinate r_3. For nonlinear molecules, the integration over $d\theta_3$ proceeds the same way. The integral

$$\int_0^\pi \exp\{-f_{\theta_3}(\theta_3 - \theta_3{}^0)^2/2kT\} \sin\theta_3 \, d\theta_3$$

is evaluated by taking $\sin\theta_3{}^0$ out of the integral sign. (The integrand functions are plotted in Fig. 10-6.) The result is

$$\sin\theta_3{}^0 \left(\frac{2\pi kT}{f_{\theta_3}}\right)^{1/2}$$

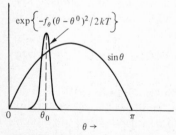

FIGURE 10-6 The functions exp $\{-f_\theta(\theta-\theta^0)^2/2kT\}$ and $\sin\theta$. The integral of the product of these functions over $d\theta$ may be evaluated approximately by taking $\sin\theta$ as a pre-integral factor $\sin\theta_0$.

FIGURE 10-7 A simple nonlinear chain molecule without branching.

Thus, the configuration integral for a nonlinear polyatomic molecule has the general form

$$Z = 8\pi^2 V(r_2^0)^2 \left(\frac{2\pi kT}{f_{r_2}}\right)^{1/2} (r_3^0)^2 \left(\frac{2\pi kT}{f_{r_3}}\right)^{1/2} \sin\theta_3^0 \left(\frac{2\pi kT}{f_{\theta_3}}\right)^{1/2} \tag{10-21}$$

$$\cdot \int e^{-U/kT} \, dx_4 \ldots dz_N$$

In particular, for a bent triatomic molecule,

$$Z = \frac{8\pi^2 V(r_2^0)^2 (r_3^0)^2 \sin\theta_3^0 (2\pi kT)^{3/2}}{(f_{r_2}f_{r_3}f_{\theta_3})^{1/2}} \tag{10-22}$$

For a molecule which is simply a nonlinear chain with no branching, as shown for example in Fig. 10-7, if the potential energy can be regarded as a sum of diagonal terms in the internal coordinates; each term being of the form

$$\tfrac{1}{2}f_R(R - R^0)^2 \tag{10-23}$$

where R can be r_n, θ_n, or ϕ_n, then Z has the form

$$Z = 8\pi^2 V(r_2^0)^2 (r_3^0)^2 \ldots (r_N^0)^2 \sin\theta_3^0 \sin\theta_4^0 \ldots \sin\theta_N^0$$

$$\times \left(\frac{(2\pi kT)^{(3N-6)/2}}{(f_{r_2}\ldots f_{r_N}f_{\theta_3}\ldots f_{\theta_N}f_{\phi_4}\ldots f_{\phi_N})^{1/2}}\right) \tag{10-24}$$

For linear molecules, the same procedure is used, except that θ_{x_N} and θ_{y_N} are used instead of θ_n and ϕ_n, resulting in

$$Z = \frac{4\pi V(r_2^0)^2 (r_3^0)^2 \ldots (r_N^0)^2 (2\pi kT)^{(3N-5)/2}}{(f_{r_2}\ldots f_{r_N}f_{\theta_{x_3}}\ldots f_{\theta_{x_N}}f_{\theta_{y_3}}\ldots f_{\theta_{y_N}})^{1/2}} \tag{10-25}$$

For chain molecules containing both linear and nonlinear parts, as shown in Fig. 10-8, serial integration can be used, beginning with

$$\frac{4\pi V(r_2^0)^2 (2\pi kT)^{1/2}}{(f_{r_2})^{1/2}} \tag{10-26}$$

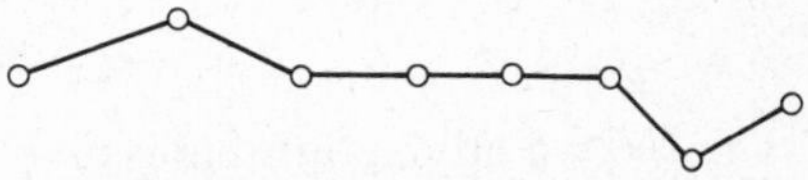

FIGURE 10-8 A nonbranched chain molecule with a linear segment.

for the first two atoms. If the third atom is collinear with respect to the first two, multiply Eq. (10-26) by

$$\frac{(2\pi kT)^{3/2}(r_3{}^0)^2}{(f_{r_3}f_{\theta_{x_3}}f_{\theta_{y_3}})^{1/2}} \tag{10-27}$$

to include the third atom. If the third atom is not collinear with the first two, multiply by

$$\frac{(2\pi)(r_3{}^0)^2 \sin\theta_3{}^0 (2\pi kT)}{(f_{r_3}f_{\theta_3})^{1/2}} \tag{10-28}$$

In either case, to include atom j which is collinear with atoms $j-1$ and $j-2$, multiply by

$$\frac{(2\pi kT)^{3/2}(r_j{}^0)^2}{(f_{r_j}f_{\theta_{x_j}}f_{\theta_{y_j}})^{1/2}} \tag{10-29}$$

To include atom k which is not collinear with atoms $k-1$ and $k-2$, multiply by

$$\frac{(2\pi kT)^{3/2}(r_k{}^0)^2 \sin\theta_k{}^0}{(f_{r_k}f_{\theta_k}f_{\phi_k})^{1/2}} \tag{10-30}$$

Upon reviewing the procedures used in deriving Eqs. (10-24) and (10-25) from Eq. (10-11), it will become clear that the serial evaluation of the configuration integral is obtained by using a Jacobian factor to transform the coordinates of each atom from cartesian to internal coordinates, and then approximately evaluating the resultant integrals. The addition of each atom α results in a factor V_α, which is a measure of the effective volume available for the atom to move in. If there were no forces acting between the atoms, each term V_α would be equal to the container volume V, and Z would be equal to V^N.

For molecules with potential fields which are harmonic in terms of a set of $3N-6$ ($3N-5$ for linear molecules) internal coordinates, it is found that

$$Z = \prod_{\alpha=1}^{N} V_\alpha \tag{10-31}$$

The volume factors have been evaluated for zig-zag and linear chain molecules. The first atom contributes a volume factor equal to the entire container volume:

$$V_1 = V$$

The second atom contributes a volume equal to a spherical shell of radius $r_2{}^0$, and thickness $(2\pi kT/f_{r_2})^{1/2}$, as illustrated in Fig. 10-9. This volume may be interpreted as resulting from the oscillations of atom 2 about its

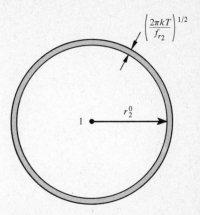

FIGURE 10-9 The volume available to atom 2 is $(2\pi kT/f_{r_2})^{1/2}$ relative to atom 1. Atom 1 can be anywhere in the container. The restricted volume of atom 2 is due to the fact that for any fixed position of atom 1, atom 2 is free to rotate about atom 1 in any orientation, but can only stretch a small amount from the equilibrium separation of atoms 1 and 2.

equilibrium position, with atom 2 free to rotate about atom 1 over an entire sphere. The volume of the shell is

$$V_2 = 4\pi (r_2^0)^2 \left(\frac{2\pi kT}{f_{r_2}}\right)^{1/2}$$

If the third atom is not collinear with atoms 1 and 2, the volume available to atom 3 is as shown in Fig. 10-10. It is a circular ring of radius $r_3^0 \sin \theta_3^0$, with thickness $(2\pi kT/f_{r_3})^{1/2}$ in the r_3 direction, and $(2\pi kT/f_{\theta_3})^{1/2}r_3^0$ in the θ_3 direction. The volume enclosed is

$$V_3 = (2\pi r_3^0 \sin \theta_3^0) \left(\frac{2\pi kT}{f_{r_3}}\right)^{1/2} \left(\frac{2\pi kT}{f_{\theta_3}}\right)^{1/2} r_3^0 = 2\pi (r_3^0)^2 \frac{(2\pi kT)}{(f_{r_3}f_{\theta_3})^{1/2}}$$

If the third atom is collinear with atoms 1 and 2, the volume available is as shown in Fig. 10-11. It follows that in this case

$$V_3 = (r_3^0)^2 \frac{(2\pi kT)^{3/2}}{(f_{r_3}f_{\theta_{x_3}}f_{\theta_{y_3}})^{1/2}}$$

Each additional atom contributes a volume term of the form given in Eqs. (10-29) and (10-30). Equations (10-24) can be rewritten as

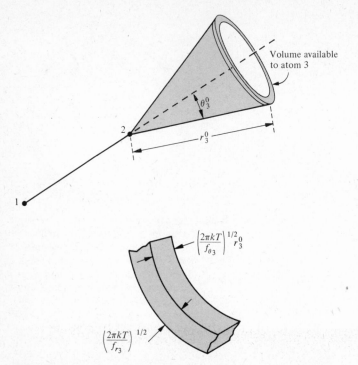

FIGURE 10-10 The volume available to the third atom of a bent triatomic molecule. Atom 3 can freely rotate about the 1–2 axis at fixed θ_3, and can stretch slightly in the θ_3 and ϕ_3 directions.

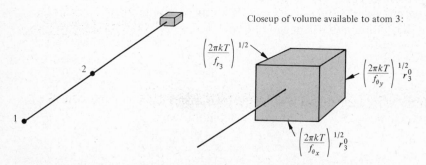

FIGURE 10-11 Volume available to atom 3 of a linear triatomic molecule. Atom 3 can stretch slightly in the r_3, θ_{x_3}, and θ_{y_3} directions, leading to a small cubical volume element.

$$Z = \prod_{\alpha=1}^{N} J_\alpha \frac{(2\pi kT)^{(3N-6)/2}}{|F|^{1/2}} \tag{10-32}$$

for nonlinear molecules, and

$$Z = \prod_{\alpha=1}^{N} J_\alpha \frac{(2\pi kT)^{(3N-5)/2}}{|F|^{1/2}} \tag{10-33}$$

for linear molecules. The J_α factors are effective Jacobian factors for each additional atom, and $|F|$ is the determinant of the force constant matrix. If F is diagonal, $|F|$ is simply the product of all force constants. The cases with branching are evaluated by specialized techniques which will not be considered here.[4]   Other, more complicated branchings can also be considered.   Typical Jacobian factors are given in Table 10-1.

TABLE 10-1   Jacobian factors for atoms[a]

Atom	Configuration	Internal Coordinates	J
1	○ 1	x, y, z	V
2	○——○ 1 2	r_{12}, θ, ϕ	$4\pi (r_{12}^0)^2$
3	○—○—○ 1 2 3	$r_{23}, \theta_x, \theta_y$	$(r_{23}^0)^2$
	(bent 1—2—3)	r_{23}, θ, ϕ	$2\pi (r_{23}^0)^2 \sin \theta^0$
	(triangle 1,2,3 with τ)	r_{13}, r_{23}, τ	$2\pi \dfrac{r_{23}^0 r_{23}^0}{r_{12}^0}$
4	○—○—○—○ 1 2 3 4	$r_{34}, \theta_x, \theta_y$	$(r_{34}^0)^2$
	(zigzag 1,2,3,4 τ)	r_{34}, θ, τ	$(r_{34}^0)^2 \sin \theta$
	(branch 1,2,3,4)	$r_{34}, \theta_x, \theta_y$	$(r_{34}^0)^2$
	(branched with 3)	$r_{24}, \theta_{124}, \theta_{324}$	$r_{24}^2/\sin \tau_{341}$
	(branched with τ)	$r_{24}, \theta_{124}, \tau$	$r_{24}^2 \cos \theta/\cos \tau$

[a]Herschbach, Johnston and Rapp, *J. Chem. Phys.* **31**, 1652 (1959).

[4]Herschbach, Johnston and Rapp, *J. Chem. Phys.* **31**, 1652 (1959).

10.4 QUANTUM CORRECTION FACTORS

The quantum correction factors apply to each of the $3N - 6$ ($3N - 5$ for linear molecules) normal modes of vibration. The ratio of Q_{qu}/Q_{cl} for a harmonic oscillator is[5] equal to

$$\Gamma = \frac{e^{-u/2}/(1 - e^{-u})}{u^{-1}} = \frac{(u/2)}{\sinh(u/2)} \qquad (10\text{-}34)$$

where $u = h\nu/kT$, and ν is the oscillator frequency. For a molecule of N atoms, from Eq. (10-3):

$$Q_{qu} = \frac{g_0{}^{el}Q_{cl}}{\sigma} \prod_{j=1}^{3N-6} \Gamma_j \qquad (10\text{-}35)$$

The quantum correction factors approach unity at high T, and become very small at low T.

At any particular temperature, the Γ_j factors are closest to unity for the lower vibrational frequencies. For example, at room temperature, $\Gamma_i \cong 0.9$ for $\nu_i \cong 300$ cm^{-1}, whereas $\Gamma_i \cong 0.008$ for $\nu_i = 3000$ cm^{-1}. Fortunately, the high vibrational frequencies of a molecule are relatively easy to estimate because they often consist mainly of localized motions such as stretching of a single bond. The low frequencies usually consist of large torsional motions, and are difficult to estimate. However, they need not be known very accurately because Γ_i is nearly unity anyway.

10.5 EFFECT ON EQUILIBRIUM CALCULATIONS

In calculating an equilibrium constant (see Chapter 7) one takes the ratio of partition functions of products and reagents raised to powers equal to the stoichiometric factors in the reaction. For example, in the process

$$aA + bB + \cdots \rightleftarrows pP + qQ + \cdots \qquad (10\text{-}36)$$

$$K_{eq} = \frac{Q_P{}^p Q_Q{}^q \cdots}{Q_A{}^a Q_B{}^b \cdots} e^{-\Delta\epsilon^0/kT} \qquad (10\text{-}37)$$

and the Q's are partition functions per unit volume[6] based on the "bottom of the well" as the zero of energy, and $\Delta\epsilon^0$ is the energy of reaction for all substances at the equilibrium rest positions.[7] According to Eqs. (10-6),

---

[5] See Chapter 1. This is based on a zero of energy as the "bottom of the well," not the lowest quantum level.

[6] That is, Q here represents $Q/V = Q'$ from Chapter 7.

[7] One could alternately use Q's based on the lowest quantum level as the zero of energy, in which case $\Delta\epsilon^0$ would be the energy of reaction for all substances in ground quantum levels, and one would then leave out the factor $e^{-u/2}$ in Eq. (10-34).

(10-7), and (10-35):

$$K_{eq} = V^{a+b-p-q}(\text{TIF})\frac{Z_P{}^p Z_Q{}^q \cdots}{Z_A{}^a Z_B{}^b \cdots}\gamma\, e^{-\Delta\epsilon 0/kT} = A\, e^{-\Delta\epsilon 0/kT} \quad (10\text{-}38)$$

where (TIF) is a temperature-independent factor:

$$\text{TIF} = \frac{\sigma_A{}^a \sigma_B{}^b \cdots (g_0{}^{el})_P{}^p (g_0{}^{el})_Q{}^q \cdots}{\sigma_P{}^p \sigma_Q{}^q \cdots (g_0{}^{el})_A{}^a (g_0{}^{el})_B{}^b \cdots}$$

and γ is the ratio of quantum correction factors for the vibrations:

$$\gamma = \frac{\underset{\substack{\text{all} \\ \text{vibrations} \\ \text{of } P}}{\prod} \Gamma_i{}^{pp} \quad \underset{\substack{\text{all} \\ \text{vibrations} \\ \text{of } Q}}{\prod} \Gamma_i{}^{Qq} \cdots}{\underset{\substack{\text{all} \\ \text{vibrations} \\ \text{of } A}}{\prod} \Gamma_i{}^{Aa} \quad \underset{\substack{\text{all} \\ \text{vibrations} \\ \text{of } B}}{\prod} \Gamma_i{}^{Bb}} \quad (10\text{-}40)$$

Equation (10-38) may be written in the form

$$K_{eq} = (\text{TIF})\gamma K_{cl} \quad (10\text{-}41)$$

where

$$K_{cl} = \frac{Z_P{}^p Z_Q{}^q \cdots}{Z_A{}^a Z_B{}^b \cdots} e^{-\Delta\epsilon 0/kT} V^{a+b-p-q} \quad (10\text{-}42)$$

PROBLEM: Show that K_{cl} for the reaction $A + BC \rightleftarrows ABC$ is given by (ABC is nonlinear)

$$K_{cl} = A_{cl}\, e^{-\Delta\epsilon 0/kT} \quad (10\text{-}43)$$

$$A_{cl} = 2\pi(r_2{}^0)^2 \left(\frac{r_3{}^0}{r_{BC}^0}\right)^2 \sin\theta_3{}^0 \left(\frac{2\pi kT}{f_{12}}\right)^{1/2} \left(\frac{2\pi kT}{f_\theta}\right)^{1/2} \left(\frac{f_{BC}}{f_{23}}\right)^{1/2} \quad (10\text{-}44)$$

The advantage of the use of Eq. (10-42) is that local bond properties, which do not change in a reaction, will cancel between reactants and products. For example, in the reaction:

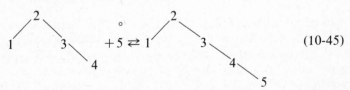

it might be assumed that the local bond properties (force constants, bond distances, and bond angles) one atom away from the reaction site do not change upon reaction. Thus, there is substantial cancellation between numerator and denominator of K_{cl}.

PROBLEM: Show that after canceling

$$8\pi^2 (r_2{}^0 r_3{}^0 r_4{}^0 r_5{}^0)^2 \sin \theta_3{}^0 \frac{(2\pi kT)^{5/2}}{(f_{12} f_{23} f_{\theta_{123}} f_{34} f_{\theta_{234}})^{1/2}} \qquad (10\text{-}46)$$

between numerator and denominator of Eq. (10-42) for reaction Eq. (10-45), what is left is

$$A_{cl} = \frac{(2\pi kT)^{3/2} (r_4{}^0)^2}{(f_{45} f_{\theta_{345}}^2)^{1/2}} \qquad (10\text{-}46a)$$

The ratio of quantum correction factors for products to reactants will involve approximate cancellation of all factors due to vibrations of molecule 1-2-3-4. Quantum correction factors are left in the numerator corresponding to the three additional vibrational modes introduced by adding atom 5 to molecule 1-2-3-4. These will include the 4-5 stretching mode, and two 3-4-5 bending modes. Fairly good approximations can be made for these modes.

10.6   ISOTOPE EFFECTS ON EQUILIBRIUM CONSTANTS[8]

Consider two separate chemical reaction equilibria where the reactions are chemically the same, but there are isotopic differences in some of the atoms in the reactions. The reactions will be written:

$$aA + bB + \cdots \rightarrow pP + qQ + \cdots \qquad (10\text{-}47)$$

$$aA' + bB' + \cdots \rightarrow pP' + qQ' + \cdots \qquad (10\text{-}48)$$

where the   following a symbol means that the substances may have isotopically substituted atoms. The equilibrium constants of reactions (10-47) and (10-48) are written as K_{eq} and K'_{eq}, respectively. According to Eq. (10-38),

$$\frac{K_{eq}}{K'_{eq}} = \frac{(\text{TIF})}{(\text{TIF})'} \frac{\gamma}{\gamma'} \frac{Z_P{}^p Z_Q{}^q \ldots Z_A'^a Z_B'^b \ldots}{Z_P'^p Z_Q'^q \ldots Z_A{}^a Z_B{}^b \ldots} \cdot \exp\{-(\Delta\epsilon^0 - \Delta\epsilon^{0'})/kT\} \qquad (10\text{-}49)$$

According to Eqs. (10-32) and (10-33), the configuration integral for a substance depends on the J_α factors and force constants. Since the J_α are mass-independent, and the force constants are independent of isotopic mass to a high degree of approximation, it follows that for any substance, Z is independent of isotopic substitution. Thus, the terms involving ratios of Z's in Eq. (10-49) cancel between numerator and

---

[8]In Chapter 11, it will be shown that similar considerations hold for isotope effects on reaction rate constants.

denominator.   Hence,

$$\frac{K_{eq}}{K'_{eq}} = \frac{TIF}{TIF'}\frac{\gamma}{\gamma'}\exp\{-(\Delta\epsilon^0 - \Delta\epsilon^{0'})/kT\} \qquad (10\text{-}50)$$

Furthermore, since $\Delta\epsilon^0$ and $\Delta\epsilon^{0'}$ are measured from the equilibrium rest positions of the molecules, $\Delta\epsilon^0 = \Delta\epsilon^{0'}$ to the extent that the molecular potential curves are independent of mass.   Thus,

$$\frac{K_{eq}}{K'_{eq}} = \frac{TIF}{TIF'}\frac{\gamma}{\gamma'} \qquad (10\text{-}51)$$

In the limit as $T \rightarrow \infty, \gamma/\gamma' \rightarrow 1$ and

$$\lim_{T\to\infty}\frac{K_{eq}}{K'_{eq}} = \frac{TIF}{TIF'} \qquad (10\text{-}52)$$

It is evident from Eq. (10-51) that those vibrational motions which are most affected by isotopic substitutions will contribute the most to K_{eq}/K'_{eq}.   For example, in the molecule CO_2, the vibrational modes are

$\leftarrow\cdot\qquad\cdot\qquad\cdot\rightarrow\qquad\qquad \nu_1$ symmetric stretch

$\leftarrow\cdot\quad\cdot\rightarrow\ \leftarrow\cdot\qquad\qquad \nu_2$ asymmetric stretch

$\uparrow\qquad\quad\qquad\qquad \nu_3, \nu_4$ bend (doubly degenerate)
$\downarrow\qquad\qquad\downarrow$

A change in the carbon isotope does not change ν_1, but it does change ν_2, ν_3, and ν_4.   A change in the oxygen mass has a relatively larger effect on ν_1 than on ν_2, ν_3, and ν_4.

Problems

1. Calculate the equilibrium constant for reaction (10-45) from the following "reasonable" molecular parameters:
   4-5 bond energy $= 70$ Kcal/mole
   4-5 bond distance $= 1.0$ Å
   $f_{45} = 5 \times 10^5$ dynes/cm
   $f_{\theta_{345}} = 5 \times 10^{-12}$ erg/radian^2
   Assume that the nine vibrational modes of molecule (12345) include the six vibrational modes of molecule (1234) plus a 4-5 stretching mode with frequency 1500 cm^{-1}, plus two degenerate bending modes of frequency 800 cm^{-1}.

Answer:
(a) Use Eq. (10-46a) to obtain

$$A_{cl} = 7.24 \times 10^{-26}\, T^{3/2}\ \text{cm}^3/\text{molecule}$$

(b) Quantum correction factors:

$$\gamma = \Gamma(1500)\Gamma^2(800)$$

(c) Exponential factor

$$70 \text{ Kcal/mole} = 4.87 \times 10^{-12} \text{ erg/molecule}$$

$$\exp(-\Delta\epsilon^0/kT) = \exp(-3.53 \times 10^4/T)$$

(d) In summary:

$$K_{eq} = 7.24 \times 10^{-26} T^{3/2} \Gamma(1500)\Gamma^2(800) \exp\left[\frac{-3.53 \times 10^4}{T}\right]$$

cm^3/molecule

2. Plot $\ln K_{eq}$ vs. $1/T$ with K_{eq} in cm^3/mole from $T = 10000°K$ to $3000°K$ from Problem 1.

3. Consider the series of reactions

$$H + D_2 \xrightarrow{1} HD + D$$

$$D + H_2 \xrightarrow{2} HD + H$$

$$H + HD \xrightarrow{3} H_2 + D$$

$$D + DH \xrightarrow{4} D_2 + H$$

(a) Calculate

$$(TIF)_1/(TIF)_n \text{ for } n = 2, 3, 4$$

(b) Using the fact that $\nu_{H_2} = 4395$ cm^{-1}, calculate ν_{HD} and ν_{D_2}.

(c) Calculate K_{eq_1}/K_{eq_n} for $n = 2, 3, 4$ at $T = 500, 1000, 2000$ and $4000°K$.

(d) Plot K_{eq_1}/K_{eq_n} vs $1/T^2$. Explain why this should be a straight line at high temperature. What is the slope?

4. Consider the reactions

$$(1) \quad O^{18} + O^{16}C^{12}O^{16} \rightarrow O^{16} + O^{18}C^{12}O^{16}$$

$$(2) \quad O^{16} + O^{18}C^{12}O^{18} \rightarrow O^{18} + O^{18}C^{12}O^{16}$$

(a) What is K_1/K_2 in the limit as $T \rightarrow \infty$?

(b) Using data on $O^{16}C^{12}O^{16}$ from the Appendix, calculate the frequency of the symmetric stretch mode of $O^{18}C^{12}O^{18}$.

(c) Assume that the ratios of the other frequencies of $O^{18}C^{12}O^{18}$ to $O^{16}C^{12}O^{16}$ are the same as for the symmetric stretch frequency. (Note: this is not actually correct. See p. 230 of G. Herzberg, *Infrared and Raman Spectra*, New York: Van Nostrand, 1945.) Calculate K_1/K_2 at $T = 2000$, 1000 and 500°K. Which frequencies determine K_1/K_2 to the greatest extent?

chapter 11

THE TRANSITION-STATE THEORY OF CHEMICAL KINETICS

11.1 INTRODUCTION

Transition-state theory is based on a model for the interaction potential between reactant molecules. First, it is assumed that the Born-Oppenheimer principle is a good approximation. Thus the rapid motions of the electrons compared to the nuclei allow the use of the electronic energy of any configuration of the nuclei at rest as a potential energy for motion of the nuclei. Second, it is assumed that the collisions are "adiabatic" so that a collision between molecules in particular electronic states leads to a collision complex in a particular electronic state, that can only form products in particular electronic states. The reactants and products are asymptotic forms of the collision complex, the entire range of allowed configurations corresponding to one grand electronic state.

For a total number of atoms in the reactants N, there are $3N-6$ coordinates[1] required to specify the relative configurations of the atoms with respect to one another. Three coordinates specify the location of the center of mass of the entire system, and three more locate the orientation in space of the principal axes of inertia for rotation of the system as a unit. These motions do not affect the potential energy and can be separated out in the beginning. If the potential energy is now plotted vs the $3N-6$ coordinates (see footnote 1) which specify the configuration of a system of atoms, a surface will be obtained representing a particular electronic state. Regions of stable configuration lie in "valleys," whereas unstable configurations are represented by "mountainous" parts of the surface. The instantaneous configuration of a system of atoms can be described by the position of a representative point on the surface. Chemical reaction is assumed to occur in a collision in which the representative point passes from the reactant valley to a product valley.

Transition-state theory requires the form of this surface to have a

---

[1] $3N-5$ for linear molecules.

mountain range separating the reactant and product valleys. The reactant valley corresponds to a range of configurations in which the atoms are grouped into reactant molecules, separated by relatively large distances. The product valley is similar, except the atoms are grouped into different molecules. It is then not unreasonable that there is some intermediate range of configurations, in which the atoms are grouped as neither product molecules nor reactant molecules, for which the repulsive energy (mountain height with respect to reactant or product valley levels) is a minimum. There is some region of the mountain range separating reactant and product valleys, which has a minimum elevation (that is, there is a pass through the mountain range). Any route drawn on the surface from reactant valley to product valley possesses a maximum in potential energy. Of all the possible routes that can be drawn, there is a class of routes which possess the smallest maxima which must be surmounted. All these routes traverse a small region of the energy surface corresponding to the top of the pass. The configuration corresponding to the lowest maximum is defined as the "equilibrium" configuration of the "transition state." Configurations of the "transition state" or "activated complex" not too different from this, can be thought of as being due to "vibrations" about the "equilibrium" configuration.

The nature of the energy surface in this region is that of a saddle point. That is, the slope of the surface at the "equilibrium" configuration in any direction is zero, and when the coordinates are transformed to principal axes of the surface, the curvatures along all but one of these axes are positive, the remaining curvature being negative.

If any set of $3N-6$ relative coordinates (measured from the "equilibrium" configuration as zero) S_i are used to define a configuration near the top of the pass, there is the condition $(\partial V/\partial S_i)^* = 0, i = 1, 2, \ldots, 3N-6$, where V is the potential energy, and the superscript asterisk means the quantity is to be evaluated at the top of the pass. For small motions about this point, the potential energy may be expanded in a power series, and only the first terms retained:

$$V(S_1, \ldots, S_{3N-6}) = V^* + \frac{1}{2} \sum_{i=1}^{3N-6} \sum_{j=1}^{3N-6} f_{ij} S_i S_j + \cdots \qquad (11\text{-}1)$$

where $f_{ij} = (\partial^2 V/\partial S_i \partial S_j)^*$. The classical kinetic energy is

$$T(\dot{S}_1, \ldots, \dot{S}_{3N-6}) = \frac{1}{2} \sum_{i=1}^{3N-6} \sum_{j=1}^{3N-6} G_{ij}^{-1} \dot{S}_i \dot{S}_j \qquad (11\text{-}2)$$

where the G_{ij}^{-1} are effective masses for the relative coordinates, and are given by Wilson, Decius, and Cross[2] for any configuration of atoms.

---

[2] E. B. Wilson, Jr., J. C. Decius and P. C. Cross, *Molecular Vibrations* (New York: McGraw-Hill Book Co., 1955).

Transition-state theory rests on the assumption that the systems and motions which predominantly determine the rate of reaction, occur in a small enough region near the "equilibrium" configuration of the transition state, that Eq. (11-1) is a good approximation. Using the standard methods of classical mechanics one can transform to (mass-weighted) normal coordinates Q_i, for which:

$$V(Q_1, \ldots, Q_{3N-6}) = V^* + \frac{1}{2} \sum_{i=1}^{3N-6} \lambda_i Q_i^2 \tag{11-3}$$

$$T(\dot{Q}_1, \ldots, \dot{Q}_{3N-6}) = \frac{1}{2} \sum_{i=1}^{3N-6} \dot{Q}_i^2 \tag{11-4}$$

An activated complex is defined as a system whose configuration is sufficiently close to the "equilibrium" configuration that Eq. (11-1) is a good approximation. The only difference between an activated complex and a normal molecule is that whereas all the λ_i are positive for a stable molecule, one of the λ_1 is negative (corresponding to an imaginary frequency) in the activated complex. The reaction coordinate is defined as that normal coordinate Q_* which corresponds to the imaginary frequency $\nu_* = (\lambda_*/2\pi)^{1/2}$. In addition to these internal coordinates, which specify the relative configuration of the system, there are three coordinates for the rotational orientation of the complex, and three coordinates to locate the center of mass of the system in space. The vibrational-rotational interaction is assumed negligible for the small range of motions involved.

The quantum-mechanical description of the system is as follows. After separating out the translations and rotations (the potential energy being independent of these quantities), the use of Eqs. (11-3) and (11-4) gives the $3N-6$ dimensional Schroedinger equation for motion of the complex about the "equilibrium" configuration.

$$-\frac{\hbar^2}{2} \sum_{i=1}^{3N-6} \frac{\partial^2 \psi}{\partial Q_i^2} + \left\{ V^* + \frac{1}{2} \sum_{i=1}^{3N-6} \lambda_i Q_i^2 \right\} \psi = E\psi \tag{11-5}$$

By using a wave function which is a product of wave functions in the normal coordinates, and an energy which is a sum of energies in the normal coordinates, this equation is factored into $3N-6$ one-dimensional equations. For the $3N-7$ normal coordinates corresponding to real classical frequencies, equations of the following form are obtained:

$$-\frac{\hbar^2}{2} \frac{\partial^2 \psi_i}{\partial Q_i^2} + \frac{1}{2} \lambda_i Q_i^2 \psi_i = E_i \psi_i \tag{11-6}$$

The solutions of these equations give ordinary harmonic oscillator wave functions and energy levels. The normal coordinate corresponding to the reaction coordinate has the equation:

$$-\frac{\hbar^2}{2}\frac{\partial^2\psi_*}{\partial Q_*{}^2}+\left\{V^*+\frac{1}{2}\lambda_* Q_*{}^2\right\}\psi_* = E_*\psi_* \qquad (11\text{-}7)$$

corresponding to a one-dimensional barrier penetration problem. This coordinate is not bound, and represents the means for getting past the region of high energy in going from reactants to products. The motion of a system in a small region near the "equilibrium" configuration is defined by Eqs. (11-6) and (11-7) regardless of transition-state theory itself. Transition-state theory uses these results, together with two basic assumptions. The first assumption is that the rate of chemical reaction is equal to the number of systems in a small region near the "equilibrium" configuration, multiplied by the average frequency with which such systems pass through this region (in the direction toward products) in the reaction coordinate. The second is that equilibrium statistical mechanics can be used to determine the number of systems in the transition-state region. It is thereby assumed that the essential motions which determine the course of reaction occur in the immediate vicinity of the transition state, and that once a system in the transition state starts to fall apart in the direction of products, it continues to do so completely. The rate of reaction must be slow enough that there is an equilibrium distribution over the excited states of reactants, which is undepleted as activated complexes fall apart in reaction.

11.2   TUNNELING AND THE REACTION COORDINATE

In discussing quantum mechanical conditions in the rate theory, one must distinguish between the quantum mechanics of a small region at the top of the barrier as expressed in Eqs. (11-6) and (11-7), and the problem of nonseparability of coordinates which occurs when a larger region is required to describe the essential motions which determine whether reaction will occur. The former problem can be treated within the framework of transition-state theory, while the latter represents a breakdown of the first assumption listed in the previous paragraph. When the range of essential motions is small enough, motion in the reaction coordinate will be orthogonal to the other coordinates of the complex, and the motions can be treated as independent of one another. In this case, one calculates the rate at which systems pass through the transition state in the reaction coordinate, allowing all the other coordinates to vary over their full range of statistically allowed values. An activated complex with particular momentum and coordinate values in the reaction coordinate has a specific probability of reaction, independent of what values the other coordinates might have in any specific case. The problem of exactly what range of motions is essential depends on how quantum-mechanical the reaction coordinate is. The required range of a purely classical reaction coordinate (very heavy masses and potentials that vary slowly with coordinates) is merely an infinitesimal distance at the top of the barrier. The range of motion required to describe the passage through the transition state increases as the effective mass in the reaction coordinate becomes smaller or the potential varies more rapidly. As this range increases, Eqs. (11-6) and (11-7) become less valid to describe the full range of motions involved, and interactions with other coordinates become more of a problem.

A more easily visualized picture of this discussion can be seen from the following discussion. Suppose there is a one-dimensional potential hill $V(x)$, and a beam of particles is fired at the potential hill from $x = -\infty$ as shown in Fig. 11-1. Suppose we wish to calculate the rate at which systems pass through the barrier and go to $x = +\infty$. If the incoming particles have a distribution of energies as shown in Fig. 11-2, we are also interested in which of these particles react.

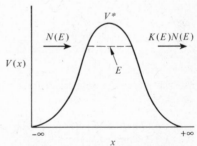

FIGURE 11-1 A beam of particles with energy distribution. $N(E)$ is incident on a one-dimensional potential hill $V(x)$. The probability of penetration is $K(E)$.

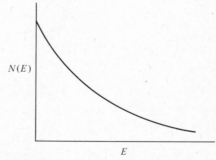

FIGURE 11-2 Distribution $N(E)$ of incident particles.

If the mass of the particles is sufficiently high, and the variation of the potential with distance sufficiently smooth, classical mechanics will be a good approximation. In this case, the exact shape of the potential barrier is unimportant. Only the maximum height V^* is of physical significance. The transmission probability per approach with energy E is given by:

$$K(E) = 1 \qquad E > V^*$$
$$K(E) = 0 \qquad E < V^* \tag{11-8}$$

The distribution of transmitted systems has the form given by the shaded area in Fig. 11-3.

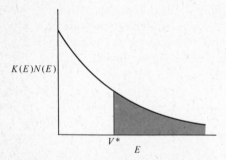

FIGURE 11-3 Distribution of transmitted particles for a classical transmission function $K(E)$. The shaded portion $(E > V^*)$ is transmitted, and the unshaded area $(E < V^*)$ is reflected.

For a system of light masses, quantum mechanics must be used to obtain the transmission function $K(E)$. As Eckart[3] has shown, this function has the form given in Fig. 11-4 for a potential of the form given in Fig. 11-1. Curves A, B, and C represent successively lighter particles, and/or sharper potentials. In the limit of a heavy particle and a broad barrier, the classical step function is obtained. The distribution of transmitted energies has the form given in Fig. 11-5.

As the system becomes more quantum mechanical, the quantum-mechanical tunneling occurs at lower energies. A detailed knowledge of $V(x)$ over a broader range of x is, therefore, required to obtain an accurate count of systems transmitted. For example, the shape of $V(x)$ is only needed near the top of the barrier in case A, whereas it is needed

---

[3]C. Eckart, *Phys. Rev.* **35**, 1303 (1930); D. Rapp, *Quantum Mechanics* (New York: Holt, Rinehart and Winston, 1971).

down to about half the barrier height for case C. In any application of activated complex theory, it is necessary to have a description of the situation which includes curves like Figs. 11-4 and 11-5 for the reaction coordinate, to determine what range of reaction coordinate is involved in tunneling. This must be compared to the range over which the reaction coordinate is separable [in the sense that Eqs. (11-6) and (11-7) are good approximations]. It is only when the main contribution toward $K(E) \cdot N(E)$ of energies E occur in the separable region, that activated complex theory is self consistent, let alone accurate. That is, the assumption that the rate can be obtained by counting the number of systems per unit of time, that are passing through the narrow region at the top of the pass, is always subject to the question of whether tunneling is important or not. However, the entire description of the activated complex as a quasi-molecule with a partition function is incorrect when tunneling becomes severe.

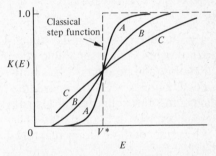

FIGURE 11-4 Transmission characteristics of a real potential barrier. Curves A, B, and C represent successively less classical systems (that is, lighter mass and/or sharper potential change with distance). For a heavy mass and a broad smooth barrier, the classical step function is approached.

It can be stated that in every case that the total number of systems transmitted in the one-dimensional hypothetical problem is greater in the quantum-mechanical case than in the classical case. That is, the areas under the curves of Fig. 11-5 are greater than the shaded area in Fig. 11-3. In the same way, the rate of reaction calculated with classical mechanics, as expressed in Eq. (11-8), is always less than the true rate.

Before going into the mathematical formulation of activated complex theory, we will point out the significance of the reaction coordinate as a separable normal mode of motion, in contrast to the picture given in Figs.

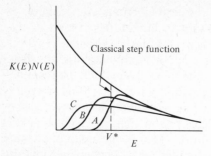

FIGURE 11-5 Energy distribution of transmitted particles for the transmission characteristics given in Fig. 11-4. As the system becomes less classical, greater penetration of the barrier at low energies, and greater reflection at high energies is produced. The areas under the curves increase from A to C.

(15) through (18) by Glasstone, Laidler, and Eyring.[4] The reaction coordinate is not a separable coordinate connecting reactants, activated complex, and products.   It is only separable in a small region at the activated complex.   Furthermore, the reaction coordinate is not the distance along the dotted line in Fig. (15) of Eyring *et al.* (see footnote 4) even for a small distance at the top of the barrier.   Assume for simplicity that there is a system of three atoms reacting in a straight line:

$$X\text{------}Y\text{------}Z$$
$$\quad r_1 \qquad r_2$$

and analogous to Fig. (15) of Eyring *et al.*, there is a potential-energy surface as shown in Fig. 11-6.   The potential energies along the lines *CFG* and *HFI* of Fig. 11-6 have the forms shown in Fig. 11-7.   Note that the distance along *CFG* is *not* the reaction coordinate as shown in Figs. (15) to (18) of Eyring *et al.*   Point *F* defines the "equilibrium" position of the transition state.   For some region *R* of the energy surface surrounding point *F*, Eq. (11-1) is a good approximation, and using Eqs. (11-3) and (11-4), normal coordinates can be found for the system.   The reaction coordinate is one of these normal coordinates, and is only separable inside region *R*.   It is not necessarily tangent to the dotted line in Fig. 11-6, although it turns out to be tangent for symmetrical complexes.

[4]S. Glasstone, K. Laidler and H. Eyring, *Theory of Rate Processes* (New York: McGraw-Hill Book Co., 1941).

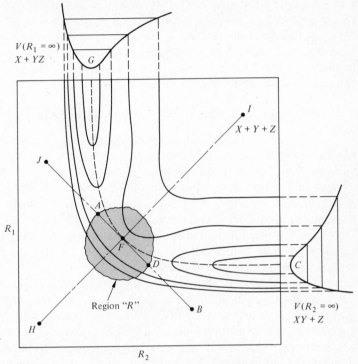

FIGURE 11-6 Energy surface for a three particle system $X + Y + Z$. The reactant and product valleys are at C and G, respectively. Point F is the top of the pass corresponding to the "equilibrium" configuration of the transition state. Line BFC defines motion along the reaction coordinate.

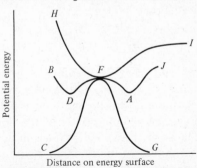

FIGURE 11-7 Sections of the energy surface shown in Fig. 11-6. Curve $BDFAC$ shows the "sidewall repulsions" which reduce tunneling along the reaction coordinate compared to what one would have guessed from the curve CFG.

Various isotopic substitutions for atoms X, Y, and Z will produce different reaction coordinates, even though the energy surface remains unchanged. The exact slope of the reaction coordinate line in any case is not obvious from Fig. 11-6 alone. It is necessary to find the quadratic approximation to the energy surface in region R and use this in a vibrational analysis of the system to obtain the reaction coordinate.

11.3 FORMULATION OF TRANSITION-STATE THEORY

In formulating activated complex theory, it is always assumed that the partition function for the transition state can be written as a product of partition functions for the reaction coordinate and the other $3N - 1$ degrees of freedom of the activated complex. This is only true in region R where Eqs. (11-6) and (11-7) are a good approximation, and the moments of inertia of the system are reasonably constant.

First the usual formulation of activated complex theory for a purely classical reaction coordinate will be derived. It will be seen later how quantum effects change the situation. The discussion given by Eyring et al.[5] will be followed. However, a subtle difference in viewpoints must be pointed out. Consider the second paragraph on p. 306 of Eyring. Eyring uses a separable reaction coordinate extending from the activated complex right down to reactants, so that his region R covers the entire energy surface; he makes use of an element of reaction coordinate dx in the "reactant valley." We, however, start with an element of reaction coordinate dx at the edge of region R, on the reactant side of the barrier, at point D in Fig. 11-6, for example.

The coordinate x is similar to Q^*, except that x is not mass-weighted. Equation (11-7) in x is

$$-\frac{\hbar^2}{2m_*}\frac{\partial^2\psi_*}{\partial x^2}+\{V^*+\tfrac{1}{2}f^*x^2\}\psi_* = E_*\psi_*$$

where f^* is the curvature along the reaction coordinate in Fig. 11-6 at point F, and m_* is the effective mass along the reaction coordinate. Eyring's dx was meant to be at point C, which is impossible if it is part of the reaction coordinate. Consider the section of the reaction coordinate shown in Fig. 11-8. Since the coordinates are separable in the region R, motion in the reaction coordinate does not affect (and is not affected by) motion in the other (orthogonal) coordinates. Thus, for purposes of calculating rate constants, we can deal with hypothetical activated complexes with specific coordinate and momentum in the reaction co-

---

[5]H. Eyring, J. Walter, and G. E. Kimball, *Quantum Chemistry* (New York: John Wiley & Sons, 1944).

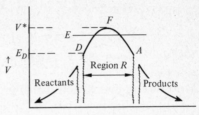

Section of reaction coordinate

FIGURE 11-8 Section of the reaction coordinate from Fig. 11-6. Outside of region R exists an abyss of interacting coordinates.

ordinate, but the other coordinates of the activated complexes are statistically distributed over the full range of allowed values. The classical partition function for an activated complex with reaction coordinate between x and $x + dx$, and conjugate momentum between p and $p + dp$, is

$$Q^{*\prime} \exp(-p^2/2m^*kT) \, dp/h \exp(-V(x)/kT) \, dx \qquad (11\text{-}8a)$$

in which $Q^{*\prime}$ is the partition function for the $3N - 1$ coordinates orthogonal to the reaction coordinate (3 translations, 3 rotations, and $3N - 7$ internal vibrations). If we choose the element of reaction coordinate at D, the concentration of activated complexes with momentum p in the direction toward the top of the barrier is

$$dC^*(p, x) = (Q^{*\prime}/Q_A Q_B \ldots)(C_A C_B \ldots) \exp(-p^2/2m^*kT) \, dp/h$$
$$\times \exp(-V_D/kT) \, dx \qquad (11\text{-}9)$$

for the reaction $A + B + \cdots \rightarrow$ products. The C's are concentrations and the Q's are partition functions per unit volume based on a zero of energy as the electronic energy of the equilibrium configuration. The energy levels determined by Eqs. (11-6) give the form of $Q^{*\prime}$ as a product of three translational functions, three rotational functions, and $3N - 7$ vibrational partition functions.

The frequency with which such systems approach the barrier is:

$$1/dt = (p/m)(1/dx) \qquad (11\text{-}10)$$

and the rate at which systems approach the barrier with momentum p in the reaction coordinate is $dC^*(p, x)/dt$. Because the energy V_D is measured relative to reactants electronic energy, the momentum p is based on point D. Thus if energies E are measured from reactants, we have $p = \{2m^*(E - V_D)\}^{1/2}$, where m^* is the "mass in the reaction coordinate." If the fraction $K(p)$ is transmitted through the barrier, the total rate of reaction is

$$\underline{k}C_A C_B \ldots = (C_A C_B \ldots)(Q^{*\prime}/Q_A Q_B \ldots) \exp{(-V_D/kT)}$$
$$\int_0^\infty \exp{(-p^2/2mkT)} K(p) p \, dp/(m^*h). \qquad (11\text{-}11)$$

The rate constant is then given by:

$$\underline{k} = (Q^{*\prime}/Q_A Q_B \ldots) \int_{V_D}^\infty \exp{(-E/kT)} K(E) \, dE/h \qquad (11\text{-}12)$$

Use in Eq. (11-12) of the classical transmission coefficient $K(E)$ from Eq. (11-8) leads to the classical rate constant:

$$(\underline{k})_{cl} = (kT/h)(Q^{*\prime}/Q_A Q_B \ldots) \exp{(-V^*/kT)}. \qquad (11\text{-}13)$$

Now turn to the case where the reaction coordinate must be treated as a quantum-mechanical degree of freedom. Equations (11-6) and (11-7) are still good approximations within region R. The problem is to obtain the proper boundary conditions for Eq. (11-7). In the classical limit an element dx at point D could be picked and the systems passing through could be counted. In the quantum-mechanical case there is Eq. (11-7) for the reaction coordinate inside region R, but the boundary conditions are not clear, since the motions outside this region can only be discussed in terms of $3N - 6$ interacting coordinates. A reasonable way to proceed, therefore, is to invent a hypothetical situation outside region R, which allows the required boundary conditions to be obtained. If the solution shows that the particular details outside region R are not important in determining the rate constant, then the choice of an arbitrary (though not highly unreasonable) situation outside R is justified.

In order to find the permeability of a potential barrier to a particle, a diagram is required like that in Fig. 11-1, in which the potential is specified for all values of the coordinate. If a wave train of unit amplitude is allowed to approach from one side, the square of the amplitude of the transmitted wave is the probability of penetration for the approach of a single particle. The potential in the reaction coordinate has the form as shown in Fig. 11-8. At the edge of region R there is an abyss of interacting coordinates. Therefore, the hypothetical situation is chosen, where the reaction coordinate may be extended indefinitely, without affecting the other $3N - 1$ degrees of freedom of the activated complex. In other words, Eq. (11-6) is assumed to hold for the other vibrational degrees of freedom, but Eq. (11-7) is replaced by the following:

$$-\frac{\hbar^2}{2} \frac{d^2 \psi_*}{dQ_*^2} + V(Q_*) \psi_* = E_* \psi_* \qquad (11\text{-}7a)$$

where $V(Q_*)$ has a form something like that in Fig. 11-1. Then there is an arrangement which coincides with Eqs. (11-6) and (11-7) inside region R, and which can be solved outside region R for a transmission coefficient. If the results show that only a negligible number of systems tunnel below energy V_D as shown in Fig. 11-8, when plotted on a graph like that of Fig. 11-5, then the results of this method should be reliable. It does not matter what the exact form of the potential below V_D is, if substantially no tunneling occurs for energies less than V_D. A possible pitfall of this method is that the hypothetical model chosen outside region R might conceivably fail in its ability to predict whether substantial tunneling occurs below V_D. This seems doubtful, however, because if one uses a potential $V(Q_*)$ which is an extension of the reaction coordinate on the energy surface beyond region R, this represents the type of barrier which must be surmounted, whether in one or more degrees of freedom. Thus, a reasonable hypothetical problem has been chosen which coincides with the real one in the region of interest. If the results show that the important motions occur in the region of interest, they can be used.

If the potential chosen has the general form shown in Fig. 11-1, then an element of "reaction coordinate" can be chosen on the flat part of the potential on the reactant side, and the number of systems passing can be counted by means of equations like Eq. (11-9) and (11-10) (except that V_D must now be taken as zero). This follows from the assumption that all along the entire hypothetical reaction coordinate, the degrees of freedom which contribute to $Q^{*\prime}$ remain unchanged. The use of the classical statistics instead of quantum statistics for the hypothetical reaction coordinate is justified because the two are identical in any region of phase space where the derivatives of the potential energy are zero. It is important to realize that an element of reaction coordinate in the "reactant valley" has *not* been chosen, but an element of a hypothetical coordinate has been chosen, which has the same energy as reactants, but which is completely different from reactants in structure and properties. The resulting analysis parallels the development of Eq. (11-9), and results in an equation like Eq. (11-12):

$$(\underline{k})_q = (Q^{*\prime}/Q_A Q_B \ldots) \int_0^\infty \exp\left(-E/kT\right) K(E)\, dE/h, \quad (11\text{-}14)$$

except the lower limit of integration is now zero. This development has been presented because Eq. (11-12) rests on a much more solid basis than Eq. (11-14), despite their similarity in form. Equation (11-14) has built into it the implicit statement that when the integrand is not negligible for energies below V_D, the theory is no longer reasonable. There is no necessity for this in Eq. (11-12), because the classical $K(E)$ must be used in it.

11.4   QUANTUM CORRECTION FOR THE REACTION COORDINATE

By dividing Eq. (11-14) by Eq. (11-13), we derive the general expression for the tunneling correction to activated complex theory for a non-classical reaction coordinate.

$$(\underline{k})_q/(\underline{k})_{cl} = e^{V^*/kT} \int_0^\infty \exp\left(-E/kT\right) K(E)\, dE/kT, \qquad (11\text{-}15)$$

where the expression is only reasonable when the integrand is negligible below V_D. The ratio of rate constants in Eq. (11-15) is always greater than unity.

This rather lengthy derivation has been presented here because there are so many instances in the literature of inadequate derivations. The derivations that write the partition function for the activated complex as $Q^{*'}$ $(kT/h\nu_*)$ are incorrect. The second term in this expression comes from treating the reaction coordinate as a weak (but bound) vibration. This factor is then multiplied by ν^*, the frequency of the "weak vibration," assuming the transition state to decompose on every vibration, to obtain the factor kT/h in Eq. (11-13). Although this derivation gives the right answer, it is physically absurd, and yet it is given in several textbooks.

In calculating rate constants from transition state theory, one usually calculates the classical rate constant from Eq. (11-13) and multiplies by a quantum correction factor for the reaction coordinate ("tunneling correction"). The tunneling corrections for the inverted parabolic barrier

$$V(Q) = V^* + \tfrac{1}{2}\lambda_* Q^2 \qquad (11\text{-}16)$$

and the Eckart barrier

$$V(A) = \frac{V_*}{\cosh^2\{|\lambda^*|Q/2V_*\}^{1/2}} \qquad (11\text{-}17)$$

have been worked out in detail. The transmission of the Eckart barrier to a wave train of energy E is[6]

$$K(\xi) = \frac{\cosh\left(2\alpha\xi^{1/2}\right) - 1}{\cosh\left(2\alpha\xi^{1/2}\right) + \cosh\left(4\alpha^2 - \pi^2\right)^{1/2}} \qquad (11\text{-}18)$$

where $\xi = E/V^*$, and $\alpha = (2\pi V^*/h|\nu_*|)$.
For $4\alpha^2 \gg \pi^2$, and for energies close to the top of the barrier ($\xi \cong 1$), this expression reduces to

$$K(\xi) = \{1 + \exp\left(\alpha[1-\xi]\right)\}^{-1} \qquad (11\text{-}19)$$

---

[6]C. Eckart, *Phys. Rev.* **35**, 1303 (1930).

which is the WKB solution for a truncated inverted parabola derived for example by Rapp.[7] These barriers are shown in Fig. 11-9. For the case where tunneling is not very severe, most of the systems that pass through the barrier have energies close to V^*. In this case, Eq. (11-19) can be used in Eq. (11-15) to obtain the tunneling correction. When $4\alpha^2 \gg 1$, and u^* is less than about 3–4, the resulting integral can be evaluated in closed form to obtain:

$$(\underline{k})_q / (\underline{k})_{cl} = \frac{u^*/2}{\sin(u^*/2)}, \qquad (11\text{-}20)$$

where $u^* = h|\nu^*|/kT$. For $u^* \ll 1$, this reduces to the Wigner correction, $1 + (u^*)^2/24$. The tunneling correction in the general case depends on the two parameters u^* and α. The result of using Eq. (11-18) in Eq. (11-15) is shown graphically in Fig. 11-10. The dependence on α disappears at low u^*.

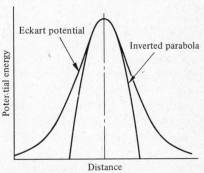

FIGURE 11-9 Eckart and inverted truncated potential barriers.

It is important to realize that all these results are based on the potential shown in Fig. 11-9, and that u^* is a constant which specifies the entire potential at any temperature. It is the entire potential which is of importance. In a case where the potential cannot be approximated by an Eckart potential, the results will be different. In using the Eckart potential for any specific problem, ν_* should be adjusted to fit the potential over a rather wide region, rather than to use the small-vibration value which corresponds to the curvature at the top of the barrier.

The transmission curves similar to Fig. 11-5 are of great interest in these calculations, since they essentially determine the range of validity of this approach. Johnston and Rapp[8] have given such curves for various

---

[7]D. Rapp, *Quantum Mechanics* (New York: Holt, Rinehart and Winston, 1971).
[8]H. S. Johnston and D. Rapp, *J. Amer. Chem. Soc.* **83**, 1 (1960).

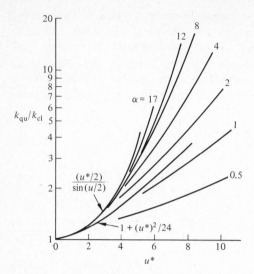

FIGURE 11-10 Tunneling correction factors for the Eckart barriers having various α. The parabolic barrier correction and the Wigner correction are shown for comparison.

parameters. The nature of the curves show that u^* is predominant in determining the shape of the transmission curve, which determines what fraction of systems pass over (rather than through) the barrier. The parameter α determines the "spread" of energies that contribute to reaction. A high α and a low (u^*) produce the most acceptable conditions: a narrow spread of energies peaked high in energy. If the energy V_D can be estimated in any case, it can be drawn as a vertical line in these figures. The validity of transition-state theory depends on the area to the left of this line being small compared to the total area under the appropriate transmission curve. In the parabolic case it can be shown that when $(u^*) = \pi$, exactly half of the transmitted systems tunnel through the barrier. Higher (u^*) result in a greater fraction tunneling, and vice versa. For very low (u^*), the transmission curve approaches the classical step function shown in Fig. 11-3.

Therefore, it is evident that a criterion for the applicability of activated complex theory is that the fraction of transmitted systems having energy less than V_D should be small. It is very possible in some cases to have a sharp high barrier with deep tunneling, which satisfies this requirement better than a lower, flatter barrier which has a lower value of α. In the case of the high barrier, a greater fraction of systems tunnel through the barrier. However, the Boltzmann factor, $e^{-E/kT}$, which enters into the distribution function, assures that the "spread" of transmitted energies is narrow, so that the systems which do tunnel, do so in a narrow range near the top of the barrier.

In actual practice tunneling effects will not be as severe as Fig. 11-10 seems to indicate. The actual potential energy along the hypothetical extended reaction coordinate $AFDB$ in Fig. 11-6 has the form shown in Fig. 11-7, causing a cutoff, below which tunneling cannot occur. This reduces the severity of the tunneling correction at low temperature. Correction for this type of effect is not possible within the framework of ordinary transition-state theory, and a more elaborate treatment is called for.[9]

11.5 ISOTOPE EFFECTS ON REACTION RATES

Before discussing the effect of isotopic substitution on chemical reaction rates, it is necessary to derive an expression analogous to Eq. (11-9), but in terms of the partition functions in local bond properties (see Chapter 10). According to Eqs. (10-3), (10-6), (10-32) [or (10-33)], and (10-35), the partition function for a nonlinear molecule can be written

$$Q = \left\{ \frac{g_0^{\text{el}}}{\sigma} \prod_{\alpha=1}^{N} \left(\frac{2\pi m_\alpha kT}{h^2} \right)^{3/2} \prod_{\alpha=1}^{N} J_\alpha \frac{(2\pi kT)^{(3N-6)/2}}{|F|^{1/2}} \right\} \prod_{j=1}^{3N-6} \Gamma_j \qquad (11\text{-}21)$$

where N is the number of atoms in the molecule. It is not immediately obvious how to apply this procedure to a transition state. According to our model, a transition state is to be treated as an ordinary molecule in every respect, except that one of the normal modes has an imaginary frequency [negative value of λ^* in Eq. (11-16)]. This results from a force constant matrix which contains off-diagonal terms. Now examine Eq. (11-8a) in more detail. The partition function per unit volume in $3N-1$ coordinates is

$$Q^{*\prime} = g_0^{\text{el}} \left(\frac{Q^*_{\text{tr}}}{V} \right) (Q^*_{\text{rot}}) (Q^{*\prime}_{\text{vib}}) \qquad (11\text{-}22)$$

The vibrational partition function is a product over $3N-7$ ($3N-6$ for linear molecules) vibrations of partition functions for the normal modes, excluding the reaction coordinate:

$$Q^{*\prime}_{\text{vib}} = \prod_{j=1}^{3N-7} (Q^*_{\text{vib}})_j = \prod_{j=1}^{3N-7} u_j^{-1} \prod_{j=1}^{3N-7} \Gamma_j \qquad (11\text{-}23)$$

---

[9]See, for example, J. D. Russell and J. C. Light, *J. Chem. Phys.* **54**, 4881 (1971), and references therein.

where $u_j^{-1} = kT/h\nu_j$ is the jth classical vibrational partition function and Γ_j is the jth quantum correction factor. The partition function for the reaction coordinate in Eq. (11-8a) is denoted as

$$Q\ddagger = \exp\left(-p^2/2mkT\right) dp/h \exp\left(-V(x)/kT\right) dx \qquad (11\text{-}24)$$

Thus, the complete partition function for the transition state is

$$Q^* = Q\ddagger Q^{*\prime} = Q\ddagger g_0{}^{\mathrm{el}}(Q^*{}_{\mathrm{tr}}/V)Q^*{}_{\mathrm{rot}} \prod_{j=1}^{3N-7} u_j^{-1} \prod_{j=1}^{3N-7} \Gamma_j \qquad (11\text{-}25)$$

This can now be multiplied and divided by $u_* = h|\nu_*|/kT$ for the reaction coordinate, and the following can be obtained:

$$Q^* = Q\ddagger u_*\left\{g_0{}^{\mathrm{el}}(Q^*{}_{\mathrm{tr}}/V)Q^*{}_{\mathrm{rot}} \prod_{j=1}^{3N-6} u_j^{-1}\right\} \prod_{j=1}^{3N-7} \Gamma_j \qquad (11\text{-}26)$$

The term in braces in this equation has the same form as $(g_0{}^{\mathrm{el}}/\sigma)$ times the classical partition function of a hypothetical bound molecule with the same structure as the transition state. But, by a purely geometrical transformation, this can be transformed to a function of the form in the braces of Eq. (11-21). Thus, the partition function for the transition state is

$$Q^* = Q\ddagger u_* \frac{g_0{}^{\mathrm{el}}}{\sigma} \prod_{\alpha=1}^{N}\left(\frac{2\pi m_\alpha kT}{h^2}\right)^{3/2} \prod_{\alpha=1}^{N} J_\alpha \frac{(2\pi kT)^{(3N-6)/2}}{|F|^{1/2}} \prod_{j=1}^{3N-7} \Gamma_j \qquad (11\text{-}27)$$

In taking the ratio $(Q^*/Q_A Q_B \ldots)$ as in Eq. (11-9), the factors $(2\pi m_\alpha kT/h^2)^{3/2}$ will cancel. The factors J_α depend only on geometry, and the force constants depend only on structure, not mass. Therefore, the ratio

$$\prod_{\alpha=1}^{N} J_\alpha \frac{(2\pi kT)^{3N-6}}{|F|^{1/2}} = S \qquad (11\text{-}28)$$

will be denoted for any substance or transition state, and this is the same for isotopically substituted molecules of the same species. Thus,

$$(Q^*/Q_A Q_B \ldots) = \frac{Q\ddagger u_*(g_0/\sigma)^*}{(g_0/\sigma)_A(g_0/\sigma)_B \ldots} \frac{S^*}{S_A S_B \ldots} \frac{\prod\limits^{3N-7} \Gamma_j{}^*}{\prod\limits^{3N-6} \Gamma_j{}^A \prod\limits^{3N-6} \Gamma_j{}^B \ldots}$$

Note that $\prod\limits^{3N-7} \Gamma_j{}^*$ for the transition state is taken only over the real vibrations. With the ratio of partition functions expressed in this form, the calculation of the rate constant proceeds exactly as in Eqs. (11-9) to

(11-15). Therefore,

$$\underline{k}_{cl} = \nu^* \frac{(g_0/\sigma)^*}{(g_0/\sigma)_A (g_0/\sigma)_B \dots} \frac{S^*}{S_A S_B \dots} \frac{\overset{3N-7}{\Pi} \Gamma_j^*}{\overset{3N-6}{\Pi} \Gamma_j^A \overset{3N-6}{\Pi} \Gamma_j^B \dots} e^{-V^*/kT}$$

instead of Eq. (11-13), and Eq. (11-15) still holds for $\underline{k}_q/\underline{k}_{cl}$. Now consider two reactions between chemical species which are the same except for isotopic substitution:

$$A + B + \dots \rightarrow (\text{transition state}) \rightarrow \text{products}$$

$$A' + B' + \dots \rightarrow (\text{transition state})' \rightarrow \text{products}'$$

The ratio of rate constants with classical reaction coordinates is now taken, assuming that V^* and S do not change for a molecule or a transition state upon isotopic substitution. The following is found:

$$\frac{k_{cl}}{k'_{cl}} = (\text{TIF}) \frac{\overset{3N-7}{\Pi} \Gamma_j^* \overset{3N-6}{\Pi} \Gamma_j^{A'} \overset{3N-6}{\Pi} \Gamma_j^{B'} \dots}{\overset{3N-7}{\Pi} \Gamma_j^{*'} \overset{3N-6}{\Pi} \Gamma_j^A \overset{3N-6}{\Pi} \Gamma_j^B \dots} \tag{11-31}$$

where the temperature independent factor is

$$\text{TIF} = \frac{\nu^*}{\nu^{*'}} \frac{(g_0/\sigma)^*}{(g_0/\sigma)^{*'}} \frac{(g_0/\sigma)_A'}{(g_0/\sigma)_A} \frac{(g_0/\sigma)_B'}{(g_0/\sigma)_B} \dots \tag{11-32}$$

and $\underline{k}/\underline{k}'$ goes to this limit at high T. If the tunneling correction for the reaction coordinate is denoted as Γ_{RC}^*, then in general,

$$\frac{\underline{k}}{\underline{k}'} = \frac{\underline{k}_{qu}}{\underline{k}'_{qu}} = (\text{TIF}) \overset{3N-6}{\Pi} \left(\frac{\Gamma_j^*}{\Gamma_j^{*'}}\right) \overset{3N-6}{\Pi} \left(\frac{\Gamma_j^{A'}}{\Gamma_j^A}\right) \overset{3N-6}{\Pi} \left(\frac{\Gamma_j^{B'}}{\Gamma_j^B}\right) \dots \tag{11-33}$$

where Γ_{RC}^* is the $(3N-6)$th Γ_j^*. It is clear that $\underline{k}/\underline{k}'$ goes to TIF as $T \rightarrow \infty$ since all the quantum correction factors tend toward 1 as $T \rightarrow \infty$. Equation (11-33) can be rewritten in the form of Eq. (10-51) if TIF and γ are defined slightly differently.

Problems

1. Consider the reaction

$$H + D_2 \rightarrow (H\text{---}D\text{---}D)^* \rightarrow HD + D$$

Assume the transition state is linear with the following properties:

$$r_{HD} = r_{DD} = 0.92 \text{ Å}$$

$$\nu_{ss} = 2800 \text{ cm}^{-1}, \nu_* = 1100i$$

$$\nu_{bend} = 950 \text{ cm}^{-1}$$

Calculate the rate constant according to transition-state theory at $T = 1000°K$ assuming the activation energy is 9 Kcal/mole, and using Eq. (11-20) for the tunneling correction. Compare with Fig. 10-1 of Johnston, *Gas Phase Reaction Rate Theory* (New York: Ronald Press, 1966).

Answer:

$$\underline{k} = \frac{kT}{h}\left(\frac{Q^{*\prime}}{Q_H Q_{D_2}}\right) \exp\left(-V^*/kT\right)(k_q/k_{c1})$$

$$Q_H = \left(\frac{2\pi m_H kT}{h^2}\right)^{3/2} = 5.91 \times 10^{23} \text{ cm}^{-3}$$

$$Q_{D_2} = \left(\frac{2\pi m_{D_2} kT}{h^2}\right)^{3/2} \left(\frac{T}{2\theta_{rot}}\right) \frac{e^{-\theta_v/2kT}}{1 - e^{-\theta_v/kT}}$$

$$Q_{D_2} = (4.73 \times 10^{24})\left(\frac{1000}{2 \times 14.4}\right)\left(\frac{e^{-2.23}}{1 - e^{-4.47}}\right)$$

$$Q_{D_2} = 4.16 \times 10^{25} \text{ cm}^{-3}$$

Note: θ_{rot} for D_2 is 1/2 of the value for H_2, and θ_v for D_2 is $1/\sqrt{2}$ of the value for H_2.

$$Q^{*\prime} = \left(\frac{2\pi MkT}{h^2}\right)^{3/2}\left(\frac{T}{\theta_{rot}}\right) Q_{vib_{ss}}(Q_{vib_{bend}})^2$$

$$M = m_H + m_D + m_D$$

$$\theta_{rot} = \hbar^2/2Ik$$

To calculate I, we first locate the center of mass:

$$2(0.93 - x) = 2x + 1(0.93 + x)$$

$$x = \frac{0.93}{5} = 0.186 \text{ Å}$$

$$I = 2(0.744)^2 + 2(0.186)^2 + 1(1.116)^2 = 3.08 \text{ amu-Å}^2$$

$$\theta_{rot} = 2.58°K$$

$$\theta_{ss} = 4030°K, \qquad \theta_{bend} = 1370°K$$

$$Q^{*\prime} = (6.62 \times 10^{24})\left(\frac{1000}{2.58}\right)\left(\frac{3^{-2.02}}{1 - e^{-4.03}}\right)\left(\frac{e^{-0.69}}{1 - e^{-1.37}}\right)^2$$

$$Q^{*\prime} = 1.89 \times 10^{26} \text{ cm}^{-3}$$

$$e^{-V^*/kT} = \exp\left(-9 \times 6.95 \times 10^{-14}/1.38 \times 10^{-16} \times 1000\right) = 0.0109$$

From Eq. (11-20), $k_q/k_{cl} \cong 1.12$ for $u^* = 1.58$.
Thus,

$$k = (2.08 \times 10^{13} \sec^{-1}) \left[\frac{1.89 \times 10^{26} \text{ cm}^{-3}}{5.91 \times 10^{23} \text{ cm}^{-3} \times 4.16 \times 10^{25} \text{ cm}^{-3}} \right] (0.0109)(1.12)$$

$k = 1.0 \times 10^{12}$ cm³/molecule-sec

2. Consider the reactions

(1) $H + H_2(o,p) \rightarrow H_2(p,o) + H$
(2) $D + D_2(o,p) \rightarrow D_2(p,o) + D$
(3) $D + H_2 \rightarrow DH + H$
(4) $H + D_2 \rightarrow HD + D$
(5) $D + DH \rightarrow D_2 + H$
(6) $H + HD \rightarrow H_2 + D$

In the first two reactions, an ortho (or para) molecule is converted to para (or ortho) by atom interchange. The product molecule from an exchange reaction probably has no "memory" of the initial rotational state of the hydrogen reactant molecule, and so the probability of the product molecule being ortho or para is probably 50%. Since only half of the atom interchange events in reactions 1 and 2 change the rotational state of H_2, these reactions are intrinsically slower by a factor of two than reactions 3 and 4. This is equivalent to the fact that the symmetry number of the transition state is 2.
The problem is:

(a) Calculate TIF for the ratios k_1/k_n for $n = 2, \ldots, 6$.
(b) Calculate $k_3 k_4/k_1(k_3 + k_4)$ in the limit of $T \rightarrow \infty$.
(c) Prepare plots of $\log(k_1/k_n)$ vs $1/T$ for $n = 2, \ldots, 6$, and also $\log[k_3 k_4/(k_3 + k_4)k_1]$ vs $1/T$.
(d) Compare with Figs. 10-16 and 10-17 of Johnston's book.
Use the following properties[10]

(linear symmetrical transition state)
Eckart tunneling correction (Fig. 11-10) with $\alpha \cong 12$.

Approximate frequencies of transition state (cm⁻¹)

Reaction	ν_{ss}	ν_*	ν_{bend}	Reactant frequency (cm⁻¹) ν
1	3250	1600i	1250	4400
2	2300	1100i	900	3100
3	2800	1500i	1200	4400
4	2800	1100i	950	3100
5	2800	1100i	950	3600
6	2800	1500i	1200	3600

[10]D. Rapp, Ph.D. Thesis, University of California (Berkeley), January, 1960.

Answer:

(a)

Reaction	$[(g_0/\sigma)^*/(g_0/\sigma)_{\text{RCTS}}]$	ν_*	$\text{TIF} = (\underline{k}_1/\underline{k}_n)_{T=\infty}$
1	1	$1600i$	1
2	1	$1100i$	1.41
3	2	$1500i$	0.53
4	2	$1100i$	0.70
5	1	$1100i$	1.41
6	1	$1500i$	1.07

(b) At $T = \infty$, $\dfrac{\underline{k}_3\underline{k}_4}{\underline{k}_1(\underline{k}_3+\underline{k}_4)} = \dfrac{1}{(0.53)(0.70)\left(\dfrac{1}{0.53}+\dfrac{1}{0.70}\right)} = 0.81$

(c) $\dfrac{\underline{k}_1}{\underline{k}_n} = (\text{TIF})_n \dfrac{[\Gamma_{\text{ss}}\Gamma^2_{\text{bend}}]_1}{[\Gamma_{\text{ss}}\Gamma^2_{\text{bend}}]_n}\dfrac{(\underline{k}_q/\underline{k}_{\text{cl}})_1}{(\underline{k}_q/\underline{k}_{\text{cl}})_n}$

See Fig. 11-11.

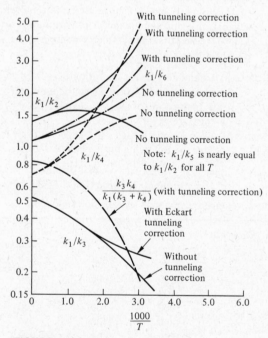

FIGURE 11-11

chapter 12

THE LIQUID STATE

12.1 PHASE DIAGRAMS

It is a matter of experimental knowledge that the phase diagrams for many substances have the general form shown in Fig. 12-1. There is a mathematical relationship between the variables p, V, and T at equilibrium, and if two of these are specified, the third becomes determined. It

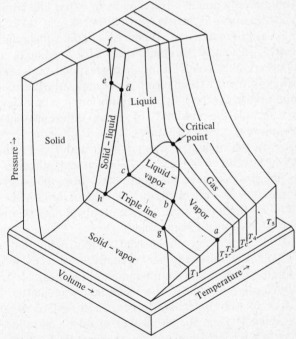

FIGURE 12-1 Surface representing the equation of state of a substance obtained by plotting p vs V and T. From F. W. Sears, *An Introduction to Thermodynamics, the Kinetic Theory of Gases, and Statistical Mechanics*, second edition, 1953, Addison-Wesley, Reading, Mass.

is perhaps easiest to visualize the diagram as a surface in which p is plotted vs V and T in three-dimensional space. Cuts through the surface with T fixed give *isotherms* of p vs V. At very high T, such as T_5, the substance is essentially an ideal gas and the isotherms are $pV = $ const. At lower temperatures such as T_4, substantial deviations from ideality occur in the gas, and the isotherms become irregularly shaped. Finally, at T_c, the *critical temperature*, the critical isotherm is reached, which separates the gas and liquid phases at high P. A gas cannot be liquified if $T > T_c$. For temperatures like T_2 and T_3, the phase diagram can be separated into several regions. Consider the isotherm $T = T_2$. At very low pressure (high volume) the substance exists as a vapor in the region along the line ab. In the region bc, liquid and vapor exist in equilibrium. If a fixed amount of substance is placed in a cylinder with a movable piston under conditions of point a, T is held constant at T_2, and the piston is steadily compressed, this corresponds to motion along the line $abcdef$. At b, liquid will begin to appear, and between b and c the ratio of amounts of liquid to gas will steadily increase. Finally, at c, the cylinder will contain only liquid. In going from c to d, the liquid will be compressed, and the curve is nearly vertical because it takes a very large pressure increase to reduce the volume of a liquid substantially. At point d, solid begins to appear in equilibrium with the liquid, and at e, the cylinder contains only solid. Line ef corresponds to compression of the solid. The *triple line* extends from g to h and is the line along which solid, liquid, and vapor can co-exist in equilibrium. The diagram in Fig. 12-1 does not include the possibility of several different solid phases corresponding to different crystal structures.

In the region between $T = T_c$ and $T = T_5$, the theory of imperfect gases developed in Chapter 9 should be applicable. This theory leads to isotherms that have the qualitative form shown in Fig. 12-2, which is a projection of part of the p-V-T surface on a plane of constant T. For $T \leqq T_c$, the liquid state can be involved, and this is beyond the scope of the discussion of Chapter 9.

It should be emphasized at this point that although Fig. 12-1 is actually found to be typical of many substances, it is not obvious that this would be the case from fundamental reasons. If a hypothetical scientist with no knowledge of the usual states of matter were asked what the effect of intermolecular forces on the state of a system might be, a reasonable guess for an answer might be as follows: "At high temperature, the forces will play only a minor role due to the large average kinetic energy of the molecules, and the gas will be essentially ideal. As the temperature is reduced, the attractive forces play a more important role, and clusters of molecules tend to form. Finally, in the limit of very low T, the substance will solidify into a lattice with the molecules oscillating

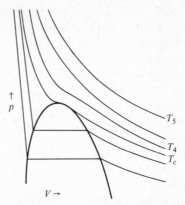

FIGURE 12-2 *Isotherms* of p vs V for fixed T. The critical isotherm is for $T = T_c$. For $T < T_c$, the intersections of the isotherms with the inverted bowl curve give the volumes of liquid and gas in equilibrium with one another in the two phase region.

about most stable distances between one another." In this description, there is no mention of the liquid state, nor of sudden phase transitions or temperatures where more than one phase can be in equilibrium. Rather, the conjecture is made that there is a continuous variation in the spatial distribution of molecules in space from a perfect gas at high T to a solid at low T. In fact, of course, this conjecture is wrong. The entity we call the liquid state does exist, discontinuities and phase transitions do take place, and the p-V-T diagram separates into regions of clearly recognized phase. One of the goals of statistical mechanics is to understand how this behavior follows from the known interactions between molecules, or more realistically, to conclude what form the unknown potentials must have in order to lead to this behavior.

12.2 THE CRITICAL REGION AND THE PRINCIPLE OF CORRESPONDING STATES

An examination of Figs. 12-1 and 12-2 shows that the critical point is defined by the requirements

$$\left(\frac{\partial p}{\partial V}\right)_T = 0$$

$$\left(\frac{\partial^2 p}{\partial V^2}\right)_T = 0 \qquad (12\text{-}1)$$

Since the critical region involves gas and liquid phases in equilibrium, the treatment given in Chapter 9 for nonideal gases would be expected to be highly approximate in this region.   Nevertheless, if it is assumed that the equation of state obtained from the procedures of Chapter 9 can be used even in the critical region, a very useful result is obtained.

In Section 9.5, the principle of corresponding states for nonideal gases was discussed.   It was found that if the intermolecular potential between two molecules is of the form in Eq. (9-113), and the total interaction potential for the substance is a sum of pairwise-additive potentials as in Eq. (9-112), then a universal relationship relates the dimensionless variables p^*, V^*, and T^*.   These variables were defined as

$$T^* = kT/\epsilon, \qquad p^* = p\sigma^3/\epsilon, \qquad V^* = V/\sigma^3 \tag{12-2}$$

where ϵ and σ are the energy and diameter parameters of the intermolecular potential:

$$u(r_{ij}) = \epsilon\phi(r_{ij}/\sigma) \tag{12-3}$$

The p^*-V^*-T^* surface corresponding to any potential $\phi(r_{ij}/\sigma)$ will usually be of the general form shown in Fig. 12-1.   Since the p^*-V^*-T^* surface is universal for all substances that behave as in Eq. (12-3), the critical point on such a surface is the same for all substances.   Let the coordinates of the critical point be p_c^*, V_c^*, and T_c^*, which are universal constants for substances which interact as in Eq. (12-3).   Then the actual values of the critical constants of any substance are

$$T_c = \epsilon T_c^*/k, \qquad p_c = \epsilon p_c^*/\sigma^3, \qquad V_c = \sigma^3 V_c^* \tag{12-4}$$

From Eq. (12-2), it follows that for any arbitrary point on the p^*-V^*-T^* surface,

$$T^* = TT_c^*/T_c, \qquad p^* = pp_c^*/p_c, \qquad V^* = VV_c^*/V_c \tag{12-5}$$

Since T_c^*, p_c^*, and V_c^* are universal constants, we define *reduced variables* of state as

$$T_r = T/T_c = T^*/T_c^*, \qquad p_r = p/p_c = p^*/p_c^* \tag{12-6}$$
$$V_r = V/V_c = V^*/V_c^*$$

It is clear that all substances that obey Eq. (12-3) will have the same p_r-T_r-V_r surface.   Hence, the more usual statement of the principle of corresponding states can be made, namely, that a universal relation exists between p_r, T_r, and V_r for all substances which interact as in Eq. (12-3). To the extent that Eq. (12-3) is a good model for the interactions in a series of molecules, the substances should have the same *reduced equation of state*.   A comparison of these parameters for a series of nearly-spherical nonpolar molecules is given in Table 12-1.   It can be seen

TABLE 12.1 Critical constants for some almost spherical nonpolar molecules reduced by means of Lennard-Jones (6–12) force constants[a]

Gas	$T_c(°K)$	$\tilde{V}_c(cm^3/mole)$	p_c (atm)	$T_c{}^*$	$V_c{}^*$	$p_c{}^*$	$p_c\tilde{V}_c/RT_c$
He	5.3	57.8	2.26	0.52	5.75	0.027	0.300
H_2	33.3	65.0	12.8	0.90	4.30	0.064	0.304
Ne	44.5	41.7	25.9	1.25	3.33	0.111	0.296
A	151	75.2	48	1.26	3.16	0.116	0.291
Xe	289.81	120.2	57.89	1.31	2.90	0.132	0.293
N_2	126.1	90.1	33.5	1.33	2.96	0.131	0.292
O_2	154.4	74.4	49.7	1.31	2.69	0.142	0.292
CH_4	190.7	99.0	45.8	1.29	2.96	0.126	0.290

[a]J. O. Hirschfelder, C. F. Curtiss, and R. L. Bird, *Molecular Theory of Gases and Liquids*, copyright© 1957. By permission of John Wiley & Sons, Inc.

that except for He and H_2 which have strong deviations due to quantum effects, the values of p_c^*, T_c^*, and V_c^* tend to be relatively constant, and p_cV_c/kT_c varies only slightly.

It would be interesting to evaluate the critical constants for substances obeying various force laws subject to Eq. (12-3). This would require the complete equation of state including all virial coefficients, for use in Eqs. (12-1). Unfortunately, this is not practical. One simple case is the Van der Waals gas, for which the approximate equation of state was shown to be

$$\left(p + \frac{\alpha N^2}{V^2}\right)(V - \beta) = NkT \tag{12-7}$$

in Chapter 9. This equation of state is based on a simple model of the potential, includes only the second virial coefficient, and is based on the assumptions $V \gg \beta$ and $p \gg \alpha N^2/V^2$. When it is used in Eqs. (12-1), two equations in p_c, V_c, and T_c are generated. These equations, together with Eq. (12-7), can be solved to yield

$$V_c = 3\beta$$

$$T_c = \frac{8}{27}\left(\frac{\alpha}{pk}\right) \tag{12-8}$$

$$p_c = \frac{1}{27}\left(\frac{\alpha}{\beta^2}\right)$$

and[1]

$$\frac{p_c \tilde{V}_c}{R T_c} = \frac{p_c V_c}{k T_c} = 3/8 = 0.375 \tag{12-9}$$

It can be seen from Table 12.1 that $p_c V_c / k T_c$ tends closer to 0.29 than to 0.375, so the equation of state Eq. (12-7) is not very adequate to describe the critical state. This is not surprising considering that liquid and gas are essentially indistinguishable in the critical state.

It is rather remarkable that the Van der Waals equation of state, which is based on a dilute-gas model, shows critical point behavior and can be qualitatively interpreted reasonably even in the liquid region. The isotherms for $T > T_c$ have the form shown in Fig. 12-3, which are of the proper form when compared with Fig. 12-2. For $T < T_c$, the isotherms have double loops in the region where liquid and vapor should coexist. The left side of an isotherm rises very rapidly compared to the right side, and this rise is qualitatively in agreement with the actual situation. Although the Van der Waals equation is not based on a model that

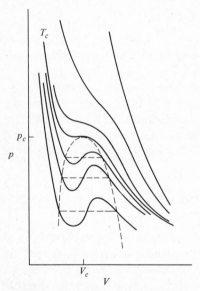

FIGURE 12-3 Isotherms predicted by the Van der Waals equation of state. The inverted bowl is obtained by drawing horizontal (dotted) lines through the region where "looping" occurs, so as to equalize the areas of the positive and negative loops.

---

[1] We use the notation: V is the volume per molecule and $\tilde{V}$ is the volume per mole.

specifically deals with the liquid phase, it does lead to two regions of vastly differing compressibility in the isotherms for $T < T_c$. This comes about mathematically because for $T < T_c$ when V is not much greater than β, it takes a rather large change in p to keep Eq. (12-7) balanced for a small change in V.   For $V \gg \beta$, the isotherms are more characteristic of a gas.   In the region of coexistence of liquid and gas, the Van der Waals equation gives a looped isotherm which is physically nonsensical. However, it is intriguing to note that the region where the looping occurs is shaped the same way as the liquid-vapor region in Fig. 12-2.

12.3 DESCRIPTION OF THE LIQUID STATE

For more detailed discussions of the liquid state, the reader is urged to read the following:

J. O. Hirschfelder, C. F. Curtiss and R. L. Bird, *Molecular Theory of Gases and Liquids* (New York: John Wiley & Son, 1954).

T. L. Hill, *Introduction to Statistical Mechanics* (Palo Alto, California: Addison-Wesley, 1960), chaps. 16, 17.

H. Eyring, et al., *Statistical Mechanics and Dynamics* (New York: John Wiley & Son, 1964), chap. 12.

J. M. Richardson and S. R. Brinkley, Jr., *Thermodynamics and Physics of Matter*, ed. Rossini (Princeton, New Jersey: Princeton University Press, 1965), section F.

The liquid state usually consists of a collection of molecules in relatively close proximity, where the interactions between neighboring molecules are not unlike the forces between molecules in a crystal lattice. The average distances between molecules in a liquid are not much larger than for the same molecules in solid form,[2] and there is considerable ordering in the liquid.   The main differences between a solid and a liquid are that the packing in a liquid is imperfect and irregular, leaving vacant volumes or "holes" throughout the volume where the ordering changes, enabling particles to slide past one another and change relative positions in the volume.   If one were to conceptually take a crystal lattice, crumble it up into parts consisting of from one to many molecules, and insert "holes" in about 10% of the internal volume, one would have a liquid. One of the difficult aspects of formulating a statistics for liquids is that the molecules are not fixed to recognizable lattice sites and in this respect behave like gas molecules, whereas the actual intermolecular forces are much more like the forces in a solid.

---

[2]In H_2O the average distance is actually less in the liquid state.

One of the best descriptions of the liquid state, is in the opening paragraphs of the chapter by Richardson and Brinkley, from which we quote:

> In a liquid the molecules are in constant interaction with one another, a given molecule being in more or less intimate association with a rather vaguely defined shell of nearest neighbors. However, the thermal motion is sufficiently violent to prohibit the intermolecular attractions from binding the molecules into a rigid, ordered system. The degree of order in a liquid is rather hard to describe in clear cut qualitative terms. However, it can be said that a liquid is sufficiently ordered to manifest rather small fluctuations in the number of molecules neighboring a given molecule, but otherwise the system is quite disordered, there being no discernible pattern in the arrangement. In a solid the thermal motion is sufficiently weak for the system to closely approach configurations of minimum potential energy. As far as we know, these configurations always correspond to a periodic structure. Thus a solid occurs in a crystalline form with a high degree of order extending over long distances. As an exception to this statement one immediately seizes upon the case of noncrystalline solids such as glass. It seems to be generally true that such substances are not in true thermodynamic equilibrium. They have become frozen in a metastable range of configurations from which the configurations of minimum potential energy are not readily accessible. These substances must be regarded as supercooled liquids.
>
> From the standpoint of a theoretician working in statistical mechanics, a liquid could be most fruitfully defined as a nonperiodic state of matter whose density is too large to allow a rapid convergence of the virial expansion, i.e., too large to allow an imperfect gas description.
>
> It is not possible to delimit the range of the liquid state by means of phase transitions. It is always possible to differentiate liquids from solids this way since it is not possible to go from the liquid to the solid state without suffering a phase transition. However, it is possible to go continuously from the liquid to the gaseous state by increasing the pressure and temperature above the critical point.

> *Thermodynamics and Physics of Matter*, ed. F. D. Rossini, Vol. I in *The High Speed Aerodynamics and Jet Propulsion Series* (copyright 1955 by Princeton University Press), from section F, "Properties of Liquids, and Liquid Solutions," by John M. Richardson and Stuart R. Brinkley, Jr. Reprinted by permission of Princeton University Press and Oxford University Press, London.

The theoretical discussions of the statistics of the liquid state tend to divide into two approaches. In the more rigorous and general approach, calculations are made from first principles of the state of a liquid based on an assumed form of the intermolecular potential. Although of great interest in the understanding of the liquid state, this approach has not yielded as many practical results as less rigorous cell theories. The cell theories treat the liquid state as a quasi-solid using various nonrigorous but intuitively reasonable models for the statistics of a molecule in the liquid lattice.   These theories can be arranged to correlate and predict many useful properties of liquids.   A brief description of these approaches is given in the next two sections.   Subjects such as liquid mixtures, solutions of electrolytes and quantum effects are not considered in this book.

12.4 RADIAL DISTRIBUTION FUNCTION APPROACH TO STATISTICAL MECHANICS OF LIQUIDS

This section gives a very brief introduction to a rigorous approach for the statistical mechanics of liquids.   Classical statistical mechanics are used throughout.   The molecules in a liquid are considered to be indistinguishable particles which roam around the container and which are subject to relatively strong intermolecular forces.   Equations (9-1) through (9-3) are then directly applicable to liquids as well as imperfect gases.   The problem is that $U(q_1, \ldots, q_{3N})$ is difficult to evaluate for liquids.   Since the intermolecular forces are assumed to be velocity-independent, the statistical distribution of molecular velocities in a liquid is the same as that in an ideal gas, and the average kinetic energy per molecule is $(3/2)kT$.

Next, the *radial distribution function*, $g(r)$, is defined in the following way.   Suppose there is a fluid of density $\rho = N/\tilde{V}$ at temperature T.[3] If there were no intermolecular forces, the substance would be an ideal gas, and the average number of molecules that would be found in a small volume element $d\tau$ at a distance r from an arbitrary molecule would be simply $\rho\, d\tau$.   This is because the molecule chosen as the origin does not affect the local density near it if no intermolecular forces exist.   In general, when forces do exist, the function $g(r)$ is defined so that $g(r)\rho\, d\tau$ is the average number of molecules in a small volume element a distance r from any particular molecule.   For an ideal gas, $g(r) = 1$ for all r.   For any real liquid or gas the intermolecular potential has a form which goes to zero at large r and which essentially goes to ∞ as $r \to 0$.   Thus, the effect of intermolecular forces should disappear at large r, and $g(r) \to 1$ as $r \to \infty$.   The probability of a second molecule being found at small r is small because the potential is very high, so $g(r) \to 0$ as $r \to 0$.   In a gas, $g(r)$ is very similar to the function $f_{ij}(r_{ij})$ defined in Chapter 9 if the gas is sufficiently dilute that complexes of more than two molecules occur rarely, and the $g(r)$ function would have the form shown in Fig. 12-4.   At low T, the molecules tend to bunch up at the distance where the intermolecular potential has a minimum.

---

[3]The nomenclature here is that $\tilde{V}$ is the volume per molecule and V is the volume of the entire collection of N particles.

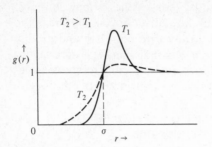

FIGURE 12-4 The function $g(r)$ at two temperatures, where $T_2 > T_1$. $g(r) = 1$ at $r = \sigma$.

In a solid, the ordering of the particles in the crystal lattice is perfect. For a close-packed cubic lattice, the radial distribution function has the form shown in Fig. 12-5. The other molecules are spaced at regular intervals about any abritrarily chosen molecule. In a liquid, the ordering varies with the p-V-T conditions. In the region of the solid-liquid equilibrium, there is considerable ordering, whereas there is less near the liquid-gas transition region. The ordering in a liquid tends to produce a strong maximum in $g(r)$ near $r =$ one molecular diameter, corresponding to a shell of molecules fairly closely packed around any molecule. Structure due to second- and third-nearest neighbors is also apparent at low T, as illustrated in Fig. 12-6 (taken from p. 899 of Hirschfelder, Curtiss, and Bird). These results were obtained by x-ray diffraction experiments.

The next task is to derive the thermodynamic properties of a liquid in terms of the function $g(r)$. It will be assumed that $g(r)$ can be obtained either from x-ray experiments or by theoretical calculation.

The total energy of the molecules in a liquid is the sum of kinetic energy plus potential energy of interaction. The kinetic energy of N molecules is $\frac{3}{2}NkT$, and the total potential energy is denoted as U_{tot}.

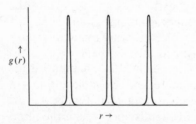

FIGURE 12-5 The function $g(r)$ for a pure crystalline solid, showing the vertical "spikes" corresponding to the lattice spacing.

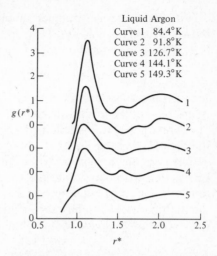

FIGURE 12-6 Experimental curves of the radial distribution function, $g(r)$. The curves are given as a function of the reduced distance $r^* = r/\sigma$, where $\sigma = 3.42$A. [This figure is taken from the review article of J. de Boer, Reports on Progress in Physics **12**, 305 (1949) and is based on the experimental measurements of A. Eisenstein and N. S. Gingrich, *Phys. Rev.* **62**, 261 (1942).] From J. O. Hirschfelder, C. F. Curtiss, and R. L. Bird, *Molecular theory of Gases and Liquids*, copyright © 1954. By permission of John Wiley & Sons, Inc.

The total potential energy is a sum over all molecular interactions of the potential energies of the molecules. It will be assumed here that this potential is a sum of pairwise spherically symmetric interactions $u(r_{ij})$ between pairs of molecules i and j. Thus, the total potential due to three molecules is $u(r_{ij}) + u(r_{ik}) + u(r_{jk})$. This was previously used for imperfect gases in Eqs. (9-8) and (9-9). The total potential energy U_{tot} depends on all the pairwise distances in the liquid. Instead of attempting to calculate this, we shall assume that U_{tot} varies only very slightly with time, and that it can be replaced by its average value, which shall be denoted $\bar{U}$. The average value $\bar{U}$ is obtained by using the average distribution function $g(r)$ and summing the molecular interactions over all r, over all molecules.[4] Consider any arbitrary molecule as an

---

[4]We shall follow the treatment given by Hill in this discussion.

origin. The average number of other molecules in a thin radial shell between r and $r + dr$ is $\rho g(r) 4\pi r^2\, dr$. If the pairwise interaction potential is $u(r)$, then the contribution to $\bar{U}$ made by the spherical shell is

$$u(r)\rho g(r) 4\pi r^2\, dr \tag{12-10}$$

If this is integrated from 0 to very large r, the net contribution to $\bar{U}$ of the interaction of the arbitrarily chosen molecule with its average atmosphere of nearest neighbors is obtained. If this is repeated for all N molecules, the total $\bar{U}$ for the liquid is obtained, except that each pairwise interaction is counted twice. When molecule j is chosen as the origin, the interaction with, say, molecule k, is considered in its peripheral interaction. When molecule k is taken as the origin, molecule j is included in the periphery of k. Thus, the j-k interaction has been included twice, and it is necessary to divide by 2 to obtain $\bar{U}$:

$$\bar{U} = \frac{N\rho}{2} \int_0^\infty u(r)\, g(r)\, 4\pi r^2\, dr \tag{12-11}$$

Since $g(r)$ depends on ρ and T, we may write

$$E = \frac{3NkT}{2} + \frac{N\rho}{2} \int_0^\infty u(r)\, g(r, \rho, T)\, 4\pi r^2\, dr \tag{12-12}$$

If $g(r, \rho, T)$ is expanded as a power series in ρ as

$$g(r, \rho, T) = g_0(r, T) + \rho g_1(r_1, T) + \cdots \tag{12-13}$$

then in limit of very low densities, $g(r, \rho, T) \cong g_0(r, T)$, and $g_0 = \exp(-u(r)/kT)$. In this limit, $g(r)$ becomes identical with $f(r) + 1$ from Chapter 9, and this is applicable only to dilute gases. For dilute gases, the density is low enough that there is only a very slight chance of more than one molecule being in the vicinity of another, and the peripheral interaction of a molecule is determined by the Boltzmann factor $\exp(-u(r)/kT)$ for one pair of molecules. For the densities encountered in liquids, there are many molecules in the vicinity of any particular molecule, and the net total interaction between all the peripheral molecules will determine the distribution function about the central molecule. It is clear from Eq. (12-12) how a knowledge of $g(r, \rho, T)$ leads to E at any T, ρ.

The remaining thermodynamic properties of a liquid can also be evaluated in terms of $g(r, \rho, T)$. Unfortunately, it is difficult to evaluate S directly, and an indirect procedure must be used. It is possible to calculate the pressure p and the chemical potential μ,[5] in terms of

---

[5]The chemical potential is defined as $(\partial A/\partial N)_{T,V}$ where A is the Helmholtz free energy $(A = E - TS)$.

$g(r, \rho, T)$. The entropy, S, may be calculated from a knowledge of p, E, and μ at fixed N, T, and V. Refer to Chapter 17 of Hill's book for this procedure. For information on the theoretical calculation of $g(r, \rho, T)$, consult the references given at the beginning of this chapter.

12.5 CELL THEORIES OF THE LIQUID STATE

The difficulties encountered in theoretical calculation and experimental measurement of the radial distribution function $g(r)$ make it desirable to formulate less rigorous approaches that lead to results which are more generally useful. The cell theories tend to achieve this aim.

In the cell theories, each molecule in a liquid is treated as if it were in a small cell or potential cage formed by the interactions with its nearest neighbors. At various intervals, a molecule can leave its cell and fill a "hole" or void to become a new cell. Most liquids contain roughly 10% voids, since the densities of liquids tend to be about 10% less than the density of the solids. To calculate the thermodynamic properties of a liquid, one would then need to estimate the average potential felt by a molecule in such a cell, and calculate the configuration integral. However, a problem immediately arises because a molecule in a cell is treated as a quasi-solid with partial localization. It is, therefore, not immediately obvious what statistics to use since the particles are not independent localized particles as in Chapter 1, nor are they completely free to wander around the container. An example will make this clearer.

It should be recalled from early chapters in this book that if one has a collection of N gas molecules in a container of volume V in the corrected Boltzmann limit, the pertinent statistical thermodynamic quantities are

$$Q_{\text{gas}} = \left(\frac{2\pi mkT}{h^2}\right)^{3/2} \tilde{V}, \qquad E = \tfrac{3}{2}NkT \qquad (12\text{-}14)$$

$$S_{\text{gas}} = Nk \ln Q_{\text{gas}} + E/T - k(N \ln N - N) \qquad (12\text{-}15)$$

$$S_{\text{gas}} = Nk \left\{ \tfrac{3}{2} \ln \left(\frac{2\pi mkT}{h^2}\right) + \ln \tilde{V} + \tfrac{3}{2} - \ln N + 1 \right\} \qquad (12\text{-}16)$$

in which the particles are treated as indistinguishable. On the other hand, consider a solid according to a model in which each molecule is confined to a cell of volume $V = \tilde{V}/N$ from which it cannot escape. The molecules establish thermal equilibrium through very weak interaction forces, and they are essentially free to move about in their respective cells. Since the cells are distinguishable, the molecules can be treated as localized gas particles in volume V.   Thus,

$$Q_{\text{solid}} = \left(\frac{2\pi mkT}{h^2}\right)^{3/2} V, \qquad E = \tfrac{3}{2}NkT \qquad (12\text{-}17)$$

$$S_{\text{solid}} = Nk \ln Q_{\text{solid}} + E/T \qquad (12\text{-}18)$$

$$S_{\text{solid}} = Nk \left\{ \tfrac{3}{2} \ln \left(\frac{2\pi mkT}{h^2}\right) + \ln \tilde{V} + \tfrac{3}{2} - \ln N \right\} \qquad (12\text{-}19)$$

On comparison of Eqs. (12-16) and (12-19), it will be seen that

$$S_{\text{gas}} = S_{\text{solid}} + Nk \qquad (12\text{-}20)$$

The extra factor Nk is often called the "communal entropy".   It is due to the fact that there are more ways to put N indistinguishable particles into translational levels characteristic of volume $\tilde{V}$, than there are to put N distinguishable particles into translational levels characteristic of cell volume $V = \tilde{V}/N$.   This is illustrated in Fig. 12-7, reproduced from Fig. 4.4-1 of Hirschfelder, Curtiss, and Bird.

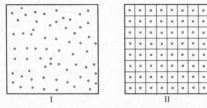

I

Perfect gas of N molecules
in volume $\tilde{V}$

II

N molecules, each one confined
to a cell of volume $\tilde{V}/N$

FIGURE 12-7 Illustration of the concept of communal entropy. From J. O. Hirschfelder, C. F. Curtiss, and R. L. Bird, *Molecular Theory of Gases and Liquids*, copyright © 1954. By permission of John Wiley & Sons, Inc.

In the cell theories of a liquid, it is assumed that each liquid molecule moves in a cell created by the forces exerted by its neighbors. Although the forces are not really constant across the cell, it is convenient to define the *free volume* v_f as the effective volume available to a liquid molecule for free motion in a cell.   In general, v_f will not be equal to $\tilde{V}/N$.   The average effect of the neighboring molecules on a molecule in a cell can be approximated by a potential $U(\mathbf{r})$ in the cell.   If it were assumed that $U(\mathbf{r})$ had the spherically symmetric constant form

$$U(\mathbf{r}) = \langle U_0 \rangle \qquad r < r_0$$
$$U(\mathbf{r}) = \infty \qquad r \geq r_0 \qquad (12\text{-}21)$$

then the free volume would simply be

$$v_f = \int_0^{r_0} 4\pi r^2 \, dr = \tfrac{4}{3}\pi r_0^3 \tag{12-22}$$

The energy of a molecule in the cell would be the sum of its kinetic energy plus $\langle U_0 \rangle / 2$. The factor of $\tfrac{1}{2}$ comes from the fact that the potential is assumed to be due to pairwise interactions, and any two molecules share the potential between them. For example, if one had two oppositely charged particles at a distance r, the potential would be $-q_1 q_2 / r$ where q_1 and $-q_2$ are the charges. Now if the particles were removed to ∞, the gain in energy would be $q_1 q_2 / r$, and each particle will share the energy equally if the masses are the same. Therefore, one might say that the energy of one of the particles in the presence of the other is the kinetic energy plus $-q_1 q_2 / 2r$. This conclusion will not change when more than two particles interact, provided the total interaction is a sum of pairwise additive potentials.

In the more general case where the potential, though spherically symmetric, varies with r as $U(r)$, the free volume is obtained by the following procedure. The point at which the potential is a minimum is defined as the origin $r = 0$, and the potential at this point is called $u_0 = U(0)$. If this potential extended across the entire cell, the probability of finding the particle in any part of the cell would be uniform. Since $U(r)$ does vary with r, it is assumed that the probability per unit volume of finding the particle in a spherical shell of volume $4\pi r^2 \, dr$ is the Boltzmann factor $\exp\left[-(U(r) - U_0)/kT\right]$, and that the contribution each spherical shell makes to the free volume is equal to probability of finding the particle in the shell. Thus,

$$v_f = \int_{\text{cell}} \exp\left[-(U(r) - U_0)/kT\right] 4\pi r^2 \, dr \tag{12-23}$$

Since the free volume so calculated effectively reduces the cell volume to include only the region over which the potential could be set equal to U_0, the average energy of a particle in such a cell must be taken as the sum of its kinetic energy plus $U_0/2$. Hence, one must set $\langle U \rangle = U_0/2$.

Assuming that an approximate free volume can be calculated, there still remains the problem of the proper statistics for a liquid. If the liquid molecules were absolutely restricted to their cells and could never escape, Eqs. (12-17) through (12-19) would be applicable, except that V would be replaced by v_f, and E would be replaced by $E + U_0$, corresponding to a new zero of energy lower by U_0. However, the liquid molecules are not restricted to their cells, and can eventually wander around the entire container. Therefore, it has been argued that an expression for the entropy of a liquid more like that of a gas should be used. According to Eq. (12-20), it might seem possible to write

$$S_{\text{liquid}} = S_{\text{solid}} + S_c \tag{12-24}$$

where S_c is the communal entropy, approximately equal to Nk. It can be shown that variations of S_c from Nk can be attributed to occupancy of cells by either zero, or more than one molecule. Although it is not yet completely resolved, it appears that the most reasonable value for S_c is slightly less than Nk. Thus, in summary, the following is used for a liquid:

$$Q_{\text{liquid}} = v_f \left(\frac{2\pi mkT}{h^2}\right)^{3/2} e^{-\langle U \rangle / kT}, \qquad E = \tfrac{3}{2}NkT + \langle U \rangle \tag{12-25}$$

$$v_f = \int_{\text{cell}} \exp\left[-(U(r) - U_0)/kT\right] 4\pi r^2 \, dr \tag{12-26}$$

$$S_{\text{liquid}} = Nk\left\{\frac{3}{2}\ln\left(\frac{2\pi mkT}{h^2}\right) + \ln v_f + \frac{3}{2} + S_c\right\} \tag{12-27}$$

$$S_c \lessgtr Nk \tag{12-28}$$

The equation of a state of a liquid is obtained from Eq. (9-66):

$$p = kT(\partial \ln Q / \partial \tilde{V})_{T,N} \tag{12-29}$$

A great many different calculations of v_f and U_0 have been made over the years, based on various models and hypotheses. Some are semi-empirical and others are completely theoretical. Here an example of each is given. In the semi-empirical approach, some of the (presumed) known properties of the substance are utilized as independent variables in the calculation. Then other properties (presumably more difficult to measure) are calculated in terms of the known properties. There is usually a certain amount of intelligent guessing involved in such calculations.

One semi-empirical approach[6] is based on the following procedures. It is known[7] that if a close-packed face-centered cubic crystal lattice is formed from hard spheres of diameter σ, the volume of N spheres is $N\sigma^3\sqrt{2}$. The volume required per sphere is $\sigma^3/\sqrt{2}$, and this exceeds the volume of a single sphere $(\pi\sigma^3/6)$ because of the small gaps between adjacent spheres. Now suppose that there is a liquid with density $\rho = N/\tilde{V}$. A volume associated with each molecule can be assigned as $V/\tilde{N} = V$. It may be assumed that each molecule moves in a rectangular cell of side $V^{1/3}$. If the hard-sphere molecules of diameter σ are placed in cells, as shown in Fig. 12-8, the free volume between adjacent mole-

[6]H. Eyring and J. O. Hirschfelder, *J. Phys. Chem.* **41**, 249 (1937).

[7]R. J. Buehler, R. H. Wentorf, J. O. Hirschfelder, and C. F. Curtiss, *J. Chem. Phys.* **19**, 61 (1951).

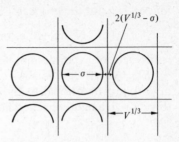

FIGURE 12-8 Illustration of cell model where the cells are cubes of volume $V^{1/3}$, and V is the molecular volume. The molecular diameter is σ, and the spacing between molecules is $3(V^{1/3} - \sigma)$.

cules depends on the linear dimension $2(V^{1/3} - \sigma)$. Since equal distances occur in directions perpendicular to that shown, the free volume is nearly equal to

$$v_f \cong [2(V^{1/3} - \sigma)]^3 = 8(V^{1/3} - \sigma)^3 \tag{12-30}$$

For purposes of calculation of the equation of state, the factor 8 in Eq. (12-30) cancels out, and all that it is necessary to assume is that $v_f \alpha (V^{1/3} - \sigma)^3$. It is now assumed that the diameter of a molecule can be obtained from the Van der Waals constant for the substance. According to the discussion following Eq. (9-86), the Van der Waals constant in the equation of state for a Van der Waals gas,

$$\left(p + \frac{\alpha}{V^2}\right)(V - \beta) = NkT \tag{12-31}$$

is given by

$$\beta = 4\left(\frac{\pi}{6}\sigma^3\right) = 2\pi\sigma^3/3 \tag{12-32}$$

Therefore,

$$\sigma = \left(\frac{3}{2\pi}\right)^{1/3}\beta^{1/3} = 0.782\beta^{1/3} \tag{12-33}$$

The free volume may, therefore, be expressed as

$$v_f \cong 8(V^{1/3} - 0.782\beta^{1/3})^3 \tag{12-34}$$

where β is assumed to be independent of V. In order to use Eqs. (12-25) through (12-28), $\langle U \rangle$ is required as well as v_f. Instead of attempting to calculate $\langle U \rangle$ from first principles, it is merely assumed that the energy of vaporization is known for whatever substance is under consideration.

Since ΔE_{vap} for the process.

$$N \text{ molecules of liquid} \rightarrow N \text{ molecules of gas}$$

is the heat of vaporization, it follows that $\langle U \rangle$ should be chosen equal to $-\Delta E_{vap}/N$. It can be shown that the energy of vaporization tends to be inversely proportional to the molecular volume V (it takes more energy to vaporize a tightly packed liquid), and $\langle U \rangle$ may be approximated by

$$\langle U \rangle \cong -\Delta E_{vap}/N \cong -a(T)/V \qquad (12\text{-}35)$$

where $a(T)$ is a different (positive) function of T for every substance. Equations (12-25) to (12-29) are now applied, obtaining

$$\left(\frac{\partial \ln Q}{\partial V}\right)_{T,N} = -\frac{a(T)}{kT^2} + \left(\frac{\partial \ln v_f}{\partial V}\right)_{T,N} \qquad (12\text{-}36)$$

$$\left(\frac{\partial \ln v_f}{\partial V}\right)_{T,N} = (V^{1/3} - 0.782\,b^{1/3})^{-1}V^{-2/3} \qquad (12\text{-}37)$$

Thus,

$$p = -\frac{a(T)}{V^2} + kT(V^{1/3} - 0.782b^{1/3})^{-1}V^{-2/3} \qquad (12\text{-}38)$$

or,

$$\left[p + \frac{\alpha(T)}{V^2}\right][V - 0.782\beta^{1/3}V^{2/3}] = kT \qquad (12\text{-}39)$$

This has the general outline of a Van der Waals equation of state, except that, due to the cell treatment, the volume of the molecules is proportional to $V^{2/3}$. The agreement with experiment on some properties, such as compressibility and coefficient of thermal expansion of the liquid, are quite good.[8]

The next example is the calculation of Lennard-Jones and Devonshire,[9] which is the predecessor of all cell theories of the liquid state. This is an *a priori* theory in the sense that it is only assumed that the intermolecular potential between any pair of molecules is known. In this theory, it is assumed that each liquid molecule is in a cell formed by its neighboring molecules in a close-packed lattice structure. In the original treatment, only the effects of nearest and second-nearest neighbors were included, but since then, the effects of neighbors further removed have also been included.[10] Only nearest neighbors are considered here. In a close-packed lattice there will be 12 nearest neighbors to any

---

[8]See Table 4.5-2 on p. 282 of Hirschfelder, Curtiss, and Bird.

[9]J. E. Lennard-Jones and A. F. Devonshire, *Proc. Roy. Soc.* (London) **A163**, 53 (1937).

[10]R. H. Wentorf, R. J. Buehler, J. O. Hirschfelder and C. F. Curtiss, *J. Chem. Phys.* **18**, 1484 (1950).

molecule. The positions of these molecules will vary, but on the average, they will tend to be distributed over a spherical surface of radius ρ, where ρ is the nearest-neighbor distance in the liquid lattice. For close-packed spheres, the volume allotted to each molecule is $\rho^3/\sqrt{2}$. Thus,

$$\rho^3 = \sqrt{2}V \qquad (12\text{-}40)$$

where V is the molecular volume per molecule. The effect of the twelve nearest-neighbor molecules on the central molecule will be taken, not as a sum of twelve point-to-point interactions, but in terms of an average interaction from a spherical surface to the central molecule. In other words, the twelve point centers of force are replaced by a uniform surface of force so as to average out the effects of the twelve particles. Since there are 12 particles to be replaced by a surface of area $4\pi\rho^2$, we hypothesize a number density of $12/4\pi\rho^2 = 3/\pi\rho^2$ particles per unit area on the surface. Each element of surface is assumed to interact with the central molecule with a Lennard-Jones potential

$$u(R) = 4\epsilon \left\{ \left(\frac{\sigma}{R} \right)^{12} - \left(\frac{\sigma}{R} \right)^{6} \right\} \qquad (12\text{-}41)$$

multiplied by the number of particles in the element of surface. This is illustrated in Fig. 12-9. Now consider a spherical cell of radius ρ, with the central molecule at point P, a distance r from the center of the cell, as shown in Fig. 12-10. The total interaction between the molecule and the particles uniformly distributed on the surface of radius ρ is obtained by summing over annular elements of the spherical surface. All

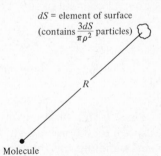

dS = element of surface
(contains $\dfrac{3dS}{\pi\rho^2}$ particles)

R

Molecule

FIGURE 12-9 An element of surface dS containing $3dS/\pi\rho^2$ particles at a distance R from the central molecule. The $3\,dS/\pi\rho^2$ molecules are assumed to interact with the central molecule with a Lennard-Jones potential.

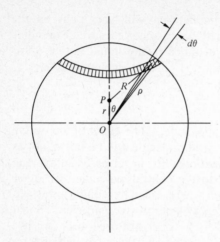

FIGURE 12-10 A molecule happens to be at point P, a distance r from the center, O, of spherical cell of radius ρ. The interaction between the molecule and the surface is obtained by summing interactions over annular rings at constant distances R from the central molecule.

regions of the annular surface are at the distance R from the molecule, and therefore, the contribution to the total potential from this area is equal to Eq. (12-41), times $3/\pi\rho^2$, times the area of the annular surface. The area of the surface is $2\pi(\rho \sin \theta)\rho\ d\theta$. The number of particles in the annular element of surface is

$$\frac{3}{\pi\rho^2}\left[2\pi\rho^2 \sin \theta\ d\theta\right] = 6 \sin \theta\ d\theta \tag{12-42}$$

The distance R is determined by the law of cosines as

$$R^2 = r^2 + \rho^2 - 2\rho r \cos \theta \tag{12-43}$$

Thus, $U(R)$ can be expressed as a function of θ for fixed ρ and r. The total interaction between the central molecule and the surface is obtained by summing the interactions from all annular elements of surface from $\theta = 0$ to $\theta = \pi$. Thus, the interaction potential between the molecule and the entire surface is

$$U(\rho, r) = \int_0^\pi u(r, \rho, \theta) \cdot 6 \sin \theta\ d\theta \tag{12-44}$$

It is more convenient to perform this integral over dR rather than $d\theta$ because $u(R)$ is easier to work with. By differentiating Eq. (12-43), we obtain (at constant r and ρ)

$$\sin \theta \, d\theta = \frac{R \, dR}{\rho r} \tag{12-45}$$

Thus,

$$U(\rho, r) = \frac{24\epsilon}{\rho r} \int_{\rho-r}^{\rho+r} \left\{ \left(\frac{\sigma}{R}\right)^{12} - \left(\frac{\sigma}{R}\right)^{6} \right\} R \, dR \tag{12-46}$$

The integrals can be easily performed, leading to

$$U(\rho, r) = \frac{24\epsilon}{\rho r} \left[\frac{-\sigma^{12}}{10 R^{10}} + \frac{\sigma^{6}}{4 R^{4}} \right]_{\rho-r}^{\rho+r} \tag{12-47}$$

Equation (12-47) cannot be used in the limit $r \to 0$ because r occurs in the denominator. Therefore, care must be taken in trying to apply Eq. (12-47) to the case where $r \to 0$ because the limit must be taken correctly. It is simpler to evaluate $U(\rho, 0)$ by summing the interactions over a sphere with a particle at the center:

$$U_0 = U(\rho, 0) = \left(\frac{3}{\pi \rho^2}\right)(4\pi\rho^2) u(\rho) = 48\epsilon \left\{ \left(\frac{\sigma}{\rho}\right)^{12} - \left(\frac{\sigma}{\rho}\right)^{6} \right\} \tag{12-48}$$

By expanding Eq. (12-47), and subtracting Eq. (12-48), the following is obtained:

$$U(\rho, r) - U_0 = 24\epsilon\sigma^{12} \left\{ \frac{1}{10\rho r} \left[\frac{1}{(\rho-r)^{10}} - \frac{1}{(\rho+r)^{10}} \right] - \frac{2}{\rho^{12}} \right\}$$

$$+ 24\epsilon\sigma^{6} \left\{ \frac{1}{4\rho r} \left[\frac{1}{(\rho+r)^{4}} - \frac{1}{(\rho-r)^{4}} \right] + \frac{2}{\rho^{6}} \right\} \tag{12-49}$$

Now calculate the free volume from Eq. (12-26). There is some uncertainty as to what to choose for the upper limit of integration in this equation. At first sight, it might seem easiest to choose the upper limit as $r = \rho$. However, it should be remembered that the nearest-neighbor *centers* are on a sphere of radius ρ away from the center of the cell, and therefore, the maximum effective radius of the cell will be somewhat *less* than ρ. A reasonable choice is afforded by assigning the cell volume equal to the molecular volume per cell. As mentioned previously, the volume per molecule in a close-packed lattice is

$$V = \rho^3/\sqrt{2} \tag{12-50}$$

if ρ is the diameter of the close-packed spheres. If the upper limit for integration in Eq. (12-26) is chosen as r_m, then

$$\tfrac{4}{3}\pi r_m^3 = V = \rho^3/\sqrt{2} \tag{12-51}$$

and

$$r_m = \rho \left(\frac{3}{4\pi\sqrt{2}}\right)^{1/3} \tag{12-52}$$

With this choice of the upper limit, the cell volume will be the molecular volume of the liquid per molecule. The free volume is then given by

$$v_f = \int_0^{r_m} \exp\left\{-\left[U(\rho, r) - U_0\right]/kT\right\} 4\pi r^2 \, dr \tag{12-53}$$

In order to make this integral more tractable, it is convenient to manipulate $U(\rho, r) - U_0$ into a simpler form. By combining fractions, it can easily be shown from the binomial theorem that

$$\frac{1}{r\rho}\left[\frac{1}{(\rho - r)^n} - \frac{1}{(\rho + r)^n}\right] = (\rho^2 - r^2)^{-n}\left[(\rho + r)^n - (\rho - r)^n\right]\frac{1}{r\rho}$$

$$= \frac{1}{r\rho}(\rho^2 - r^2)^{-n}\left[2\left(n\rho^{n-1}r + \frac{n(n-1)(n-2)}{1\cdot 2\cdot 3}\rho^{n-3}r^3 + \cdots\right)\right] \tag{12-54}$$

In terms of the variable $y = (r/\rho)^2$, this becomes

$$2\rho^{-(n+2)}(1-y)^{-n}\left(n + \frac{n(n-1)(n-2)}{1\cdot 2\cdot 3}y + \cdots\right)$$

For $n = 4$, this becomes

$$8\rho^{-6}(1-y)^{-4}(1+y)$$

For $n = 10$, it becomes

$$20\rho^{-12}(1-y)^{-10}(1 + 12y + \tfrac{252}{10}y^2 + 12y^3 + 4^4)$$

Therefore,

$$U(\rho, r) - U_0 = 48\epsilon\left\{\frac{\sigma^{12}}{\rho^{12}}\overbrace{\left[(1 + 12y + \tfrac{252}{10}y^2 + 12y^3 + y^4)(1-y)^{-10} - 1\right]}^{l(y)}\right.$$

$$\left. + \frac{\sigma^6}{\rho^6}\underbrace{\left[(1+y)(1-y)^{-4} - 1\right]}_{m(y)}\right\} \tag{12-55}$$

The functions in braces are defined as $l(y)$ and $m(y)$. According to Eq. (12-50), the reduced molecular volume is

$$V^* \equiv \frac{V}{\sigma^3} = \frac{\rho^3}{\sqrt{2}\sigma^3} \tag{12-56}$$

Therefore, $\sigma^3/\rho^3 = (V^*\sqrt{2})^{-1}$, and Eq. (12-55) may be rewritten

$$U(\rho, r) - U_0 = 12\epsilon\left\{\frac{l(y)}{V^{*4}} - \frac{2m(y)}{V^{*2}}\right\} \tag{12-57}$$

Furthermore, the variable of integration in Eq. (12-53) for v_f can be changed to y as follows:

$$dy = 2r \, dr/\rho^2$$
$$r^2 \, dr = \rho^3 y^{1/2} \, dy/2 \tag{12-58}$$

The integral for v_f may then be expressed as

$$v_f = 2\pi\rho^3 \int_0^{y_m} \exp\left\{-\left[12\left(\frac{l(y)}{V^{*4}} - \frac{2m(y)}{V^{*2}}\right)\right]T^{*-1}\right\} y^{1/2}\, dy \qquad (12\text{-}59)$$

where $T^* = kT/\epsilon$ and $y_m = r_m^2/\rho^2 = (3/4\pi\sqrt{2})^{2/3}$. *The integral in this equation is usually denoted as* $g(V^*, T^*)$. The equation of state of the liquid according to this model is obtained from Eqs. (12-25) and (12-29).

$$\frac{p}{kT} = -\frac{1}{kT}\left(\frac{\partial\langle U\rangle}{\partial V}\right) + \left(\frac{\partial \ln v_f}{\partial V}\right) \qquad (12\text{-}60)$$

Since $\langle U\rangle = U_0/2$, and $V = \rho^3/\sqrt{2}$, it follows that

$$\frac{\partial\langle U\rangle}{\partial V} = \frac{1}{2}\frac{\partial}{\partial V}\left\{12\epsilon\left[\frac{\sigma^{12}}{V^4} - \frac{2\sigma^6}{V^2}\right]\right\} = 24\epsilon\left[\frac{\sigma^6}{V^3} - \frac{\sigma^{12}}{V^5}\right]$$

$$\frac{\partial\langle U\rangle}{\partial V} = \frac{24\epsilon}{V}\left[\frac{1}{V^{*2}} - \frac{1}{V^{*4}}\right] \qquad (12\text{-}61)$$

and

$$\frac{\partial \ln v_f}{\partial V} = \frac{\partial \ln}{\partial V}\{2^{3/2}\pi V g\} = \frac{1}{V} + \frac{1}{g}\frac{\partial g}{\partial V} \qquad (12\text{-}62)$$

From the definition of g,[11] it follows that

$$\frac{\partial g}{\partial V} = \left\{\frac{g_l(V^*, T^*)}{V^{*4}} + \frac{g_m(V^*, T^*)}{V^{*2}}\right\} \cdot \frac{48}{VT^*} \qquad (12\text{-}63)$$

where

$$g_l = \int_0^{y_m} \exp\left[-12l(y)/V^{*4}T^*\right]l(y)y^{1/2}\, dy \qquad (12\text{-}64)$$

$$g_m = \int_0^{y_m} \exp\left[-24m(y)/V^{*2}T^*\right]m(y)y^{1/2}\, dy \qquad (12\text{-}65)$$

Thus,

$$\frac{\partial \ln v_f}{\partial V} = \frac{1}{V} + \frac{48}{VT^*}\left\{\frac{g_l}{g}\cdot\frac{1}{V^{*4}} + \frac{g_m}{g}\cdot\frac{1}{V^{*2}}\right\} \qquad (12\text{-}66)$$

and the equation of state is

$$\frac{pV}{kT} = 1 + \frac{24}{T^*}\left\{\frac{1}{V^{*4}}\left(1 + \frac{2g_l}{g}\right) + \frac{1}{V^{*2}}\left(-1 + \frac{2g_m}{g}\right)\right\} \qquad (12\text{-}67)$$

The integrals g, g_l, and g_m have been extensively evaluated by Lennard-Jones and Devonshire and extended by others.[12] In addition to the

---

[11]Given below Eq. (12-59).

[12]I. Prigogine, *J. Chim. Physique* **45**, 273 (1948); T. L. Hill, *J. Phys. Colloid Chem.* **51**, 1219 (1947).

equation of state, other thermodynamic properties can be calculated.[13]
A comparison (see footnote 12) of the *modified theory* taking into account
effects of second- and third-neighbor interactions, with experiment shows
only semiquantitative agreement.   It is to be expected that a cell model
would work best at high densities, and begin to fail badly near the critical
region.   The critical constants predicted by the modified cell theory are

$$T_c{}^* = kT_c/\epsilon = 1.30$$
$$V_c{}^* = V_c/\sigma^3 = 1.77$$
$$p_c{}^* = p_c\sigma^3/\epsilon = 0.434$$
$$\frac{p_cV_c}{kT_c} = 0.591$$

These may be compared with the average values $T_c{}^* \cong 1.3$, $V_c{}^* \cong 3.0$,
$p_c{}^* \cong 0.13$, and $p_cV_c/kT_c \cong 0.292$ taken from Table 12-1.   The agreement is far from satisfactory.

12.6   CELL THEORIES INCLUDING "HOLES"

The cell theories of the liquid state tend to work best at moderately
high densities where the model of a central molecule surrounded by close-
packed neighbors is most reasonable.   At lower densities, such as occur
at the critical point, the liquid will contain a substantial fraction of
vacancies or "holes" due to the relatively high average kinetic energy of
the molecules which tends to break up the packing.   Therefore, even a
moderately good cell theory, which describes the liquid fairly well for
$V \ll V_c$, might be expected to fail near the critical region.   To obtain a
theory that works well for a wider range of densities, one must introduce
the concept of vacant cells or *holes* into the cell model.[14]

In the hole theories of a liquid, the molecules are assumed to be in a
fixed lattice of cells, and the communal entropy term is *not* added *ad hoc*
as in the cell theories.   Each cell is assumed to be either occupied by a
molecule or vacant; in the latter case it is said that the site is occupied by a
hole.   In the range of ordinary substances at nonextreme conditions, a
typical value for the fraction of cells occupied by holes is 0.1.   At high
densities this is reduced, and at low densities it increases, ending ultimate-
ly in being nearly 1.0 for the gas phase.   It will be shown that the

---

[13]R. H. Wentorf, R. J. Buehler, J. O. Hirschfelder, and C. F. Curtiss, *J. Chem. Phys.*
18, 1484 (1950); J. O. Hirschfelder, C. F. Curtiss, and R. L. Bird, *Molecular Theory
of Gases and Liquids* (New York: J. Wiley & Son, 1964), pp. 296–310.
[14]For further details than contained in this chapter, the reader is referred to F.
Cernuschi and H. Eyring, *J. Chem. Phys.* **7**, 547 (1939); H. M. Peek and F. L. Hill,
J. Chem. Phys. **18**, 1252 (1950), J. S. Rowlinson and C. F. Curtiss, *J. Chem. Phys.*
19, 1519 (1951) and J. de Boer, *Proc. Roy. Soc.* **A215**, 4 (1952).

statistics of a mixture of molecules and holes in a fixed lattice of cells leads automatically to the communal entropy factor S_c which goes to $N_m k$ when the total number of holes (N_h) greatly exceeds the total number of molecules N_m. This entropy is due to random mixing of holes and molecules in the lattice.

12.6.1 One-Dimensional Lattice Ising Model

Before dealing with real substances, a one-dimensional version of the Ising[15] model will be explored, which is very important in statistical mechanics.   This model can be used very effectively in the understanding of adsorption on surfaces,[16] and the theory of ferromagnetism[17] of metals, as well as the liquid state.[18]   Later, this will be extended to more than one dimension, and compared with properties of real substances.

Consider a one-dimensional lattice where each site is occupied by either a hole or a molecule.   There are a large number of possible configurations involving N_h holes and N_m molecules, one of which is partly shown in Fig. 12-11.   The total number of sites is N_t.   For any particular configuration, N_{11} is defined as the number of pairs of adjacent sites filled by two molecules, N_{00} is defined as the number of pairs of adjacent vacant sites, and N_{10} is defined as the number of pairs of sites filled with one molecule and one hole.

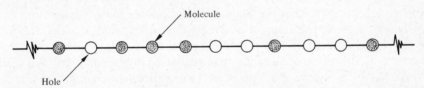

FIGURE 12-11 One possible arbitrary configuration of a linear lattice where each lattice point can contain either a molecule or a hole.

The relation between N_m and N_h, and N_{11}, N_{00}, and N_{10} is determined by the following argument.[19]   Draw a line from each site occupied by a molecule to its nearest-neighbor sites.   Thus, $2N_m$ lines are drawn. If a nearest-neighbor site is filled with a molecule, two lines connect the sites, as shown in Fig. 12-12.   If the nearest-neighbor site has a hole, only a single line connects the sites, as shown in Fig. 12-12.   The total

---

[15]E. Ising, *Z. Physik* **31**, 253 (1925).

[16]See T. L. Hill, *Statistical Thermodynamics* (Palo Alto, California: Addison-Wesley, 1960), Chapters 7, 14.

[17]G. F. Newell and E. W. Montroll, *Revs. Mod. Phys.* **25**, 353 (1953).

[18]C. N. Yang and T. D. Lee, *Phys. Rev.* **87**, 404 (1952).

[19]Given in Chapter 14 of Hill's book.

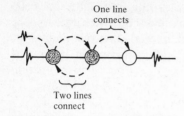

One line
connects

Two lines
connect

FIGURE 12-12 Illustration of method for calculating N_m and N_h. A line is drawn from each molecule to each adjacent site. Hole-hole pairs have no lines connecting them, hole-molecule pairs have one line, and molecule-molecule pairs are connected by two lines.

number of lines is $2N_{11} + N_{10}$. Thus,

$$2N_m = 2N_{11} + N_{10} \qquad (12\text{-}68)$$

Similarly, if we draw two lines from each empty site, we obtain

$$2N_h = 2N_{00} + N_{10} \qquad (12\text{-}69)$$

It is assumed here that the ends of the chain are joined to form an extremely large closed loop. For fixed values of N_m and N_h, Eqs. (12-68) and (12-69) constitute two relations between N_{11}, N_{00}, and N_{10}. Therefore, only one of these quantities can be chosen independently. We shall take N_{10} as the independent variable. For various different arrangements of the holes and molecules with N_h and N_m fixed, N_{00} and N_{11} can be determined from N_{10}. Let us now calculate $g(N_h, N_m, N_{10})$, the number of different arrangements of N_h holes and N_m molecules such that there are N_{10} adjacent pairs consisting of a hole and a molecule. We evaluate g by writing out a typical arrangement:

$$\ldots \mid m\,m \mid h \mid m \mid \underbrace{h\,h\,h}_{\substack{\text{a group} \\ \text{of } h\text{'s}}} \mid \underbrace{m\,m\,m\,m}_{\substack{\text{a group} \\ \text{of } m\text{'s}}} \mid h \mid m\,m\,m \mid h\,h \mid m \mid \ldots \qquad (12\text{-}70)$$

Imagine that the ends of the chain are joined. Since an m-h interface occurs at the boundary of an m-group and an h-group, it follows that

$$n_h = n_m = N_{10}/2 \qquad (12\text{-}71)$$

where n_h is the number of h-groups and n_m is the number of m-groups. It is desired to calculate the number of ways to put N_h molecules into n_h h-groups, and N_m molecules into n_m m-groups. Each such arrangement corresponds to N_{10} *m-h* interfaces. Consider the holes. If N_h were equal to n_h, there would be just one hole to go around to each *h*-group. The number of ways to do this would be unity and each *h*-group would contain one hole. If N_h were $n_h + 1$, there would be one extra hole which could be put in any of the *h*-groups, so there would be n_h different arrangements. In general, for $N_h > n_h$, there is at least one hole per *h*-group. There are $N_h - n_h$ indistinguishable holes remaining, which must be distributed among the n_h *h*-groups. The problem is now analogous to the Bose-Einstein problem of putting n_i indistinguishable particles into g_i distinguishable compartments [see Eq. (3-1)]. There are

$$\rho(h) = \frac{(N_h - 1)!}{(n_h - 1)!(N_h - n_h)!} \tag{12-72}$$

ways to put $N_h - n_h$ indistinguishable particles into n_h distinguishable compartments. For example, if $N_h = n_h + 1$, $\rho(h) = n_h$. Similarly, for the molecules, there are

$$\rho(m) = \frac{(N_m - 1)!}{(n_m - 1)!(N_m - n_m)!} \tag{12-73}$$

ways to arrange N_m indistinguishable molecules into n_m groups with at least one molecule in each group. Since any arrangement of the molecules can be combined with any arrangement of the holes, the total number of ways to arrange the molecules and holes is

$$g(N_h, N_m, N_{10}) = \rho(m)\rho(h) \cong \frac{N_h! N_m!}{(N_h - N_{10}/2)!(N_m - N_{10}/2)![(N_{10}/2)!]^2} \tag{12-74}$$

in which we have assumed $N_h \gg 1$ and $N_m \gg 1$.

It may be reasonably assumed that the total interaction potential between molecules is the sum of nearest-neighbor interactions between pairs of sites occupied by molecules. If the average interaction between two adjacent molecules is denoted u, then the total interaction energy, U, is the product of u times the number of adjacent pairs, N_{11}. Thus,

$$U = N_{11}u = (N_m - N_{10}/2)u \tag{12-75}$$

The partition function for the ensemble of lattice sites containing N_h holes and N_m *localized* molecules is obtained by the following arguments. First suppose that there are no intermolecular forces between molecules, and that $N_h = 0$. In this case, $g(0, N_m, 0) = 1$ and there is only one possible arrangement of indistinguishable molecules on the lattice sites. The partition function for a single molecule would be that corresponding to an *independent localized particle*, as in Chapter 1. If the model used is that each localized molecule is free to move over a volume v_f, the one-dimensional partition function for the molecule is

$$q = \left(\frac{2\pi mkT}{h^2}\right)^{1/2} v_f e^{-u/2kT} \tag{12-76}$$

where u is the average potential in a cell. In the Einstein approximation (Chapter 1), of independent harmonic oscillators, one would have

$$q = \frac{e^{-h\nu/2kT}}{1 - e^{-h\nu/kT}} \tag{12-77}$$

Regardless of the particular form of q, the partition function for the entire collection of N_m molecules in N_m lattice sites is

$$Q = q^{N_m} \tag{12-78}$$

Now consider the case where there are again no intermolecular forces, but where $N_h \neq 0$. The partition function for the entire ensemble of N_m molecules and N_h holes will be determined from the fact that each energy level of the ensemble will be the sum of energy levels of all the molecules. Therefore, for any arbitrary arrangement of the molecules and holes, the contribution to Q is still q^{N_m}. However, there are many possible arrangements of the holes and molecules. Therefore, there is a *configurational degeneracy*, g_c, which is the number of such configurations, and

$$Q = g_c q^{N_m} \tag{12-79}$$

since all the states of the individual molecules can be repeated for all arrangements of the holes and molecules. The configurational degeneracy is simply the number of ways to arrange N_m molecules among $N_m + N_h$ distinguishable lattice sites:

$$g_c = \frac{(N_m + N_h)!}{N_m! N_h!} \tag{12-80}$$

and therefore,

$$Q = \frac{(N_m + N_h)!}{N_m! N_h!} q^{N_m} \tag{12-81}$$

When intermolecular forces exist, and $N_h \neq 0$, it follows that the partition function *for the ensemble* is q^{N_m}, multiplied by the sum over all arrangements of holes and molecules, of $g(N_h, N_m, N_{10})$ times the interaction energy term $\exp(-U/kT)$ corresponding to N_{11} pairs of adjacent molecules. Note that N_{11} is uniquely determined from N_h, N_m, and N_{10}. Since all arrangements with the same N_{10} have the same interaction potential [see Eq. (12-75)], it follows that the sum needs to be taken over all possible N_{10} consistent with fixed N_h and N_g. We have

$$Q = q^{N_m} \sum_{N_{10}} g(N_h, N_m, N_{10}) \exp\left[-(N_m - N_{10}/2)\frac{u}{kT}\right] \tag{12-82}$$

$$Q = q^{N_m} e^{-N_m u/kT} \sum_{N_{10}} \frac{N_h! N_m! \, e^{N_{10}u/2kT}}{(N_h - N_{10}/2)!(N_m - N_{10}/2)![(N_{10}/2)!]^2} \tag{12-83}$$

At this point, N_m and N_h are presumed to be very large numbers. If a plot of $g(N_h, N_m, N_{10})$ vs N_{10} was made, it would be found that a curve with a maximum would be generated, and the maximum would correspond to a large value for N_{10}. The larger N_h and N_m are, the sharper will be the maximum in g. Since all thermodynamic properties depend on $\ln Q$, not Q itself, it will be a good approximation to replace $\ln Q$ by the value obtained by replacing the sum expression for Q by the largest term in the sum. The argument here is the same as in Section 1.7. Thus, it is assumed that

$$\ln Q \cong N_m \ln q - N_m u/kT + N_{10}^* u/2kT + \ln g(N_h, N_m, N_{10}^*) \tag{12-84}$$

where N_{10}^* is the value of N_{10} that maximizes $g(N_h, N_m, N_{10}) \exp[N_{10}u/kT]$. To find N_{10}^*, treat N_{10} as if it was a continuous variable, and use Stirling's approximation for factorials of $(N_h - N_{10}/2)$, $(N_m - N_{10}/2)$, and $N_{10}/2$. Then

$$\frac{\partial \ln\{g\, e^{N_{10}u/2kT}\}}{\partial N_{10}} = 0 \tag{12-85}$$

After defining $n = N_{10}/2$, it is found that

$$\ln\{g\, e^{nu/kT}\} \cong n^*u/kT - (N_h - n^*)\ln(N_h - n^*) + (N_h - n^*)$$
$$- (N_m - n^*)\ln(N_m - n^*) + (N_m - n^*)$$
$$- 2n^*\ln n^* + 2n^* \tag{12-86}$$

and therefore,

$$\left\{ u/kT + \frac{(N_h - n^*)}{(N_h - n^*)} + \ln(N_h - n^*) - 1 + \frac{(N_m - n^*)}{(N_m - n^*)} + \ln(N_m - n^*) \right.$$
$$\left. -1 - 2\ln n^* - \frac{2n^*}{n^*} + 2 \right\} = 0 \tag{12-87}$$

Thus,

$$\ln[(N_h - n^*)(N_m - n^*)/n^{*2}] = -u/kT \tag{12-88}$$

which may be written

$$\frac{N_{10}^{*2}}{N_{11}N_{00}} = 4\, e^{u/kT} \tag{12-89}$$

from Eqs. (12-68), (12-69), and the relation $n^* = N_{10}^*/2$. This has the form of a chemical equilibrium for the "reaction"

$$N_{00} + N_{11} \rightleftarrows N_{10}^* \tag{12-90}$$

The change in energy for this reaction is $-u$. The factor of 4 is due to the fact that there is a configurational degeneracy of 2 attached to N_{10}^* because two forms ($-\bullet\!\!-\!\!\circ-$ and $-\circ\!\!-\!\!\bullet-$) can exist. Thus, Eq. (12-89) can be written

$$\frac{N_{10}^{*2}}{N_{11}N_{00}} = \frac{q_{10}^2}{q_{11}q_{10}}\, e^{-\Delta\epsilon/kT} = \frac{(2)^2}{(1)(1)}\, e^{u/kT} \tag{12-91}$$

According to Eq. (12-88),

$$\theta n^{*2} = n^{*2} - (N_m + N_h)n^* + N_h N_m \tag{12-92}$$

where $\theta = \exp(-u/kT)$. Since the total number of sites is $N_t = N_m + N_h$, this becomes

$$n^{*2}(\theta - 1) + n^* N_t - N_h N_m = 0 \tag{12-93}$$

The solution is

$$2n^* = N_{10}^* = \frac{-N_t \pm \sqrt{N_t^2 + 4N_h N_m(\theta - 1)}}{\theta - 1} \tag{12-94}$$

When $u \to 0$, $\theta \to 1$. For small values of u, θ will be $1 + \delta + \cdots$ where δ is small.[20] Thus, in this case,

$$N_{10}^* = \frac{-N_t \pm \sqrt{N_t^2 + 4N_h N_m \delta}}{\delta} = \frac{N_t}{\delta}\{-1 \pm \sqrt{1 + 4f(1-f)\delta}\} \tag{12-95}$$

where $f = N_m/N_t$. When the square root function is expanded in a power series, the result is

$$N_{10}^* = \frac{N_t}{\delta}\{-1 + (1 + 2f(1-f)\delta + \cdots)\} \tag{12-96}$$

But, Eq. (12-93) can be solved directly for $\theta = 1$, giving $N_{10}^* = 2f(1-f)N_t$. Equation (12-96) will only go to this proper limit if the plus sign in front of the square root is chosen. Thus, in general,

$$N_{10}^* = \frac{N_t}{(\theta - 1)}\{-1 + \sqrt{1 + 4f(1-f)(\theta - 1)}\} \tag{12-97}$$

With this value, $\ln Q$ can be evaluated and the thermodynamic properties evaluated. The equation of state is evaluated from Eq. (9-66). However, since a one-dimensional lattice model is being dealt with, care must

---

[20]It is assumed here that $u < 0$ corresponding to an attractive potential between neighbours.

be taken with the definitions of p and V. The one-dimensional pressure has the dimensions of a force and the one-dimensional volume is a length. The volume per molecule, V, is equal to $\tilde{V}/N_t$, where $\tilde{V}$ is the total volume of the lattice, and $N_t = N_h + N_m$ is the number of lattice sites.   Equation (9-66) may be written

$$p = RT \left(\frac{\partial \ln Q}{\partial \tilde{V}} \right)_{N_m, T} \tag{12-98}$$

A cell volume can be defined as

$$\omega = \tilde{V}/N_t \tag{12-98a}$$

The cell volume probably expands slightly as the density is lowered. However, it is assumed here that it is independent of density and temperature.   Changes in volume, $d\tilde{V}$, may be regarded as due to a change in the total number of sites, dN_t, since ω is constant.   Thus

$$d\tilde{V} = \omega \, dN_t \tag{12-98b}$$

By this, it is implied that space is divided into increments of width ω, and changes in volume imply changes in the total number of lattice sites for fixed N_m.   Since $V = \tilde{V}/N_m$, it follows that

$$d\tilde{V} = V(N_m/N_t) \, dN_t = Vf \, dN_t \tag{12-98c}$$

where, as before, $f = N_m/N_t$ is the fraction of sites occupied by molecules. Therefore, Eq. (12-98) can be re-expressed as

$$\Phi = \frac{pV}{kT} = \frac{1}{f} \left(\frac{\partial \ln Q}{\partial N_t} \right)_{N_m, T} \tag{12-98d}$$

From Eqs. (12-82) and (12-84), write

$$\ln Q = N_m (\ln q - u/kT) + \ln G^* \tag{12-99}$$

where G^* is defined by

$$\ln G^* = \ln [g(N_h, N_m, N_{10}^*)] + N_{10}^* u/2kT \tag{12-100}$$

Note that $\ln G^*$ is a maximum for fixed values of N_t and N_m; that is,

$$\left(\frac{\partial \ln G^*}{\partial N_{10}^*} \right)_{N_m, N_t} = 0 \tag{12-101}$$

since N_{10}^* is the value of N_{10} that maximizes G.   Furthermore, since $N_t = N_h + N_m$, any *two* of the variables N_t, N_m, and N_h suffice as independent parameters of the system.   Since N_{10}^* is determined by Eq. (12-97) in terms of *any two* of the variables N_t, N_m, N_h, it follows that

$$\left(\frac{\partial}{\partial N_t} \right)_{N_m} = \left(\frac{\partial}{\partial N_t} \right)_{N_{10}^*, N_m} + \left(\frac{\partial N_{10}^*}{\partial N_t} \right)_{N_m} \left(\frac{\partial}{\partial N_{10}^*} \right)_{N_m, N_t} \tag{12-102}$$

When Eqs. (12-98), (12-99), and (12-102) are combined, the result is

$$\Phi = \frac{1}{f}\left(\frac{\partial \ln G^*}{\partial N_t}\right)_{N_m} = \frac{1}{f}\left\{\left(\frac{\partial \ln G^*}{\partial N_t}\right)_{N_{10}^*,N_m} + \left(\frac{\partial N_{10}^*}{\partial N_t}\right)_{N_m}\left(\frac{\partial \ln G^*}{\partial N_{10}^*}\right)_{N_m,N_t}\right\}$$

(12-103)

From Eq. (12-101), it may be concluded that

$$\Phi = \frac{1}{f}\left(\frac{\partial \ln G^*}{\partial N_t}\right)_{N_{10}^*,N_m}$$

(12-104)

After Eq. (12-100) is combined with Eq. (12-104), the following is obtained:

$$\Phi = \frac{1}{f}\left(\frac{\partial \ln g}{\partial N_t}\right)_{N_{10}^*,N_m}$$

(12-105)

assuming u is independent of N_t. When Stirling's approximation is used for the factorials in g, the result is

$$\ln g = N_h \ln N_h + N_m \ln N_m - (N_h - N_{10}^*/2)\ln(N_h - N_{10}^*/2)$$
$$- (N_m - N_{10}^*/2)\ln(N_m - N_{10}^*/2) - N_{10}^*\ln N_{10}^* + N_{10}^*\ln 2$$

(12-106)

Thus, after replacing N_h by $N_t - N_m$, the result is

$$\left(\frac{\partial \ln g}{\partial N_t}\right)_{N_{10}^*,N_m} = 1 + \ln(N_t - N_m) - 1 - \ln(N_t - N_m - N_{10}^*/2)$$

$$= \ln\left[\frac{(1-f)}{1-f-N_{10}^*/2N_t}\right]$$

(12-107)

$$\Phi = \frac{1}{f}\ln\left\{\frac{(1-f)}{(1-f)-\frac{1}{2}(\theta-1)^{-1}[-1+\sqrt{1+4f(1-f)(\theta-1)}]}\right\}$$

(12-108)

In the limit as $u \to 0$, $\theta \to 1$, and $N_{10}^* \to 2f(1-f)N_t$, so

$$\lim_{\theta\to 1}\Phi = \frac{1}{f}\ln\left(\frac{1}{1-f}\right) = \lim_{\theta\to 1}(p\tilde{V}/RT)$$

(12-109)

For a very dilute substance, where $f \ll 1$,

$$-\ln(1-f) \cong f + \cdots$$

and

$$\lim_{\substack{\theta\to 1\\f\to 0}}\Phi = 1$$

(12-110)

Equation (12-110) is an ideal-gas equation since $\Phi = p\tilde{V}/RT = pV/kT$. When $\theta = 1$, but f is not very small, the effect of Eq. (12-108) is to increase p above the ideal-gas value for moderate f. A typical isotherm for $\theta = 1$ is shown in Fig. 12-13, with T and N_m constant for the figure. At low f, Eq. (12-109) reduces to Eq. (12-110). When $\theta \neq 1$, the only effect is to make the isotherms rise very rapidly as $f \to 1$. It can be shown that the isotherms do not show any indication of a phase transition and there is no evidence of a liquid state. The one-dimensional lattice model is clearly inadequate to explain the liquid state.

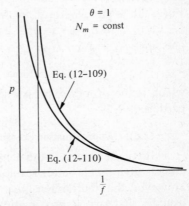

FIGURE 12-13 Comparison of equations of state for a one-dimensional lattice with no interacting forces between particles. Equation (12-109) gives the exact equation of state for $u = 0$, whereas (12-110) is only approximate for low densities.

The entropy predicted by the one-dimensional lattice model is obtained as follows:[21]

$$S = k \ln Q + E/T$$

$$E = \frac{1}{2} N_m kT + N_{11} u = \frac{N_m kT}{2} + (N_m - N_{10}^*/2) u$$

$$\ln Q = N_m (\ln q - u/kT) + \ln g (N_h, N_m, N_{10}^*) + N_{10}^* u/2kT \qquad (12\text{-}111)$$

$$S = N_m k [\tfrac{1}{2} + \ln q] + k \ln g \qquad (12\text{-}112)$$

This can be written as

$$S = S_{\text{int}} + S_{\text{config}} \qquad (12\text{-}113)$$

where

$$S_{\text{int}} = N_m k [\tfrac{1}{2} + \ln q] \qquad (12\text{-}114)$$

---

[21]Note that no communal entropy term is arbitrarily introduced here.

is the internal entropy associated with a single configuration of the ensemble of holes and molecules, and

$$S_{\text{config}} = k \ln g_c \cong k \ln g(N_h, N_m, N_{10}^*) \qquad (12\text{-}115)$$

where g_c is the total number of configurations of the holes and molecules.[22] In the calculation here, $\ln g_c$ has been replaced by $\ln [g(N_h, N_m, N_{10}^*)]$ corresponding to the maximum term in the sum over all configurations. For a dilute gas where $N_m \ll N_h$, it follows from Eq. (12-97) that $N_{10}^* \cong 2f(1-f)N_t \cong 2N_m$. In this case, Eq. (12-74) reduces to Eq. (12-80), and since $N_h \cong N_h + N_m$,

$$S_{\text{config}} \cong -k \ln (N_m!) \qquad (12\text{-}116)$$

which is the proper expression for an ideal gas of N_m indistinguishable particles.

12.6.2   Approximation Procedures for the One-Dimensional Lattice Model

It will be shown in the next section that the multidimensional lattice models give much better results than the one-dimensional lattice model. However, the multidimensional lattice models can not usually be solved exactly, and two approximation methods are often employed. These approximation procedures will be illustrated for the one-dimensional lattice where an exact solution is available for comparison. The approximate procedures are, of course, not necessary in the one-dimensional lattice model, but considerable insight is gained by examination of their use in this case. In the next section, the approximate procedures will be applied to a multidimensional lattice, where they are much more necessary.

First, consider the *Bragg-Williams approximation*. It is assumed that it is not known that Eq. (12-83) exists, and an approximate result is desired for the partition function of an ensemble of N_h holes and N_m molecules from Eq. (12-82). In the Bragg-Williams approach, it is assumed that $g(N_h, N_m, N_{10})$ has a sharp maximum at $N_{10} = N_{\text{BW}}$, and that $\exp[-(N_m - N_{10}/2)u/kT]$ does not vary greatly across the region where g is appreciable, as shown in Fig. 12-14. In this case, the exponential function can be taken out of the sum in Eq. (12-82) as $\exp[-(N_m - N_{\text{BW}}/2)u/kT]$ and the following approximation can be made:

$$Q \cong q^{N_m} \exp[-(N_m - N_{\text{BW}}/2)u/kT] \sum_{N_{10}} g(N_h, N_m, N_{10}) \qquad (12\text{-}117)$$

Note that N_{BW}, the value of N_{10} which maximizes g, is *not* equal to N_{10}^*,

---

[22]Given by Eq. (12-80).

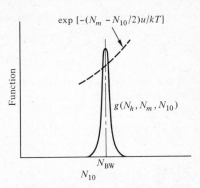

FIGURE 12-14 The functions exp $[-(N_m - N_{10}/2)u/kT]$ and $g(N_h, N_m, N_{10})$ vs N_{10}. In the Bragg-Williams approximation, it is assumed that g has a sharp maximum at $N_{10} = N_{BW}$, and the exponential function does not vary greatly across the width of g.

which is the value of N_{10} which maximizes the function under the sum sign in Eq. (12-82). The sum in Eq. (12-117) is the sum over all arrangements of N_h holes and N_m molecules. Thus,

$$Q \cong q^{N_m} \exp\left[-(N_m - N_{BW}/2)u/kT\right] \frac{(N_m + N_h)!}{N_m! N_h!} \quad (12\text{-}118)$$

To calculate N_{BW}, the further assumption is made that since the potential energy u plays an insignificant role in determining the relative importance of various configurations, and therefore, it can be assumed that the placement of molecules in lattice sites is purely random. That is, the probability of arrangements with different values of N_{11} is neither enhanced nor reduced by u. Thus, the most probable value of N_{10}, N_{BW} is approximated as follows. There are N_m randomly sprinkled molecules in the lattice. The fraction of sites unoccupied is N_h/N_t. Thus, the number of adjacent molecule-hole combinations is $2N_m(N_h/N_t)$, and the number of molecule-molecule combinations is $\frac{1}{2}(2N_m)(N_m/N_t) = N_m^2/N_t$. Therefore, since $N_{BW} = 2N_mN_h/N_t$,

$$Q \cong q^{N_m} \exp\left[-N_m^2 u/N_t kT\right] \left(\frac{N_t!}{N_m! N_h!}\right) \quad (12\text{-}119)$$

Clearly, the Bragg-Williams approximation is based on the assumptions that all configurations are equally probable regardless of the value of $N_{11}u$, and that the total potential energy is $\underline{N}_{11}u$, where $\underline{N}_{11}$ is the value of N_{11} that maximizes $g(N_h, N_m, N_{10})$.

The thermodynamic properties may then be calculated:

$$\ln Q = N_m \ln q - N_m^2 u/N_t kT + N_t \ln N_t - N_m \ln N_m - N_h \ln N_h \quad (12\text{-}119\text{a})$$

$$E = \tfrac{1}{2} N_m kT + N_m^2 u/N_t kT \quad\quad\quad\quad\quad\quad\quad\quad\quad (12\text{-}119\text{b})$$

$$S = k \ln Q + E/T = N_m \ln q + \tfrac{1}{2} N_m k + \ln \left(\frac{N_t!}{N_m! N_h!} \right) \quad (12\text{-}119\text{c})$$

$$\Phi = \frac{1}{f} \left(\frac{\partial \ln Q}{\partial N_t} \right)_{N_m, T} = \frac{1}{f} \left[\frac{f^2 u}{kT} - \ln (1-f) \right] \quad (12\text{-}119\text{d})$$

where $f = N_m/N_t$ as before. The entropy is the same as one would have from Eq. (12-112) if the molecules were randomly distributed on the lattice [that is, if $u = 0$ in Eq. (12-112)]. The equation of state is qualitatively different[23] than the exact result for a one-dimensional lattice. These results will not be discussed further in the one-dimensional case, but the Bragg-Williams approximation for multidimensional lattices will be returned to.

The *quasi-chemical* approximation is only slightly more complicated than the Bragg-Williams approximation, and is a somewhat better procedure. The term "quasi-chemical" comes from a derivation of this model due to Fowler and Guggenheim[24] in which the procedure starts with Eq. (12-91). Of course, in a one-dimensional lattice, Eq. (12-91) is exact, and the quasi-chemical approximation turns out to be exact for a one-dimensional lattice. For a multidimensional lattice, Eq. (12-91) is not generally correct, and the quasi-chemical approximation differs from the exact result. Here, a simple derivation of the quasi-chemical approximation is given, based on Bethe,[25] and it leads to the same result. In the quasi-chemical model, it is assumed that there is a collection of N_t pairs of adjacent sites[26] composed of a total of N_h holes and N_m molecules, and that *the pairs are independent of one another*. This statement is the crux of the whole approach, and it is important that it be understood. If there is the lattice arrangement

molecule　　　　　　　　　　　　hole

$i-2$　$i-1$　i　$i+1$　$i+2$

then if pair $(i-1, i)$ of sites contains, for example, two molecules, this puts a constraint on what pair $(i, i+1)$ can possibly consist of. Evidently,

[23]The Bragg-Williams approximation incorrectly leads to a liquid phase in one dimension, whereas the exact solution does not.

[24]R. H. Fowler and E. A. Guggenheim, *Proc. Roy. Soc.* (London) **A174**, 189 (1940).

[25]H. Bethe, *Proc. Roy. Soc.* (London) **A150**, 552 (1935).

[26]The number of adjacent pairs is equal to the total number of sites.

pair $(i, i+1)$ can consist of either –●—○– or –●—●–. Furthermore, if pair $(i+1, i+2)$ happened to be –○—○–, this, coupled with $(i-1, i)$ being –●—●–, would require that $(i, i+1)$ be –●—○–. Therefore, in actuality, the pairs are not *independent of one another*. Assuming the pairs are independent leads to the quasi-chemical approximation. Let each pair be one of the four types:

–●—○–   –○—●–

–●—●–   –○—○–

Let there be $N_{●○}$ pairs of type (●○), and so forth. Then, for a large number of sites, it may be assumed that

$$N_{●○} = N_{○●} = N_{10}/2 \qquad (12\text{-}120)$$

since $N_{●○} + N_{○●} = N_{10}$.

From Eqs. (12-68) and (12-69),

$$N_{●●} = N_{11} = N_m - N_{10}/2$$
$$N_{○○} = N_{00} = N_h - N_{10}/2 \qquad (12\text{-}121)$$

Now if the pairs of adjacent sites were *independent* of one another, the number of ways to arrange the N_t pairs into $N_{●○}$, $N_{○●}$, $N_{●●}$, and $N_{○○}$ pairs of the various types would be calculable as follows. Imagine that there are N_t objects to be divided into four categories, with $N_{●○}$, $N_{○●}$, $N_{●●}$, and $N_{○○}$ objects in each of the respective categories. Since it is assumed that the numbers in each category are independent of one another, the total number of arrangements is $W(N_h, N_m, N_{10})$, given by

$$\frac{N_t!}{N_{●●}! N_{○○}! N_{●○}! N_{○●}!} = \frac{N_t!}{(N_m - N_{10}/2)!(N_h - N_{10}/2)![(N_{10}/2)!]^2} \qquad (12\text{-}122)$$

This formula evidently leads to an overestimate of the total number of arrangements because it includes arrangements where, for example, pair $(i, i+1)$ is (●●) and pair $(i+1, i+2)$ is (○○), which are really impossible. The correct total number of arrangements is the number of ways to divide N_t sites into N_h holes and N_m molecules,

$$\frac{N_t!}{N_h! N_m!}$$

In order to keep the quasi-chemical approximation normalized to the proper total number of arrangements, write

$$g_{QC}(N_h, N_m, N_{10}) = C(N_h, N_m) W(N_h, N_m, N_{10}) \qquad (12\text{-}123)$$

where C is a normalizing constant such that

$$\sum_{N_{10}} g_{QC}(N_h, N_m, N_{10}) = \frac{N_t!}{N_h! N_m!} = C(N_h, N_m) \sum_{N_{10}} W(N_h, N_m, N_{10})$$
$$(12\text{-}124)$$

With this normalized value of g_{QC}, we have from Eq. (12-82),

$$Q \cong q^{N_m} e^{-N_m u/kT} \sum_{N_{10}} g_{QC}(N_h, N_m, N_{10}) e^{N_{10}u/2kT} \qquad (12\text{-}125)$$

The $\ln Q$ is evaluated, as usual, by replacing the sum in Eq. (12-125) by the maximum term. The value of N_{10}, N_{10}^*, that maximizes the term under the summation in Eq. (12-125) is obtained from

$$C \frac{\partial}{\partial N_{10}} [W e^{N_{10}u/2kT}] = 0 \qquad (12\text{-}126)$$

Using Stirling's approximation for the factorials in g_{QC},

$$\begin{aligned}
\ln (W e^{N_{10}u/2kT}) \cong\ & N_t \ln N_t - (N_m - N_{10}/2) \ln (N_m - N_{10}/2) \\
& - (N_h - N_{10}/2) \ln (N_h - N_{10}/2) \qquad (12\text{-}127) \\
& - N_{10} \ln (N_{10}/2) + N_{10}u/2kT
\end{aligned}$$

Therefore, Eq. (12-126) may be evaluated as

$$0 = \tfrac{1}{2} \ln (N_m - N_{10}^*/2) + \tfrac{1}{2} \ln (N_h - N_{10}^*/2) - \ln (N_{10}^*/2) + u/2kT \qquad (12\text{-}128)$$

Thus,

$$N_{10}^{*2} = 4(N_m - N_{10}^*/2)(N_h - N_{10}^*/2) e^{u/kT} \qquad (12\text{-}129)$$

or

$$\frac{N_{10}^{*2}}{N_{11}N_{00}} = 4 e^{u/kT} \qquad (12\text{-}130)$$

Thus, N_{10}^* satisfies the chemical equilibrium relation, Eq. (12-91). The value of N_{10}^* in Eq. (12-130) maximizes $g_{QC} \exp(N_{10}u/2kT)$. For comparison, it should be noted that in the Bragg-Williams approximation,

$$\frac{N_{BW}^2}{N_{11}N_{00}} = \frac{(2N_m N_h/N_t)^2}{(N_m^2/N_t)(N_h^2/N_t)} = 4 \qquad (12\text{-}131)$$

showing that the Bragg-Williams approximation is based on the assumption that u has no effect on determining the most probable value of N_{10}. In the special case of a one-dimensional lattice, N_{10}^* is the same in the quasi-chemical approximation as in the exact treatment. Now evaluate $W(N_h, N_m, N_{10}^*)$. Upon comparing Eqs. (12-122) and (12-74), it is seen that

$$W(N_h, N_m, N_{10}) = \left(\frac{N_t!}{N_h! N_m!}\right) g(N_h, N_m, N_{10})$$

Therefore, since

$$\sum_{N_{10}} g(N_h, N_m, N_{10}) = \frac{N_t!}{N_h! N_m!}$$

it follows from Eq. (12-124) that

$$C = \left(\frac{N_t!}{N_h! N_m!}\right)^{-1} \tag{12-132}$$

With this choice of C, it follows that

$$g_{QC}(N_h, N_m, N_{10}) = g(N_h, N_m, N_{10}) \tag{12-133}$$

and the thermodynamic properties are the same in the one-dimensional lattice, whether calculated by the quasi-chemical, or exact methods. This equality does not hold for a multidimensional lattice.

12.6.3   Multidimensional Ising Lattice Models

The one-dimensional lattice model fails to give a reasonable description of the liquid state. A two-dimensional square lattice problem can be solved exactly using the same sort of model as in the one-dimensional case.[27] Molecules are assumed to interact only with nearest neighbors along a square lattice, as shown in Fig. 12-15. The total energy is assumed to be a sum of pairwise interactions between nearest neighbors. The mathematics involved is very abstract, and it is very difficult to formulate an expression for $g(N_h, N_m, N_{10})$. We shall only briefly mention a few results. The most striking result of the solution of the two-dimensional square lattice is the possibility of existence of two phases, one of which is dilute and may be interpreted as "gas," and the

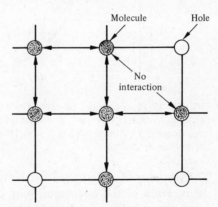

FIGURE 12-15 Illustration of the simple model of a two-dimensional square lattice with only nearest neighbor interactions.

[27]G. F. Newell and E. W. Montroll, *Revs. Mod. Phys.* **25**, 353 (1953). In this comprehensive review, many important references are given. Also, see C. N. Yang and T. D. Lee, *Phys. Rev.* **87**, 404, 410 (1952).

other of which has a much higher value of $f = N_m/N_t$ and may be interpreted as "liquid." The importance of this model is that it is the only exactly soluble problem in statistical mechanics that leads to a discontinuous phase boundary. The equation of state has the form[28] shown in Fig. 12-16. A critical point where the two-dimensional pressure satisfies the critical relations

$$\left(\frac{\partial p}{\partial V}\right)_{T,N_m} = \left(\frac{\partial^2 p}{\partial V^2}\right)_{T,N_m} = 0 \qquad (12\text{-}134)$$

is called the critical temperature, T_c, and is found to be the solution of the equation[28]

$$e^{u/2kT_c} = \sqrt{2} - 1 = 0.414 \qquad (12\text{-}135)$$

Evidently, u must be negative to have a critical point. The horizontal axis in Fig. 12-16 is $1/f = N_t/N_m$, and is proportional to the volume for fixed N_m. The critical volume is found to satisfy the relation $f = 1/2$.

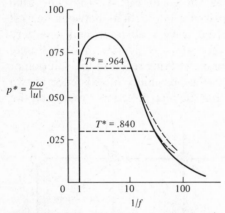

FIGURE 12-16 Equation of state for the two-dimensional lattice gas. The fraction of sites occupied by molecules is f, and $1/f$ is the molecular volume.

The isotherms for various T are shown as dotted lines. For $T < T_c$, the intersection of the horizontal part of an isotherm with the solid curve gives the densities of liquid and gas phases in equilibrium with one another. At high T, and moderate density, the isotherms are similar to "ideal-gas" isotherms. At high densities where f approaches 1, the isotherms become nearly vertical. It can be shown[27,28] after much

---

[28]T. D. Lee and C. N. Yang, *Phys. Rev.* **87**, 410 (1952).

arduous calculation, that the pressure (force/length) at which two phases co-exist at any T is given by

$$\frac{p^*}{T^*} = \frac{p\omega}{kT} = \ln(1+\mu) + \frac{1}{2\pi}\int_0^\pi \tag{12-136}$$

$$\times \ln\{\tfrac{1}{2}[1+(1-k_1^2\sin^2\phi)^{1/2}]\}\,d\phi$$

where $\mu = e^{u/kT}, p^* = p\omega/|u|, T^* = kT/|u|$, and

$$k_1 = \frac{4\mu^{1/2}(1-\mu)}{(1+\mu)^2} \tag{12-137}$$

If f_L and f_G are defined as the values of f corresponding to the liquid and gas phases in equilibrium, it can be shown that

$$f_G(\mu) = \frac{1}{2}\left\{1 - \left[\frac{(1+\mu)(1-6\mu+\mu^2)^{1/2}}{(1-\mu)^2}\right]^{1/4}\right\} \tag{12-138}$$

$$f_L(\mu) = 1 - f_G(\mu) \tag{12-139}$$

If the total number of sites occupied by the liquid phase is N_{tL} and the number occupied by the gas phase is N_{tG}, then it can be shown that if the overall average value of $f = \frac{1}{2}$,

$$f_G N_{tG} + f_L N_{tL} = \tfrac{1}{2}N_t \tag{12-140}$$

Since $N_{tG} + N_{tL} = N_t$, it follows from Eqs. (12-139) and (12-140) that

$$f_G N_{tG} + (1-f_G)(N_t - N_{tG}) = N_t/2$$

and therefore,

$$N_{tG} = N_{tL} = N_t/2$$

Thus, the volumes of both phases are equal if the overall fraction of sites occupied is $\frac{1}{2}$.

There are many other interesting aspects of the exact solution of the two-dimensional lattice that cannot be considered here. Instead, consider the approximation methods discussed in Section 12.6.2 applied to a general *three*-dimensional lattice composed of N_t sites, N_m of which are filled. Let the interaction energy between any pair of nearest-neighbor molecules be u, and let the total interaction energy be a sum over nearest-neighbor pairs. Assume that the lattice is packed in a regular array such that each site is surrounded by z nearest-neighbor sites, where, for example,

> $z = 2$ for a linear lattice
> $z = 4$ for a planar square lattice
> $z = 12$ for a close packed cubic lattice

If a suitable expression for $g(N_h, N_m, N_{11})$ could be written, the partition function could be evaluated from

$$Q = q^{N_m} \sum_{\substack{\text{all} \\ N_{11}}} g(N_h, N_m, N_{11})\, e^{-N_{11}u/kT} \qquad (12\text{-}141)$$

Unfortunately, g is generally an extremely complicated function. Even for the case of a two-dimensional square lattice the exact calculation of g is beyond the scope of this book. Therefore, consider the Bragg-Williams approximation first. It will be recalled from the discussion of one-dimensional lattices that in the Bragg-Williams approach, it is assumed that the interaction energy does not determine the maximum term in Eq. (12-141), and the molecules are distributed randomly in the lattice. Therefore,

$$Q_{\text{BW}} = q^{N_m}\, e^{-\underline{N}_{11}u/kT} \sum_{N_{11}} g(N_h, N_m, N_{11}) \qquad (12\text{-}142)$$

where $\underline{N}_{11}$ is the most probable value of N_{11} corresponding to the molecules being distributed on the sites purely randomly without regard to the interactions generated. A typical molecule at a site has z nearest-neighbor sites, of which N_m/N_t are occupied on average, for a random distribution. Thus, the average number of molecule-molecule pairs produced by the molecule and its nearest-neighbor molecules is zN_m/N_t. Since there are N_m such molecules together, there must be a total of $\frac{1}{2}(N_m)(zN_m/N_t) = zN_m^2/2N_t = \underline{N}_{11}$ occupied nearest neighbor pairs. The factor of $\frac{1}{2}$ is required because otherwise, each occupied pair would be considered twice, once with each molecule as the central one. The total number of ways to arrange N_t sites as N_m molecules and N_h holes is

$$\sum_{N_{11}} g(N_h, N_m, N_{11}) = \frac{N_t!}{N_m! N_h!} \qquad (12\text{-}143)$$

Thus, in the Bragg-Williams approximation,

$$Q_{\text{BW}} = q^{N_m}\, e^{-zN_m^2 u/2N_t kT}\, \frac{N_t!}{N_m! N_h!} \qquad (12\text{-}144)$$

This equation is the same as Eq. (12-119), except for the factor z. The various thermodynamic properties can then be derived as in Eqs. (12-119a) to (12-119d) except there are now 3 dimensions:

$$\ln Q = N_m \ln q - zN_m^2 u/2N_t kT + N_t \ln N_t \\ - N_m \ln N_m - N_h \ln N_h \qquad (12\text{-}145)$$

$$E = \tfrac{3}{2} N_m kT + zN_m^2 u/2N_t kT \qquad (12\text{-}146)$$

$$S = N_m \ln q + \tfrac{3}{2} N_m k + \ln\left[\frac{N_t!}{N_h! N_m!}\right] \qquad (12\text{-}147)$$

$$\Phi = \frac{pV}{kT} = \frac{1}{f}\left[\frac{zf^2u}{2kT} - \ln(1-f)\right] \tag{12-148}$$

Note that Eq. (12-147) is independent of u because the distribution is assumed random. When $z = 2$, these equations reduce to the one-dimensional case. The isotherms of p vs f^{-1} are illustrated in Fig. 12-17 for the case $z = 4$, corresponding to a square-plane lattice. It can be seen that a critical temperature exists, below which "looped" isotherms are obtained. These looped isotherms are an artifact of the model used which assumes only one phase exists. As in the case of the Van der Waals equations, the loops occur in the region where gas and liquid can co-exist. By drawing horizontal lines in such a way as to equalize the areas of the positive and negative loops (see Fig. 12-17), the isotherms shown in Fig. 12-18 are obtained. The curves are qualitatively similar to those obtained by exact solution of a square lattice in two dimensions. The critical point occurs where Eqs. (12-134) hold true. Since N_m is fixed, and $\tilde{V}$ is proportional to N_t, it follows that $d\tilde{V}$ is proportional to dN_t. Furthermore by defining $s = f^{-1}$ it follows that dN_t is proportional to ds at constant N_m. Since $V = N_t\omega/N_m = \omega/f$ where $\omega = \tilde{V}/N_t$, and $\omega =$

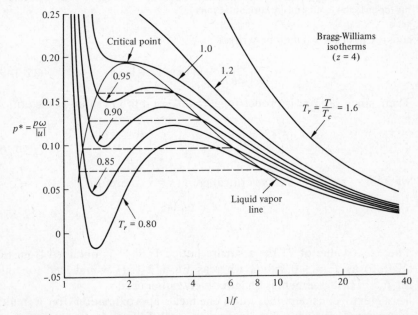

FIGURE 12-17 Isotherms for the two-dimensional lattice gas in the Bragg-Williams approximation. The horizontal dashed lines are drawn to equalize areas of positive and negative loops in the region where "looping" occurs.

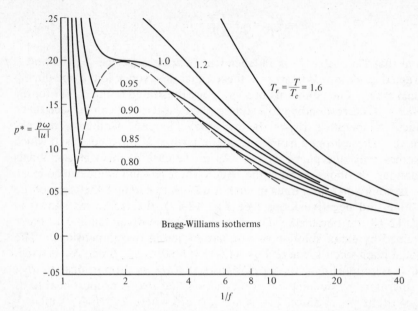

FIGURE 12-18 Equation of state for the two-dimensional square lattice gas according to the Bragg-Williams approximation obtained from Fig. 12-14 by replacing the loops in the isotherms by horizontal lines.

const, Eq. (12-148) may be written

$$p = \frac{kT}{\omega}\left[\frac{zs^{-2}u}{2kT} - \ln{(s-1)} + \ln s\right] \qquad (12\text{-}149)$$

Then, since $(\partial p/\partial s)$ is proportional to $\partial p/\partial V$, it follows that at the critical point,

$$\left(\frac{\omega}{kT}\right)\frac{\partial p}{\partial s} = 0 = \frac{-B}{s^3} - \frac{1}{s-1} + \frac{1}{s} \qquad (12\text{-}150)$$

where $B = zu/2kT$.   At the critical point, $s = \frac{1}{2}$; therefore,

$$T_c = -zu/4k$$
$$f_c = \tfrac{1}{2} \qquad (12\text{-}151)$$

The exact value of T_c for a square lattice $(z = 4)$ is obtained from Eq. (12-135) as $T_c = -0.568u/k$ whereas Eq. (12-151) would lead to $-u/k$ for T_c.   The agreement is quantitatively rather crude.

Next, consider the quasi-chemical approximation for a multi-dimensional lattice.   The procedure will follow the quasi-chemical procedure developed for a one-dimensional lattice, but takes into account the existence of z nearest neighbors.   The number of pairs of various types are found to be

$$N_{\bullet\circ} = N_{\circ\bullet} = N_{10}/2$$
$$N_{\bullet\bullet} = N_{11} = (zN_m - N_{10})/2 \tag{12-152}$$
$$N_{\circ\circ} = N_{00} = (zN_h - N_{10})/2$$
$$N_p = (z/2)(N_m + N_h)$$

instead of Eqs. (12-120) and (12-121).[29] The calculation proceeds as before. Assume the pairs are independent, and in analogy to Eq. (12-122), write

$$W(N_h, N_m, N_{10}) = \frac{N_p!}{N_{\bullet\bullet}! N_{\circ\circ}! N_{\bullet\circ}! N_{\circ\bullet}!} \tag{12-153}$$

as the number of ways to arrange N_p pairs of adjacent sites. As in the one-dimensional calculation, set

$$g_{QC}(N_h, N_m, N_{10}) = C(N_h, N_m) W(N_h, N_m, N_{10}) \tag{12-154}$$

such that $\Sigma\, g_{QC}$ is normalized to the total number of ways to arrange N_t sites:

$$\sum_{N_{10}} g_{QC}(N_h, N_m, N_{10}) = \frac{N_t!}{N_m! N_h!} = C(N_h, N_m) \sum_{N_{10}} W(N_h, N_m, N_{10}) \tag{12-155}$$

To evaluate C, take the natural log of each side of Eq. (12-155), and assume that

$$\ln\left(\sum_{N_{10}} W\right) \cong \ln W_{\max} \tag{12-156}$$

where $W_{\max}$ is the maximum term in the sum. Thus N_{10}^* is required, the value of N_{10} that maximizes W. Set

$$\left(\frac{\partial \ln W}{\partial N_{10}}\right)_{N_m, N_h} = 0 \tag{12-157}$$

When Stirling's approximation is used for the factorials in Eq. (12-153) the following is found:

$$\ln W = N_p \ln N_p - \tfrac{1}{2}(zN_m - N_{10}) \ln\left[\tfrac{1}{2}(zN_m - N_{10})\right] \tag{12-158}$$
$$- \tfrac{1}{2}(zN_h - N_{10}) \ln\left[\tfrac{1}{2}(zN_h - N_{10})\right] - N_{10} \ln N_{10}/2$$

Therefore, Eq. (12-157) becomes

$$\tfrac{1}{2}\ln\left[\tfrac{1}{2}(zN_m - N_{10}^*)\right] + \tfrac{1}{2}\ln\left[\tfrac{1}{2}(zN_h - N_{10}^*)\right] - \ln\left(N_{10}^*/2\right) = 0$$

---

[29]Note that there are N_p pairs in the general case, and N_p only reduces to N_t if $z = 2$.

or,

$$N_{10}^* = \frac{zN_mN_h}{N_t} \tag{12-160}$$

The maximum value of W is, therefore, obtained from

$$\ln W_{\max} = \frac{z}{2} N_t \ln \left(\frac{zN_t}{2}\right) + \frac{N_{10}^*}{2} \ln \left[(N_{10}^*)^2/4\right] - N_{10}^* \ln (N_{10}^*/2)$$
$$- \frac{zN_m}{2} \ln \left[\frac{z}{2} \frac{N_m^2}{N_T}\right] - \frac{zN_h}{2} \ln \left[\frac{z}{2} \frac{N_h^2}{N_t}\right] \tag{12-161}$$

The second and third terms cancel, and the following is obtained:

$$\ln W_{\max} = zN_t \ln N_t - zN_m \ln N_m - zN_h \ln N_h \tag{12-162}$$

Using Stirling's approximation in reverse, it can be inferred that

$$W_{\max} = \left(\frac{N_t!}{N_m!N_h!}\right)^z \tag{12-163}$$

Therefore, from Eqs. (12-155), (12-156), and (12-163),

$$C(N_h, N_m) = \left(\frac{N_t!}{N_m!N_h!}\right)^{1-z} \tag{12-164}$$

To obtain the partition function, Eq. (12-125) is replaced by

$$Q \cong q^{N_m} e^{-zN_m u/2kT} \sum_{N_{10}} g_{QC} e^{N_{10}u/2kT} \tag{12-165}$$

The same procedure is used.   Instead of Eq. (12-127), there results

$$\ln \{W e^{N_{10}u/2kT}\} \cong N_p \ln N_p - \tfrac{1}{2}(zN_m - N_{10}) \ln \left[\tfrac{1}{2}(zN_m - N_{10})\right]$$
$$- \tfrac{1}{2}(zN_h - N_{10}) \ln \left[\tfrac{1}{2}(zN_h - N_{10})\right] \tag{12-166}$$
$$- N_{10} \ln (N_{10}/2) + N_{10}u/2kT$$

Therefore, Eq. (12-126) can be evaluated to find the value of N_{10}, N_{10}^*, which maximizes $W \exp [N_{10}u/2kT]$ in Eq. (12-166):

$$0 = \tfrac{1}{2} \ln \left[\tfrac{1}{2}(zN_m - N_{10}^*)\right] + \tfrac{1}{2} \ln \left[\tfrac{1}{2}(zN_h - N_{10}^*)\right] - \ln (N_{10}^*/2) + u/2kT \tag{12-167}$$

Thus,

$$N_{10}^{*2} = (zN_m - N_{10}^*)(zN_h - N_{10}^*) e^{u/kT} \tag{12-168}$$

The procedure for solving this is analogous to that used in Eqs. (12-91) to (12-97).   With $\theta = \exp (-u/kT)$ and $f = N_m/N_t$, the following is obtained

$$N_{10}^{*2}(\theta - 1) + N_{10}^*(2N_p) - z^2N_mN_h = 0 \tag{12-169}$$

$$N_{10}^* = \frac{-N_p \pm \sqrt{N_p^2 + z^2N_mN_h(\theta - 1)}}{\theta - 1}$$

$$N_{10}^* = \frac{zN_t}{2(\theta - 1)} \{-1 + \sqrt{1 + 4f(1-f)(\theta - 1)}\} \tag{12-170}$$

It can be seen that Eq. (12-170) reduces to Eq. (12-97) when $z = 2$. The $+$ sign is chosen in front of the $\sqrt{}$ to make $N_{10}^* \to zf(1-f)N_t$ as $\theta \to 1$. The quasi-chemical calculation is now carried out using Eq. (12-170) in Eq. (12-165), with

$$\ln\left\{\sum g_{QC} \, e^{N_{10}u/2kT}\right\} \cong \ln g_{QC}^* + N_{10}^* u/2kT \qquad (12\text{-}171)$$

Therefore,

$$\ln Q = N_m \left(\ln q - zu/2kT\right) + \ln G^* \qquad (12\text{-}172)$$

where

$$\ln G = \ln g_{QC}(N_h, N_m, N_{10}) + N_{10}u/2kT \qquad (12\text{-}173)$$

and G^* is G for N_{10} equal to N_{10}^*. Since $\ln G$ is a maximum at $N_{10} = N_{10}^*$, it follows that

$$\left(\frac{\partial \ln G}{\partial N_{10}}\right)_{N_m, N_t} = 0 \qquad (12\text{-}174)$$

The equation of state is calculated from

$$\Phi = \frac{pV}{kT} = \frac{1}{f}\left(\frac{\partial \ln Q}{\partial N_t}\right)_{N_m, T} \qquad (12\text{-}175)$$

where, again, V is the volume per molecule. From the form of Eq. (12-172), it follows that

$$\Phi = \frac{1}{f}\left(\frac{\partial \ln G^*}{\partial N_t}\right)_{N_m, T} \qquad (12\text{-}176)$$

But from Eqs. (12-174) and (12-102), it can be shown that this reduces to

$$\Phi = \frac{1}{f}\left(\frac{\partial \ln G^*}{\partial N_t}\right)_{N_{10}^*, N_m} = \frac{1}{f}\left(\frac{\partial \ln g_{QC}}{\partial N_t}\right)_{N_{10}^*, N_m} \qquad (12\text{-}177)$$

in analogy to Eqs. (12-104) and (12-105). The following is found:

$$\ln g_{QC} = \frac{z}{2} N_t \ln\left(\frac{z}{2} N_t\right) - \frac{1}{2}(zN_m - N_{10}) \ln\left[\frac{1}{2}(zN_m - N_{10})\right]$$
$$- \frac{1}{2}(zN_h - N_{10}) \ln\left[\frac{1}{2}(zN_h - N_{10})\right] - N_{10} \ln(N_{10}/2) \qquad (12\text{-}178)$$
$$+ (1-z)\left[N_t \ln N_t - N_m \ln N_m - N_h \ln N_h\right]$$

and

$$\left(\frac{\partial \ln g_{QC}}{\partial N_t}\right)_{N_{10}^*, N_m} = \frac{z}{2} \ln\left(\frac{zN_t}{2}\right) - \frac{z}{2} \ln\left[\frac{1}{2}(zN_t - zN_m - N_{10}^*)\right]$$
$$+ (1-z) \ln N_t - (1-z) \ln(N_t - N_m)$$
$$= -\frac{z}{2} \ln\left[1 - f - N_{10}^*/zN_t\right] - (1-z) \ln(1-f) \qquad (12\text{-}179)$$
$$= \ln\left\{(1-f)^{z-1}(1-f - N_{10}^*/zN_t)^{-z/2}\right\}$$

This reduces to Eq. (12-107) when $z = 2$.   Therefore, in general,

$$\frac{pV}{kT} = \frac{1}{f} \ln \left\{ \frac{(1-f)^{z-1}}{\{1-f-\frac{1}{2}(\theta-1)^{-1}[-1+\sqrt{1+4f(1-f)(\theta-1)}]\}^{z/2}} \right\} \quad (12\text{-}180)$$

which reduces to Eq. (12-108) when $z = 2$.   The qualitative behavior of p vs f^{-1} is similar to that obtained in the Bragg-Williams approximation. The critical point occurs at $f = \frac{1}{2}$, such that

$$0 = \left(\frac{\partial p}{\partial f^{-1}}\right)_{N_m, f=1/2} = -\left[f\left(\frac{\partial p}{\partial f}\right)\right]_{N_m, f=1/2}$$

or simply

$$\left(\frac{\partial p}{\partial f}\right)_{f=1/2} = 0 \quad (12\text{-}181)$$

From Eq. (12-180), the result is

$$0 = \left\{ \frac{(z-1)}{(1-f)} \right.$$

$$\left. -\frac{(z/2)\{-1-\frac{1}{2}(\theta-1)^{-1}[-1+4f(1-f)(\theta-1)]^{-1/2}4(2f-1)(\theta-1)\}}{1-f-\frac{1}{2}(\theta-1)^{-1}[-1+\sqrt{1+4f(1-f)(\theta-1)}]} \right\}_{f=1/2}$$

$$(12\text{-}182)$$

Since $2f - 1 = 0$, this simplifies to

$$\frac{-(z-1)}{1/2} + \frac{z/2}{\frac{1}{2}[1-(\phi-1)/(\phi^2-1)]} = 0$$

where $\phi = \theta^{1/2} = \exp[-u/2kT]$.   Therefore,

$$-2(z-1)\phi(\phi-1) + (\phi^2-1)z = 0$$

$$(2-z)\phi^2 + 2(z-1)\phi - z = 0$$

$$\phi = \frac{2(1-z) + \sqrt{4(z-1)^2 + 4z(2-z)}}{2(2-z)} = \frac{(1-z) \pm 1}{(2-z)} \quad (12\text{-}183)$$

$$\theta^{1/2} = \phi = \frac{z}{z-2} = e^{-u/2kT} \quad (12\text{-}184)$$

where the $+$ sign must be chosen in Eq. (12-183).   The critical point occurs at

$$T_c = \frac{u}{2k \ln\left(\frac{z-2}{z}\right)} = \frac{|u|}{2k \ln\left(\frac{z}{z-2}\right)} \quad (12\text{-}185)$$

This result is compared with the exact and Bragg-Williams approximation results for a square lattice, where $z = 4$ in Table 12-2.

Table 12-2

	$kT_c/|u|$
Exact	0.568
Quasi-chemical	0.722
Bragg-Williams	1.000

A substantial improvement over the Bragg-Williams results is obtained in the quasi-chemical approximation. A plot of T_c vs z in the quasi-chemical and Bragg-Williams approximations is given in Fig. 12-19.

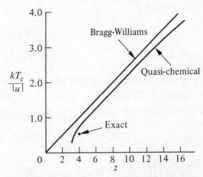

FIGURE 12-19 The dependence of critical temperature on the number of nearest neighbors for a lattice structure in the Bragg-Williams and quasi-chemical approximations.

PROBLEM: Show that the equation of state in the Bragg-Williams and quasi-chemical approximations (for $z = 4$) can be put in the following forms:

(a) Bragg-Williams ($z = 4$):

$$T_r = \frac{T}{T_c} = \frac{kT}{|u|} \tag{12-186}$$

$$p^* = \frac{p\omega}{|u|} = -2f^2 - T_r \ln (1 - f) \tag{12-187}$$

(b) Quasi-chemical ($z = 4$):

$$T_r = \frac{T}{T_c} = \frac{kT}{|u|} (2 \ln (4/2)) = \frac{kT}{0.722|u|} \tag{12-188}$$

$$p^* = \frac{p\omega}{|u|} = 0.722 T_r \{3 \ln (1-f) - 2 \ln [1 - f - \tfrac{1}{2}(\theta - 1)^{-1}(\beta - 1)]\} \tag{12-189}$$

$$\theta = e^{-u/kT} = e^{|u|/kT} = \exp\left[1/0.722T_r\right] \qquad (12\text{-}190)$$

$$\beta = \sqrt{1 + 4f(1-f)(\theta - 1)} \qquad (12\text{-}191)$$

Now compare the Bragg-Williams and quasi-chemical approximations with the exact results for $z = 4$. The BW isotherms are plotted in Figs. 12-17 and 12-18. The QC isotherms are shown in Figs. 12-20 and 12-21. In each case, the looped isotherms have been interpreted as containing the liquid-vapor region, and horizontal dotted lines are drawn to balance the areas under the positive and negative loops. The intersections of these lines with the isotherms are used to draw the envelope of the region of liquid and vapor coexistence. The critical pressure, p_c^*, is obtained by putting $T^* = 1$ and $f = 1/2$ in the appropriate expression for p^*. It is found that

$$(p_c^*)_{QC} = 0.0853$$

$$(p_c^*)_{BW} = 0.1932$$

It is interesting to compare the BW and QC approximations at the same values of T, not T_r. Since T_c is different in the two methods, T_r will be different at the same T. Thus, at the same value of T,

$$(T_r)_{BW} = 0.722(T_r)_{QC}$$

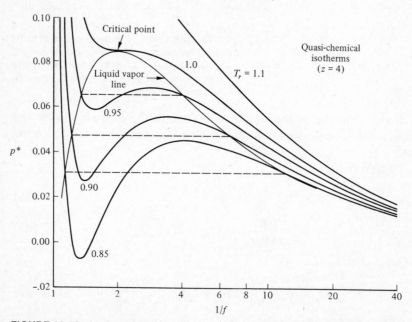

FIGURE 12-20 Isotherms for the two-dimensional square lattice gas in the quasi-chemical approximation.

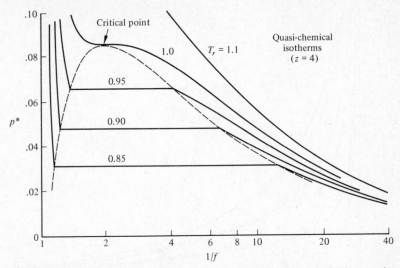

FIGURE 12-21 Equation of state of the two-dimensional square lattice gas in the quasi-chemical approximation obtained replacing the loops in the looped region by horizontal lines.

Hence, when Eq. (12-189) at $(T_r)_{QC}$ is compared with Eq. (12-187) at $(T_r)_{BW} = 0.722(T_r)_{QC}$, it follows that the isotherms are at the same absolute temperature. The result of such a calculation is given in Fig. 12-22. It can be seen that at temperatures well above the critical point, the isotherms for the same T are very similar. For lower T, the curves tend to diverge in the region between $1 < 1/f < 4$. At very low f, the two procedures tend to agree at all T.

Now compare these approximations with the exact treatment for $z = 4$. Equation (12-136) may be rewritten

$$\frac{p^*}{T^*} = \ln (1+\mu) + \frac{1}{2\pi} \int_0^\pi \ln \{\tfrac{1}{2}[1 + (1-k_1^2 \sin^2 \phi)^{1/2}]\} \, d\phi \qquad (12\text{-}192)$$

where $p^* = p\omega/|u|$, $T^* = kT/|u|$, and $T_r = T/T_c = T^*/0.568$. When Eqs. (12-192), (12-138), and (12-139) are plotted, the curve shown in Fig. 12-13 is obtained for the region of liquid-vapor coexistence. The critical pressure is $p^* = 0.027$. It can be seen that the QC approximation fits the exact results much better than the BW approximation. The isotherms in the exact calculation are obtained from expansions given by Yang and Lee.[30] The results are also given in Fig. 12-16. It can be seen that the results are quite similar to the QC results. The results of the three methods are summarized in Fig. 12-23 and in Table 12-3.

---

[30]T. D. Lee and C. N. Yang, *Phys. Rev.* **87**, 410 (1952).

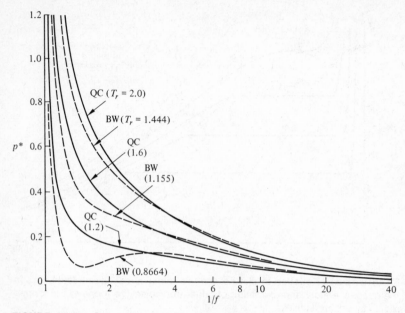

FIGURE 12-22 Comparison of the Bragg-Williams and quasi-chemical isotherms for a two-dimensional square lattice gas the same values of T.

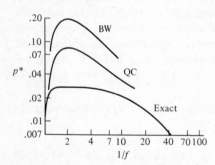

FIGURE 12-23 Comparison of liquid-vapor lines predicted by exact, Bragg-Williams, and quasi-chemical treatments of the two-dimensional square lattice.

Table 12-3

	$T_c^* = \dfrac{kT_c}{|u|}$	$p_c^* = \dfrac{p_c\omega}{|u|}$	$V_c^* = \dfrac{V_c}{\omega}$	$\dfrac{p_c^* V_c^*}{T_c^*} = \dfrac{p_c V_c}{kT_c}$
Exact	0.568	0.027	2	0.095
QC	0.722	0.0853	2	0.236
BW	1.000	0.1932	2	0.386

Note that since $V = \tilde{V}/N_m = N_t\omega/N_m = \omega/f$, it follows that $V^* = V/\omega = 1/f$ is 2 at the critical point.

The hole theory of liquids in three dimensions will now be considered. In this case, no exact solution is available, and only an approximate treatment based on the QC approximation must suffice. Imagine that the volume occupied by a liquid is subdivided into N_t sites, each of which is a sphere of diameter σ. If the spheres are close-packed in a cubic array, it follows that the volume occupied by N_t sites is[31]

$$V = N_t\sigma^3/\sqrt{2}$$

For looser forms of packing, the volume will be slightly greater. If there are a total of N_m molecules distributed over the N_t sites, the volume per molecule is

$$V = \tilde{V}/N_m = \sigma^3/(f\sqrt{2})$$

At the critical point, $f = 1/2$. Therefore, for close-packed sites,

$$V_c = \sigma^3\sqrt{2}$$

For looser forms of packing V_c will be greater than this.

The critical temperature can be obtained from Eq. (12-185). For close-packed sites, $z = 12$, and

$$T_c^* = \frac{kT_c}{|u|} = 2.74$$

The critical pressure is obtained by using Eq. (12-180) with $z = 12$, $f = 1/2$, $V_c = \sigma^3/\sqrt{2}$, and $T_c = 2.74|u|/k$. The result is

$$p^* = \frac{p\sigma^3}{|u|} = 0.663$$

The critical ratio p_cV_c/kT_c is also obtained from Eq. (12-180) with $f = 1/2$. It is found that $\theta_c = [z/(z-2)]^2$ and

$$\frac{p_cV_c}{kT_c} = 2\left[\ln 2 - \frac{z}{2}\ln\left(\frac{z}{z-1}\right)\right]$$

with $z = 12$, thus

$$\frac{p_cV_c}{kT_c} = 0.342$$

It can readily be seen that this model leads to an equation of state in conformity with the principle of corresponding states. It might be argued that the packing in an actual liquid lattice is much looser than a

---

[31]See discussion above Eq. (12-30).

close-packed cubic lattice, and therefore, a smaller value of z might be more appropriate.   In Table 12-3, the critical constants are presented for several choices of z.

TABLE 12-3

z	$\dfrac{p_c V_c}{kT_c}$	$\dfrac{kT_c}{|u|}$
12	0.342	2.74
10	0.332	2.24
8	0.306	1.74
6	0.294	1.24

According to Table 12-1, an average value for $p_c V_c / kT_c$ for a selection of roughly spherical molecules is ~ 0.29.   The agreement is best for $z \cong$ 6.   If it were hypothesized that the average interaction energy u is comparable to $-\epsilon$, where ϵ is the Lennard-Jones parameter, then the best experimental value (taken from Table 12-1) for $kT_c / |u|$ is ~ 1.29.   The same value of z (~ 6) that gives a best value of $p_c V_c / kT_c$ also gives a best value for T_c.   A number of refinements to this theory have been made[32] which bring it into closer agreement with experiment.   One of the principal difficulties with the hole theories is that the vacant volume in a liquid tends to exist as many small crevices between imperfectly packed spheres rather than as a relatively few entirely empty spherical sites.

The reader is referred to the articles by de Boer and Henderson for more details.

---

[32]J. de Boer, *Proc. Roy. Soc.* **A215**, 4 (1952); D. Henderson, *J. Chem. Phys.* **37**, 631 (1962).

APPENDICES

Appendix I

Energy Unit Conversion Factors

	$ergs^*$	eV^*	$cm^{-1}*$	$kcal/Mole$
1 erg* =	1.0	6.242×10^{11}	5.035×10^{15}	1.440×10^{13}
1 eV* =	1.602×10^{-12}	1.0	8.066×10^3	23.07
1 cm^{-1}* =	1.986×10^{-16}	1.240×10^{-4}	1.0	2.860×10^{-3}
1 Kcal/Mole =	6.945×10^{-14}	4.335×10^{-2}	3.496×10^2	1.0

*per molecule

Appendix II

Table of Fundamental Constants

$h = 6.625 \times 10^{-27}$ erg-sec

$\hbar = h/2\pi = 1.0545 \times 10^{-27}$ erg-sec

$k = 1.3804 \times 10^{-16}$ ergs/°K

$e = 4.8029 \times 10^{-10}$ esu

$m_e = 9.108 \times 10^{-28}$ g

$N_{av} = 6.0248 \times 10^{23}$ molecules/mole

$R = 8.3167 \times 10^7$ erg/mole-°K

$c = 2.9979 \times 10^{10}$ cm/sec

$\dfrac{m_p}{m_e} = 1836.13$

Appendix III

Some Atomic Energy Levels

Atom	Spectroscopic notation	Energy, eV
H	$n = 1$	0
	$n = 2$	10.20
	$n = 3$	12.08
	$n = 4$	12.75
$H^+ + e^-$	—	13.60
He	$1s^2\ {}^1S$	0
	$2s\ {}^3S$	19.80
	$2s\ {}^1S$	20.61
	$2p\ {}^3P$	20.97
	$2p\ {}^1P$	21.21
$He^+ + e^-$	—	24.58
Li	$2s\ {}^2S$	0
	$2p\ {}^2P$	1.85
	$3s\ {}^2S$	3.38
$Li^+ + e^-$	—	5.39
C	${}^3P_0, {}^3P_1, {}^3P_2$	0.000, 0.002, 0.005
	${}^1D$	1.26
	${}^1S$	2.69
$C^+ + e^-$	—	11.26
N	${}^4S$	0.0
	${}^2D$	2.38
	${}^2P$	3.58
$N^+ + e^-$	—	14.53
O	${}^3P_2, {}^3P_1, {}^3P_0$	0.000, 0.020, 0.028
	${}^1D$	1.97
	${}^1S$	4.19
$O^+ + e^-$	—	13.61
Na	$3s\ {}^2S$	0.0
	$3p\ {}^2P$	2.10
	$4s\ {}^2S$	3.19
$Na^+ + e^-$	—	5.14

Appendix IV

Properties of Some Diatomic Molecules

Molecule	E_{el}^a	Electronic State spectroscopic designation	no.[b]	Vib. Freq. (cm^{-1})	Bond Dist. (Å)	θ_{vib} (°K)	θ_{rot} (°K)	E_{diss}^c (eV)
H_2	0.0	$X\ ^1\Sigma_g^+$	0	4395	0.742	6320	28.8	4.48
	11.38	$B\ ^1\Sigma_u^+$	1	1357	1.293	1950	8.74	3.56
N_2	0.0	$X\ ^1\Sigma_g^+$	0	2360	1.094	3390	2.89	9.76
	6.24	$A\ ^3\Sigma_u^+$	1	1460	1.293	2100	2.07	3.7
	7.40	$B\ ^3\Pi_g$	2	1734	1.212	2490	2.35	4.9
NO	0.0	$X\begin{cases}^2\Pi_{3/2}\\ ^2\Pi_{1/2}\end{cases}$	0	1904	1.151	2740	2.45	6.50
	0.015							
O_2	0.0	$X\ ^3\Sigma_g^-$	0	1580	1.207	2270	2.08	5.08
	0.98	$a\ ^1\Delta_g$	1	1509	1.216	2170	2.05	4.2
	1.64	$b\ ^1\Sigma_g^+$	2	1433	1.227	2060	2.01	3.6
CO	0.0	$X\ ^1\Sigma^+$	0	2170	1.128	3120	2.78	9.35
		$A\ ^1\Pi$	1	1515	1.235	2180	2.32	—
		$B\ ^1\Sigma^+$	2	(2080)	1.120	(2990)	2.82	—

[a]E_{el} denotes the energy of the "bottom of the well" measured from the bottom of the well of the ground state.
[b]State number is ordered as 0 for ground state, 1 for first excited state, etc.
[c]E_{diss} is the dissociation energy measured from the lowest quantum state.

Appendix V

Properties of Some Polyatomic Molecules

1. *Carbon Dioxide*

$$\text{ground electronic state} = {}^1\Sigma$$

1.162 Å   1.162 Å
O———C———O

vibrational frequencies:

symmetric stretch   ←O   O O→ $\nu_1 = 1351\ cm^{-1}$
asymmetric stretch   O→ ←O O→ $\nu_2 = 2396\ cm^{-1}$

x-bend   O   O O $\nu_3 = 672\ cm^{-1}$

y-bend   ⊙   ⊕ ⊙ $\nu_4 = 672\ cm^{-1}$

$\theta_{rot} = 0.562°K$

2. *Nitrous Oxide*

ground electronic state $= {}^1\Sigma$

N———N———O

vibrational frequencies:

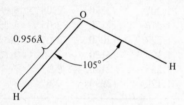

symmetric stretch $\leftarrow$O $\leftarrow$O   O$\rightarrow$ $\nu_1 = 1288$ cm^{-1}
asymmetric stretch $\leftarrow$O O$\rightarrow$ $\leftarrow$O $\nu_2 = 2237$ cm^{-1}

x-bend $\nu_3 = 588$ cm^{-1}

y-bend   ⊙   ⊕   ⊙ $\nu_4 = 588$ cm^{-1}
$\theta_{rot} = 0.601°$K

3. Water — ground electronic state = singlet

0.956Å

—105°

O

H

H

vibrational frequencies:

symmetric stretch

$\nu_1 = 3825$ cm^{-1}

asymmetric stretch

$\nu_2 = 1654$ cm^{-1}

bend

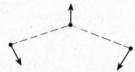

$\nu_3 = 3936$ cm^{-1}

rotations:

$\theta_{rot_1} = 39.3$ $\theta_{rot_2} = 13.6°$K $\theta_{rot_3} = 21.0°$K

Appendix VI

Volume Elements and Jacobian Determinants

It can be shown[1] that if one has a volume element in N dimensions of cartesian coordinates $\xi_1, \xi_2, \ldots, \xi_N$.

$$d\tau = d\xi_1 \, d\xi_2 \ldots d\xi_N$$

that on transforming to generalized coordinates $q_1, q_2, \ldots, q_N$, the volume element becomes

$$d\tau = ||J|| \, dq_1, dq_2, \ldots, dq_N$$

where $||J||$ is the absolute value of the *Jacobian determinant*, given by

$$|J| = \begin{vmatrix} \dfrac{\partial \xi_1}{\partial q_1} & \dfrac{\partial \xi_1}{\partial q_2} & \cdots & \dfrac{\partial \xi_1}{\partial q_N} \\ \vdots & & & \vdots \\ \dfrac{\partial \xi_N}{\partial q_1} & \dfrac{\partial \xi_N}{\partial q_2} & \cdots & \dfrac{\partial \xi_N}{\partial q_N} \end{vmatrix}$$

Furthermore, the volume element in a second set of generalized coordinates $r_1, r_2, \ldots, r_N$ is

$$d\tau = ||J'|| \, dr_1 \, dr_2 \ldots dr_N$$

where

$$|J'| = \begin{vmatrix} \dfrac{\partial q_1}{\partial r_1} & \cdots & \dfrac{\partial q_N}{\partial r_1} \\ \vdots & & \\ \dfrac{\partial q_N}{\partial r} & \cdots & \dfrac{\partial q_N}{\partial r_N} \end{vmatrix}$$

Some simple examples of the use of Jacobians in coordinate transformations are:

(1) (x, y) cartesian to (r, θ) polar:

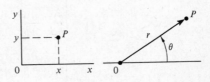

[1] J. L. Franklin, *A Treatise on Advanced Calculus* (New York: J. Wiley and Sons, 1940), p. 371.

$$dx\,dy = |J|\,dr\,d\theta = \begin{vmatrix} \dfrac{\partial x}{\partial r} & \dfrac{\partial x}{\partial \theta} \\ \dfrac{\partial y}{\partial r} & \dfrac{\partial y}{\partial \theta} \end{vmatrix} dr\,d\theta$$

$$x = r\cos\theta$$

$$y = r\sin\theta$$

$$\frac{\partial x}{\partial r} = \cos\theta, \qquad \frac{\partial y}{\partial r} = \sin\theta$$

$$\frac{\partial x}{\partial \theta} = -r\sin\theta, \qquad \frac{\partial y}{\partial \theta} = r\cos\theta$$

$$|J| = \begin{vmatrix} \cos\theta & -r\sin\theta \\ \sin\theta & r\cos\theta \end{vmatrix} = r$$

$$dx\,dy = n\,d\theta\,dr$$

(2) x, y, z cartesian to ρ, ϕ, z cylindrical

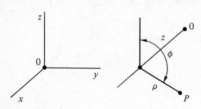

$$x = \rho\cos\phi$$
$$y = \rho\sin\phi$$
$$z = z$$

$$|J| = \begin{vmatrix} \dfrac{\partial x}{\partial \rho} & \dfrac{\partial x}{\partial \phi} & \dfrac{\partial x}{\partial z} \\ \dfrac{\partial y}{\partial \rho} & \dfrac{\partial y}{\partial \phi} & \dfrac{\partial y}{\partial z} \\ \dfrac{\partial z}{\partial \rho} & \dfrac{\partial z}{\partial \phi} & \dfrac{\partial z}{\partial z} \end{vmatrix} = \begin{vmatrix} \cos\phi & -\rho\sin\phi & 0 \\ \sin\phi & \rho\cos\phi & 0 \\ 0 & 0 & 1 \end{vmatrix}$$

$$|J| = \rho$$

$$dz\,dy\,dz = \rho\,d\rho\,d\phi\,dz$$

(3) x, y, z cartesian to r, θ, ϕ spherical polar

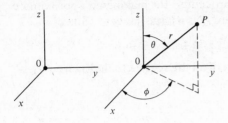

$x = r \cos \phi \sin \theta = r\,c\phi\,s\theta$
$y = r \sin \phi \sin \theta\ = r\,s\phi\,s\theta$
$z = r \cos \theta \qquad = r\,c\theta$

$$|J| = \begin{vmatrix} \dfrac{\partial x}{\partial r} & \dfrac{\partial x}{\partial \phi} & \dfrac{\partial x}{\partial \theta} \\[2mm] \dfrac{\partial y}{\partial r} & \dfrac{\partial y}{\partial \phi} & \dfrac{\partial y}{\partial \phi} \\[2mm] \dfrac{\partial z}{\partial r} & \dfrac{\partial z}{\partial \phi} & \dfrac{\partial z}{\partial \theta} \end{vmatrix} = \begin{vmatrix} c\phi\,s\theta & -r\,s\phi\,s\theta & r\,c\phi\,c\theta \\[2mm] s\phi\,s\theta & r\,c\phi\,s\theta & r\,s\phi\,c\theta \\[2mm] c\theta & 0 & -r\,s\theta \end{vmatrix}$$

$$|J| = c\phi\,s\theta \begin{vmatrix} r\,c\phi\,s\theta & r\,s\phi\,c\theta \\[2mm] 0 & -r\,s\theta \end{vmatrix} + r\,s\phi\,s\theta \begin{vmatrix} s\phi\,s\theta & r\,s\phi\,c\theta \\[2mm] c\theta & -r\,s\theta \end{vmatrix}$$

$$+ r\,c\phi\,c\theta \begin{vmatrix} s\phi\,s\theta & r\,c\phi\,s\theta \\[2mm] c\theta & 0 \end{vmatrix}$$

$|J| = c\phi\,s\theta(-r^2\,s^2\theta\,c\phi) + r\,s\phi\,s\theta(-r\,s^2\theta\,s\phi - r\,s\phi\,c^2\theta) + r\,c\phi\,c\theta(-r\,c\phi\,s\theta\,c\theta)$

$\qquad = r^2\,s\theta[-s^2\theta - c^2\theta] = -r^2\,s\theta$

$\|J\| = r^2 \sin \theta, \qquad d\tau = r^2\,dr \sin \theta\,d\theta\,d\phi$

Appendix VII

Stirling's approximation is a procedure for making an approximate estimate of the function $\ln (N!)$, where N is a large integer.   Since

$$\ln (N!) = \ln 2 + \ln 3 + \ln 4 + \cdots + \ln N$$

it follows that $\ln (N!)$ is the total area of all the rectangles in Figure A-1.

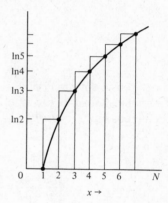

Figure A-1

Now consider the integral

$$I = \int_{1}^{N} \ln (x) \, dx$$

This can be integrated by parts, noting that

$$\int \ln x \, dx = x \ln x - \int dx = x \ln x - x$$

The definite integral is therefore

$$I = N \ln N - N + 1$$

This integral is the area under the solid curve drawn through the points $\ln x$.   This is clearly less than $\ln (N!)$, but should be a good approximation for large N.

A more exact analysis leads to the result

$$\ln (N!) \cong (N + \tfrac{1}{2}) \ln N - N + \tfrac{1}{2} \ln (2\pi) + \text{terms of order } (1/N)$$

If $N \gg 1$, we may use the approximate form

$$\ln N! \cong N \ln N - N$$

These functions are given below for N up to 1000.

N	$\ln(N!)$	$N \ln N - N$	$(N+\tfrac{1}{2})\ln N - N + \tfrac{1}{2}\ln(2\pi)$
10	15.104	13.026	15.096
20	42.336	39.915	42.331
50	148.48	145.60	148.48
100	363.9	360.8	363.9
500	2608	2604	2608
1000	5912	5908	5912

INDEX